AF802826

Aquaculture, Innovation and Social Transformation

The International Library of Environmental, Agricultural and Food Ethics

VOLUME 17

Editors

Michiel Korthals, *Dept. of Applied Philosophy, Wageningen University, Wageningen, The Netherlands*
Paul B. Thompson, *Dept. of Philosophy, Michigan State University, East Lansing, U.S.A.*

Editorial Board

Andrew Brennan, *La Trobe University, Victoria, Australia*
Lawrence Busch, *Dept. of Sociology, Michigan State University, East Lansing, U.S.A.*
Avner de-Shalit, *Hebrew University, Jerusalem, Israel*
Richard Haynes, *Dept. of Philosophy, University of Florida, Gainesville, U.S.A.*
Darryl Macer, *The Eubios Ethics Institute, University of Tsukuba, Ibaraki, Japan*
Clare Palmer, *Washington University, St Louis, U.S.A.*
Doris Schroder, *University of Central Lancashire, Preston, United Kindgom*

For other titles published in this series, go to
www.springer.com/series/6215

Aquaculture, Innovation and Social Transformation

Edited by

Keith Culver
Fredericton
Canada

and

David Castle
Ottawa
Canada

Springer

Editors
Keith Culver
P.O. Box 4400 Stn A.
Fredericton, New Brunswick
E3B 5A3
University of New Brunswick
Canada
kculver@unb.ca

David Castle
70 Laurier Avenue East
Ottawa K1N 6N5
University of Ottawa
Canada
dcastle@uottawa.ca

ISBN: 978-1-4020-8834-6 e-ISBN: 978-1-4020-8835-3

Library of Congress Control Number: 2008931596

© 2008 Springer Science+Business Media B.V.
No part of this work may be reproduced, stored in a retrieval system, or transmitted in any form or by any means, electronic, mechanical, photocopying, microfilming, recording or otherwise, without written permission from the Publisher, with the exception of any material supplied specifically for the purpose of being entered and executed on a computer system, for exclusive use by the purchaser of the work.

Printed on acid-free paper

9 8 7 6 5 4 3 2 1

springer.com

Contents

Contributors

Colin J. Barrow
Executive Vice-President of Research and Development, Ocean Nutrition

Kenneth D. Black
Head, Ecology Department, Scottish Association for Marine Science, Dunstaffnage Marine Laboratory

Kevin G. Butterworth
Director, Marine and Environmental Research, British Columbia Centre for Aquatic Health Sciences

David Castle
Canada Research Chair in Science and Society, and Director, Institute for Science, Society and Policy, University of Ottawa

Thierry Chopin
Professor of Biology, University of New Brunswick

Steve Clark
Department of Marketing and Consumer Studies, University of Guelph

Barry A. Costa-Pierce
Professor of Fisheries and Aquaculture, and Director, Rhode Island Sea Grant, University of Rhode Island

Keith Culver
Associate Professor and Director, Centre for Social Innovation Research, University of New Brunswick

Larry Felt
Professor of Sociology, Memorial University of Newfoundland

Karen Finlay
Associate Professor of Marketing and Consumer Studies, University of Guelph

K. Fiona Cubitt
Research Associate, UBC-DFO Centre for Aquaculture and Environmental Research, University of British Columbia

Gilly Griffin
Guidelines Program Director, Canadian Council for Animal Care

Brad Hicks
Vice-President, Taplow Feeds

Felicity Huntingford
Professor of Functional Ecology, University of Glasgow

Sunil Kadri
Honorary Research Associate, Aquaculture Innovation & Fish Biology Group, Division of Environmental & Evolutionary Biology, Institute of Biomedical & Life Sciences, University of Glasgow

Dan Lane
Professor of Management, University of Ottawa

Paul Lyon
Policy Analyst, Aquaculture Managament Directorate, Fisheries and Oceans Canada

Kenny McCaffrey
Editor, *Fish Farming International*

R. Scott McKinley
Canada Research Chair in Aquaculture and the Environment and Director, UBC-DFO Centre for Aquaculture and Environmental Research, University of British Columbia

Wojtek Michalowski
Professor of Management, University of Ottawa

Frode Nilssen
Senior Scientist, Department for Economics and Marketing Research, Norwegian Institute of Fisheries and Aquaculture Research

Fred Page
Director, Centre for Integrated Aquaculture Research, St. Andrews Biological Station, Fisheries and Oceans Canada

Jacques Paynter
Consultant, AMEC Earth and Environmental Services

Jeremy Rayner
Associate Professor of Political Science, University of Regina

Teresa Ryan
Member of the Gitlan, Tsimshian First Nation and PhD candidate Institute of Resources, Environment and Sustainability, University of British Columbia

Marc A. Saner
Director of Research, Regulatory Governance, School of Public Policy and Administration Carleton University

Tom Sephton
Division Manager, Environmental Science Division, Fisheries and Oceans Canada Maritimes Region

Marcel Shepert
Member of the Carrier-Sekani First Nation, Executive Director of the Fraser River Aboriginal Fisheries Secretariat (FRAFS) and Coordinator of the Upper Fraser Fisheries Conservation Alliance.

Jamey Smith
Executive Director, New Brunswick Salmon Growers Association

Robert Stephenson
Director, St. Andrews Biological Station, Fisheries and Oceans Canada

Editors' Introduction

Keith Culver and David Castle

Introduction

Aquaculture is at the leading edge of a surprisingly polarized debate about the way we produce our food. According to the United Nations Food and Agriculture Organization, aquaculture production has increased 8.8% per year since 1970, far surpassing productivity gains in terrestrial meat production at 2.8% in the same period (FAO 2007). Like the 'green revolution' before it, the 'blue revolution' in aquaculture promises rapidly increased productivity through technology-driven intensification of aquaculture animal and plant production (Costa-Pierce 2002; The Economist 2003). Proponents of further aquaculture development emphasize aquaculture's ancient origins and potential to contribute to global food security during an unprecedented collapse in global fisheries (World Fish Center; Meyers and Worm 2003; Worm et al. 2006). For them, technology-driven intensification is an ordinary and unremarkable extension of past practice. Opponents counter with images of marine and freshwater environments devastated by intensive aquaculture practices producing unsustainable and unhealthy food products. They view the promised revolution as a scam, nothing more than clever marketing by profit-hungry fish farmers looking for ways to distract the public from the real harms done by aquaculture. The stark contrast between proponents and opponents of modern aquaculture recalls decades of disputes about intensive terrestrial plant and animal agriculture, disputes whose vigor shows that the debate is about much more than food production (Ruse and Castle 2002). The debate is at least emblematic of, if not literally about, the struggle over the place of technology in human life and the life of the planet. There is little middle ground to be had, and even those who simply advocate a fair hearing for aquaculture may find themselves attacked by one side or the other.

Aquaculture's blue revolution suddenly seems less promising. Aquaculture's 'innovation' may be associated with the introduction of new, undesirable risks.

K. Culver
Associate Professor and Director, Centre for Social Innovation Research,
University of New Brunswick
e-mail: kculver@unb.ca

K. Culver, D. Castle (eds.), *Aquaculture, Innovation and Social Transformation*,

© Springer Science+Business Media B.V. 2008

'Transformation' of communities may not be positive if technological change challenges core community values to the point that social cohesion is threatened. This situation is further complicated by the fact that social aversion to innovation is not automatic or easily predicted. It does seem clear, however, that resistance to technological change may be especially strong when the potential beneficiaries of innovation have no pressing need for those benefits. Northern cold water aquaculture countries may lead aquaculture innovation, but they are wealthy enough to simply walk away from the debate over aquaculture's future. Yet doing nothing would actually turn out to be doing something, since withdrawal from aquaculture innovation might undermine both local coastal communities and broader aspirations for global food security built on sustainable industries. To the extent that the politics of food are inseparable from global food webs, aquaculture, innovation, and social transformation are inescapably both local and global issues.

Definitional starting points:

Aquaculture: "The farming of aquatic organisms in inland and coastal areas, involving intervention in the rearing process to enhance production and the individual or corporate ownership of the stock being cultivated" (FAO 2007).

Governance: "the traditions and instititutions by which authority in a country is exercised for the common good. This includes (i) the process by which those in authority are selected, monitored and replaced, (ii) the capacity of the government to effectively manage its resources and implement sound policies, and (iii) the respect of citizens and the state for the instititutions that govern economic and social interactions among them (World Bank 2007).

Innovation: An innoviation is the implementation of a new or significantly improved product (good or service), or process, a new organisational method in business practices, workplace organisation or external relations (OECD 2007).

This volume arose from an attempt to find a new way to approach the debate over aquaculture's future, facing up to the central insight that a global, technology-driven blue revolution will require new forms of governance to match the technological and social changes brought by innovative aquaculture practices. Our contribution to the debate focuses on Canada. We have adopted this focus partly because we, and the majority of our contributors, live and work in Canada, and partly because Canada's experience and leadership in aquaculture may provide valuable comparative lessons to other states and to global aquaculture development. The chapters in this volume are the result of a process started by a workshop in 2004 sponsored by AquaNet, Canada's Network of Centres of Excellence in Aquaculture, and Genome Canada, Canada's federal funder of genomics research. The workshop was intentionally organized to bring together divergent perspectives from Canadian and international university, government, and industry sectors whose members were nonetheless still willing to talk with one another. As we put it in our letter of invitation to participants:

> Aquaculture's role in national economics and coastal communities is a fact. The questions facing aquaculture are not 'whether' aquaculture should be permitted, but 'how' to develop sustainable aquaculture practices which strike an ethically acceptable balance between risk and opportunity. This workshop will investigate the state and future of Canadian aquaculture from the perspective of policy-oriented social science and humanities research. What is required is not an arid academic exercise but rather a frank discussion in which participants exchange their experiences and insights gained through academic analysis, practical governance, and the business of aquaculture. While the focus of this workshop is on Canadian aquaculture and its social implications, the scope is broadened by the addition of international perspectives on Canada's domestic and international issues.

Our authors heeded our request to provide insight into aquaculture's innovative capacity and potential for altering its own social context. Indeed, there is much to be learned from each of our contributors about aquaculture, innovation, and social transformation. At the same time, however, our contributors display passively many of the discordances and tensions that arise in the debate about aquaculture. This result is not entirely accidental, as we knew we were engineering a clash of cultures when we invited participants from quite different points of view to discuss governance practices guiding environmentally and socially sustainable aquaculture. Perhaps the greatest contribution of this book is its presentation of the result of this interaction between contributors and ideas, laying bare disputes and divisions where contributors simply disagree with one another, or talk past one another. In the remainder of this introduction, we focus on how we engineered the 'contextual' elements of our expert discussion, leaving most of the 'factual' elements for detailed discussion in each chapter.

Localising the Issues: Aquaculture, Innovation, and Social Transformation in Canada

If, as the popular saying goes, 'all politics is local,' it is certainly also true that social contexts of industries and the use of science and technology are equally local. This insight is certainly borne out in social acceptance of aquaculture, which varies not only from place to place, but within particular cultures. Canada serves as a particularly complex example of the opportunity and challenge aquaculture poses for its host communities. Canada might seem, at first glance, to be the ideal aquaculture state. Canada's coastline is the world's largest, extending nearly a quarter million kilometers, including over 50,000 islands. Within Canada's borders, the Great Lakes are the world's biggest freshwater system. Over 6 million Canadians, nearly a fifth of the population, live on its coasts and many of those coast-dwellers are employed in use of ocean resources (DFO 2007a). Yet, despite these seemingly obvious advantages, aquaculture remains in many ways an afterthought in Canada's ocean and freshwater policies. Except where voluble opponents raise public awareness of aquaculture, it is otherwise invisible to most Canadians. How could such an underperforming but simultaneously controversial industry arise in Canada?

Cultural variations between Canada's east and west coasts are part of the explanation. The aquaculture industry on the east coast of Canada often enjoys broad social support and tends to be viewed as offering a livelihood to younger workers otherwise likely to be forced to leave their hometowns to seek work. The introduction of aquaculture is often credited with vast changes to social and economic well-being. Prior to the introduction of aquaculture in the 1980s, Charlotte County was one of the poorest areas of New Brunswick. Yet in the period 1986–2001 Charlotte County enjoyed the largest drop in unemployment in New Brunswick, as aquaculture grew to provide 25% of all full-time employment in Charlotte County. Of the nearly 3000 persons employed in aquaculture, 75% were beneath the age of 40 (DFO 2001). Many of these jobs are better paid than the employment which preceded aquaculture not just because aquaculture is financially successful, but because aquaculture required a higher basic standard of education. Higher average education standards are generally associated with greater social mobility, better personal health and social cohesion (OECD 2001).

The situation is much different in British Columbia on Canada's west coast, where aquaculture is a symbolic and actual collision point of a diverse set of social interests. The act of naming those interests is itself hazardous, since the debate over aquaculture has accumulated nuanced positions which do not map neatly to simple categories of 'pro' or 'con.' Even proponents of aquaculture may disagree regarding key business and policy matters such as choice of species, aquaculture sites, and the balance between foreign and local ownership, much as opponents such as the Coastal Alliance for Aquaculture Reform vary from those proposing reform to those demanding the end of aquaculture altogether (CAAR 2007). Further differences emerge among rural and urban dwellers, and recreational and commercial users of the nearshore marine environment. The many attempts to find common ground, for example the British Columbia Pacific Salmon Forum (2007), have largely failed, for a variety of reasons discussed by many of our contributors. At the other end of the debate, where exchange collapses into fiery one-sided rhetoric and worse, hopes of consensus have sometimes collapsed into legally actionable defamation. In 2007, for example, prominent anti-aquaculture activist, Don Staniford, acting on behalf of environmental group Friends of Clayoquot Sound, was found by the British Columbia Supreme Court to have defamed aquaculture firm Creative Salmon in press releases making false statements regarding Creative Salmon's use of antibiotics (CBC News 2007; British Columbia Court of Appeal 2007).

Disputes over aquaculture in British Columbia are further complicated by the varying reception of aquaculture by First Nations peoples. In eastern Canada, First Nations are involved in the capture fisheries, with a particular interest in the lobster fishery, and there is some First Nations involvement in shellfish aquaculture and arctic charr. Yet there is little First Nations involvement in discussions of the industry-dominating issues in salmon aquaculture and broader finfish aquaculture policy. First Nations in British Columbia play a much larger role in the debate over aquaculture. Positions range from outright opposition to salmon farming by the Nine

Allied Tsimshian Tribes of Lax Kw'alaams (NATT) to First Nations such as the Kitasoo Band who have collaborated with Marine Harvest Canada to bring salmon aquaculture to the remote village of Klemtu (The Tyee 2007; Kitasoo 2007). As First Nations continue to negotiate treaties, and attempt to re-build societies fragmented by economic, political, and social oppression, there are disputes among those First Nations regarding their best interests in the longer term, and just what their traditional and local ecological knowledge should mean for First Nations' fisheries and aquaculture. At the time of writing it appears that the First Nations arguing for reform, reduction, or elimination of aquaculture, particularly net-pen aquaculture of salmon, may win over public opinion and the support of legislators struggling to decide what to do next after the lifting of a moratorium on new aquaculture sites in British Columbia (Turtle Island Native Network 2006). Yet the continued existence of the Aboriginal Aquaculture Association (Aboriginal Aquaculture 2007) perhaps indicates that divergent interests in aquaculture persist.

Other disputes regarding aquaculture stand above variations in perspectives on aquaculture in the East and West of Canada. Many of aquaculture's proponents and opponents will agree that no matter what is done about aquaculture, any further steps require improved cooperation between provincial and federal departments who share jurisdiction over Canadian aquaculture. The federal department Fisheries and Oceans Canada admits the deficiencies of the current approach in admirably clear terms in its 2005–2006 Departmental Performance Report: "The governance framework for aquaculture is highly complex and in major need of renewal. None of the federal legislation currently used to regulate the aquaculture industry was designed with aquaculture in mind (DFO 2006a)." One outstanding issue is the completion of an Aquaculture Framework Agreement between the federal government and the provinces, on the model of the existing Agriculture Framework Agreement which undergirds an integrated regulatory, policy and program approach for the benefit of Canada's agriculture. In the absence of an integrated approach enabled by a framework agreement, proponents of aquaculture believe it will be difficult to make progress on issues such as an aquaculture-specific equivalent to crop insurance. Opponents worry that the absence of an integrated governance approach to aquaculture will hamper the reform they advocate – whether relatively minor reforms or elimination of aquaculture. Still further concerns are raised regarding difficulties in implementation of an Aquaculture Framework Agreement, assuming an agreement can eventually be reached. Aquaculture sector members often complain informally that the federal department of Fisheries and Oceans Canada, which is the lead among 17 other departments and agencies related to aquaculture (DFO 2006b), is fundamentally a natural resource conservation and extraction regulator, poorly adapted to managing aquaculture. Some aquaculture advocates would prefer to see aquaculture placed with the federal ministry Agriculture and Agri-Food Canada. Shifting responsibility to another ministry is not without precedent, to the extent that in the province of New Brunswick, for example, aquaculture is governed by the Department of Agriculture and Aquaculture, while the capture fisheries are governed by a separate Department of Fisheries. It appears that the federal department of Fisheries and Oceans Canada recognizes tensions in its broad mandate, as it has sought to ensure careful focus on aquaculture via measures including designation of its St Andrews Biological Station as the location of the Secretariat for a national virtual centre of aquaculture expertise, the Centre for Integrated Aquaculture Science (DFO 2007b). Critics, however, take this as further evidence of a mandate divided between capture and culture, and do not anticipate harmony in the governance of the two in Canada.

All of the factors we have discussed are implicated in the governance of new aquaculture technologies in Canada and their reception by the public. The first and most important governance debate concerns the nature of aquaculture itself. The debate ranges from the assertion that aquaculture is simply farming the sea in a way analogous to familiar terrestrial raising of cattle, and on to the opposite conclusion, that aquaculture – particularly intensive finfish aquaculture – represents the horrible nadir of intensive food production in its use of unsustainable inputs and far-reaching environmental harms. As our contributors demonstrate, modern intensive aquaculture is likely neither of these things. Intensive aquaculture is no longer simply a theoretical possibility, and while its governance is certainly imperfect, it is evolving rapidly. Federal and provincial regulators have in fact responded to growing knowledge of the conditions of aquaculture, including matters such as use of antibiotics and other drugs, diet formulations, efficient feed techniques, and bay management approaches which reduce demands on inputs and limit effects on the marine environment. Still newer developments promise even greater control as marine 'integrated aquaculture' or polyculture takes advantage of cycles of interaction between aquaculture animals and plants, and other aquaculture technologies enable isolation of aquaculture from the marine environment via techniques for closed containment aquaculture in land-based tanks and marine bags. These developments are not yet in widespread use, but they are far from hypothetical in Canada. Scotian Halibut (www.halibut.ns.ca), for example, is producing and marketing halibut grown in land-based closed containment facilities in rural Nova Scotia. In light of these developments we are perhaps on the cusp of finding that aquaculture's transformative *social* effects are even greater and less predictable than its environmental effects.

Facing the Problem of Aquaculture Governance

From the facts discussed above it might seem reasonable to suppose that while the situation of aquaculture in Canada is complex, the future is bright for aquaculture advocates who are prepared to use new technologies in ways which stand a good chance of satisfying some of the critics of aquaculture. Yet there is no boom in Canadian aquaculture, and the tone of the debate has changed very little. Why is this?

In organizing a workshop to examine the relationship between technical innovation in aquaculture and transformation of aquaculture communities and consumers, we thought initially that we were facing a problem about governance of new technologies. Our thinking about how to investigate the problem can be expressed in the form of two hypotheses we planned to test in the course of the workshop. Our first hypothesis recognized the complexity of the issues and the impossibility of facing them from any single disciplinary or sectoral standpoint:

> Hypothesis 1: A fair hearing for aquaculture as a contributor to global food security requires an integrated understanding of the social consequences of innovation in aquaculture.

Our second hypothesis recognized that inter-disciplinary approaches to complex problems are methodologically perilous. Resolution of problems with several social and scientific dimensions tends to require much more than a simply summary list of the perspectives of the multiple insights and solutions offered by the disciplines and interests which 'own' a part of the problem.

> Hypothesis 2: A fair hearing requires a novel way of talking with one another.

We will discuss further below the fate of these hypotheses, leaving them for the moment as the backdrop to some further reflections about innovation, and what innovation means in Canada's strategy for support of bio-based industries.

Whatever else might be said about the aquaculture industry, it is bio-based and technology-intensive, willingly adopting technological innovations and accepting that their social consequences are often disruptive. We mean 'disruptive' in a particular way here, to capture the full range of social effects which upon evaluation appear to be good, bad, neutral, and sometimes of indeterminate value. This sense of disruption is evident in the two ways 'innovation' is used in talk of social consequences of technological innovation. The first refers simply to new products and services enabled by a recently developed or newly applied technology. This use of 'innovation' is related to strategies to ensure that new, incremental improvements to products and services are fully transferred from laboratory to marketplace. It seems innocuous to use 'innovation' this way, and objections to it seem muddle-headed in the face of our long history of improving our living conditions. Yet once this use of innovation is paired with its second, less common use, the reasons why innovation in aquaculture has provoked such diverse responses become clearer.

The second use of innovation comes from 18th century political theorist Edmund Burke, who cautioned that "[t]o innovate is not to reform (Burke 1963)." By contrasting innovation with reform, Burke showed that the change involved in innovation is an introduction of something entirely new replacing the old, as opposed to the gentler path of reform in which change leaves origins recognizable and present in the reformed thing. In Burke's words, "[c]hange is novelty, and whether it is to operate any one of the effects of reformation at all, or whether it may not contradict the very principle upon which reformation is designed, cannot be certainly known beforehand. Reform is not a change in the substance or in the primary modification of the object, but a direct application of a remedy to the grievance complained of (Burke 1963)." This second use of innovation illustrates what makes introduction of innovative technology so controversial – it is not the newness of technological innovation which is alarming, but the prospect that its adoption might mean the end of something else. Modern intensive aquaculture as a whole might represent an innovation which does away with some portion of the skills and community practices associated with the capture fisheries, for example. The cultural icon of the traditional fisher eking an existence from the harsh sea is very familiar – in historically Christian societies, it is of course a familiar biblical metaphor. By contrast, aquaculture offers no attractive and specific image as it threatens to displace familiar icons with tremendous nostalgic value. Whether aquaculture is better paying, safer for workers and more sustainable may be beside the point if the public will not give up the romantic ideal in which capture fishers represent a more natural and noble kind of human relation to fish and their environment.

Talk of aquaculture's threat to cultural icons is just one dimension of the question of how to govern technological innovation to produce acceptable or justified social change. While Canadian citizens try to decide just what they think of news stories providing competing views of aquaculture's social, economic, and environmental effects, successive Canadian governments have invested in biotechnology R&D with the explicit aim of creating scientific and technological innovation that brings social and economic wealth to Canadians. The 2007 federal science R&D strategy announced by the Conservative minority government follows closely the goals of prior strategies implemented by the preceding Liberal government, varying mainly in a greater faith in the efficacy of private sector involvement (Government of Canada 2007). The Canadian Prime Minister, Stephen Harper, announced the strategy's aim to bring research scientists and entrepreneurs closer together to drive technological progress which will result in greater prosperity. "Canada's New Government is charting a new direction, one that links the competitive energy of Canada's entrepreneurs to the creative genius of our scientists," Harper claimed "Our new strategy will strengthen Canada's economy by tapping our deep well of entrepreneurial energy to fuel our technological progress (Office of the Prime Minister 2007)." The commitments of the new strategy build on a stream of commitments which have had a substantial and direct effect on aquaculture.

The now defunct AquaNet Network of Centres of Excellence in Aquaculture (1999–2006) invested $3.6 million per year in university-government-private sector aquaculture R&D collaborations. Much of that investment was leveraged by co-funding from partners, reaching 50% leveraging prior to the network's non-renewal by the federal government. Since the AquaNet era and to the present, the federal department of Fisheries and Oceans Canada operates a $4.5 million per year Aquaculture Collaborative Research and Development Program (ACRDP), pairing aquaculture producers and their issues in production, animal health, and environmental performance with researchers in federal department of Fisheries and Oceans Canada laboratories (DFO 2007c). Most recently, Genome Canada and its regional centres and collaborators have invested $40 million since 2001 in projects focusing on salmonids, halibut, and cod (Genome Canada 2004). In addition to these specific, targeted investments, various opportunities exist to leverage or simple add further research to those projects using regional development funds available through federal agencies such as Western Economic Development Canada, and the Atlantic Canada Opportunities Agency.

In light of these investments, it is beyond argument that the Government of Canada values transformative scientific and technological innovation generally and aquaculture innovation specifically, even in the second sense of aquaculture as a disruptive innovation replacing old ways with new. Yet aquaculture enjoys at best variable social acceptance across Canada (DFO 2005), and despite support for aquaculture research and development, its governance remains flawed. Consequently, in our development of the workshop and this book we think our first hypothesis is confirmed: a fair hearing for aquaculture requires an integrated understanding of its challenges and opportunities. A comprehensive integrated approach requires bringing together the interests and perspectives of NGOs, provincial and federal governments, academics, and the aquaculture private sector from marine to freshwater, finfish to shellfish and aquatic plants. The trouble remains, however, that the comprehensive, integrated discussion Canada needs has yet to occur.

Previous attempts to engage government, universities and industry have not succeeded in identifying and resolving core problems, leading us to formulate our second hypothesis, that we in Canada need a new way of talking to one another about aquaculture if we are to get the integration that is needed. The need for a novel way of talking together perhaps helps to make sense of the otherwise somewhat surprising fact that this volume is edited by a legal theorist and a philosopher of science. If nothing else, philosophers of science and law are very familiar with identification and integration of insights from competing, and often mutually exclusive world views. Perhaps in light of the depth of disagreement over aquaculture we cannot hope for Canadians' favorite value, 'consensus,' yet we might hope for respect for the diversity of views, and acceptance of the need for ongoing discussion of issues while recognizing no endpoint of 'solution' to the evolving governance needs of an innovative and evolving practice. In reaching for this state we are asking for what the German philosopher Jurgen Habermas has called a 'fusion of horizons,' enabling mutual understanding even through profound disagreement. By testing our hypotheses in an active engagement between various perspectives on aquaculture, we are attempting to get past one-dimensional conceptions of innovation and single conceptions of the meaning of innovation in aquaculture that strive for simple truths regarding the right view about aquaculture. For reasons we have just discussed, the problems and their solutions are much more complex, and should direct us toward an account of aquaculture innovation with which all can live, one which can serve as the basis for widely accepted governance arrangements. It might be said that we are advocating in practice what Stuart Hampshire has argued in political theory: that justice and fairness are not end-states which establish norms to strive for or from which we have somehow fallen (Hampshire 2001). Rather, conflict is the normal situation when free people live together, and justice consists in having mechanisms to manage conflict.

Transdisciplinarity in Theory and Practice

At the workshop we tested our hypotheses in a trial version of a method we call 'contested exchange.' This method attempts to surpass typical conference and workshop approaches which presume that rational people hearing the views of others will reach the same conclusions about those views. Other approaches such as the 'consensus conference' for example, may force false consensus when the method limits the outcome of group deliberation to single statements, indistinguishable from a tyranny of the most vocal minority or the minority most willing to risk social disapproval while advancing its view. Our approach does not presume consensus is possible or desirable. Instead we take seriously the possibility that aquaculture represents a 'wicked policy problem' (Rittel and Webber 1973), one whose nature and resolution conditions change while the problem is investigated.

Three features of the structure of the workshop fostered contested exchange:

- defining seven focal topics representing particular areas of conflict in aquaculture
- structuring a conflict in each topic area by asking commentators from separate interest groups to respond to the main 'position paper' offered by an initial speaker, and
- using a facilitator to carry key disputes and ideas forward from focal topic to focal topic

Since we recognized that consensus on resolution of issues might be impossible, we were careful to avoid presupposing any specific number or kind of conclusions as the acceptable endpoint to discussion. In operating by this method we sought to get beyond conflicts of disciplinary points of view where each discipline or interest implicitly claims that its way of understanding the situation of aquaculture is the primary or best way to understand it. In the conflicts between various perspectives we have sought to build *trans-disciplinary* understanding of aquaculture and social transformation, working from understanding of one another's vocabularies, social priorities, and so on, to eventually incorporate in our individual views some of the perspectives and arguments of others. An understanding, for example, of environmental impact might come to recognize the importance of including effects of change on human communities and not just non-human animal communities.

Our method of contested exchange begins with a particular set of tensions between disputants in the aquaculture debate. Knowing we needed to start with manageable gaps in mutual comprehension, we made use of the existing tensions between university-based researchers, the aquaculture industry, and members of governance organizations from federal department to First Nations. Academics are often thought by industry to be too keen on abstract knowledge and long-term issues at the expense of real problems facing us today. Governance organizations often sympathise, charging academics with a general failure to recognize that aquaculture governance and aquaculture communities are not merely part of pilot projects to be tweaked for interest's sake. Industry, for its part, is often spoken of by academics as excessively focused on the near-term bottom line, and governance organizations are sometimes frustrated by industry's wish for unqualified support even when industry haste might result in unacceptable risks to public values or the public purse. Governance organizations are in turn subject to suspicion and sometimes outright scorn, as civil servants are sometimes seen by academics as timid do-nothing bureaucrats hiding an inability to face change behind a pretence to deference to elected officials.

Industry may tend to overtly support governance organizations, yet in discussion with them industry can be quick to criticize the inefficiency and slow speed of governance reforms which tend to hamper industry's ability to grow because of the need to demonstrate accountability to Canadians for the social, economic, and environmental consequences of aquaculture.

It would be easy to object that these three groups represent too little of the diversity of views on aquaculture. This objection only holds weight, however, if we are careless about our claims regarding the scope of our conclusions. We recognize that we have involved only some of the relevant groups, yet urge our readers to consider the value found in examining their interaction – an interaction whose hit-and-miss nature tends to show that working with any wider range of views might have resulted in a complete failure of mutual understanding, rather than the partial exchange we saw. Within these limits, then, we invited an academic researcher to provide a position paper on the following focal topics:

1. Animal welfare in aquaculture
2. Knowledge management and intellectual property issues
3. The environmental sustainability of aquaculture
4. The interaction between traditional knowledge and modern aquaculture
5. Messages, consumers and aquaculture: new products; new worries
6. The final frontier: integrated coastal zone management
7. New practices for global competitiveness: alternate species, alternate uses, and value-added aquaculture.

Beginning with an academic perspective was not motivated by undue respect for academics, but out of awareness that of our contributors, academics are most accustomed to writing position papers for abstract workshop events. We expected this approach might help governance and private sector contributors to be able to serve as commentators with the academics' papers as a ready target. We asked each academic to write on his or her topic in light of our overall objectives for the workshop. As our workshop guide for participants stated the objectives, we hoped to emerge from the workshop with:

- a comprehensive identification of emerging social issues arising because of innovation in aquaculture;
- a forecast regarding responsible governance models for aquaculture innovation.

As the event progressed, we found that we achieved much of our first objective, as you will see in contributions to each chapter. Yet we stalled as we looked to the second objective, for reasons which bear on the future for not just aquaculture, but nearly any discussion of the governance of technological innovation. Despite our making our goals clear to participants, in the pre-circulated participant guide and in our opening remarks at the workshop, it was very difficult to generate and sustain the kind of sharp, contested exchange encouraged by our model. This observation is by no means an indictment of our participants, who certainly spoke up and provided insights. Those insights were, however, closely connected with disciplinary or interest-based points of personal departure. Where governance options were identified, they often re-identified already known or enacted options or offered speculations about new directions with little consideration of how they might be implemented.

Why this was the case is not clear to us still, and conjectured explanations are given only limited support by our experience. At the close of the workshop some participants suggested that it was a valuable activity which should be conducted again. Perhaps another workshop would have expanded participants' knowledge of one another's positions, and gradual building of trust among participants might have enabled a freer dialogue and eventual focus on governance rather than individual departure-points and competences. This suggestion certainly has some merit, yet it is notoriously difficult in a culture which likes the idea that when we face a problem we should simply solve it, to return to a coalition of funding partners to say 'we're nearly there, but not quite, so we'd like to do this workshop again, more money please.' Moreover, enthusiasm for an iterative process is not a substitute for coming to an endpoint, even if that endpoint appears to be disagreement. As we have indicated above, we did not presuppose that consensus or some other form of agreement could or would result from our approach, and we accepted the possibility that disagreement might be ongoing, and at best managed or mitigated. In the workshop, our efforts to engage our two guiding goals left us quite distant from group agreement on a list of governance mechanisms to be given further consideration as options from which final choices might be made. We conclude now that the problem of aquaculture governance is even more complex than we initially thought. Our progress has come up short not just against the question of the right governance mechanisms, but the prior and broader question of a common list or 'menu' of mechanisms. This outcome can be attributed to the sheer diversity of the views regarding the nature of aquaculture, aquaculture innovation, and what counts as sustainable aquaculture environments and communities. We expected this conclusion to the extent that transdisciplinary approaches often excel at clarifying questions even when solutions to those questions remain out of reach. Yet we hoped to clarify the question of the menu of viable governance options somewhat further than we did. This volume nonetheless represents the progress we made, and leaves a foundation for further discussion.

When and where that discussion will be held remains unclear, as the demise of AquaNet has left Canada without any single organization to coordinate integrated, inter-disciplinary, multi-stakeholder collaborative research in aquaculture. Compared to ten years ago, we are now further behind because we lack institutions that support the process of contested exchange. At this point, it seems unlikely that commonality between worldviews about aquaculture will arrive at agreement on at least the parameters of the 'wicked policy problem' posed by aquaculture. A wicked problem can be controlled if one learns about its contours and adapts accordingly, but aquaculture is becoming a worst-case wicked problem now that public institutions supporting its development have been dissolved. From our academic point of view it is difficult to feel much immediate optimism regarding development of new governance options for aquaculture. The efforts of federal and provincial departments are now apparently focused on the Aquaculture Framework Agreement with what we fear might be too little regard for its connection to innovation and its social consequences. Meanwhile the federal department of Fisheries and Oceans Canada finds its attention drawn to other policy initiatives including Canada's renewed determination to assert its Arctic sovereignty. As these policy initiatives press on, we expect that much will continue as before: research will in fact be translated from laboratories to end-users to develop new aquaculture products and services. Likewise, we anticipate continued contributions of private sector funds in research partnerships, all without any corresponding increase in work to understand whether and how the resulting innovations can be managed and governed to provide promised social and economic benefits for Canadians and the citizens of developing states. We look forward nonetheless to being proven wrong about our forecast, since our being wrong will be evidence that aquaculture governance has at last received the attention it deserves.

From the Workshop to the Book, and Beyond

The process of converting a workshop into a book is rarely straightforward. It is even more difficult when what is billed as a frank and open exchange is formalized into chapters and commentaries. Three participants from the workshop declined our invitation to develop their papers further for contribution to the book, but we were able to bring new contributors to join the dialogue. On balance we remained very near our original intentions, maintaining the model of contested change even as participants varied.

In addition to the contributions from authors of principal paper writers and commentators, we have included as an appendix the report of the workshop facilitator, Mr Jacques Paynter of AMEC, Inc. His report provides a snapshot of the discussion on the day of the workshop. The report offers readers a useful contrast with the extended papers collected in this volume. We have also solicited an epilogue from Dr Barry Costa-Pierce, University of Rhode Island. We approached Dr Costa-Pierce to serve as epilogist in light of his familiarity with the Canadian context, due in part to service on AquaNet's International Advisory Board, and the special perspective he offers as an American operating in a very similar yet slightly different innovation system and society.

In closing, we are mindful that no interdisciplinary project can grow toward transdisciplinary insight without the good will and contributions of a very diverse group of persons for whom reaching beyond the ordinary can be a thankless task. We owe thanks to each of the workshop participants and book contributors for having agreed to join, and for having completed, a sometimes difficult process.

Without the research funding support of AquaNet, neither the workshop nor the subsequent development of the project would have happened. The same can be said for Genome Canada, although the story there is somewhat more complex. Castle led a research project on transgenic fish in aquaculture, supported by the first round of Genome Canada, administered through the Ontario Genomics Institute. More recently, Culver's research on social issues within the Cod Genome Project has been supported through the third round of Genome Canada funding, administered by Genome Atlantic. Similarly, Castle continues work on innovation systems, including aquaculture, supported through the third round of Genome Canada funding administered by the Ontario Genomics Institute. Each of Castle and Culver's research has greatly benefitted from interaction with investigators in the Genome Canada-funded Pleurogene Project (http://pleurogene.ca/) and Consortium for Genomic Research in All Salmonids Project – cGRASP (http://web.uvic.ca/cbr/grasp/) who we thank for sharing their views on the relation of aquaculture, innovation, and social transformation.

References

Aboriginal Aquaculture Association. 2007. http://www.aboriginalaquaculture.com/home.htm. Accessed August 17, 2007.

British Columbia Court of Appeal. 2007. Creative Salmon Company Ltd. v. Staniford. http://www.courts.gov.bc.ca/jdb-txt/sc/07/00/2007bcsc0062.htm. Accessed August 17, 2007.

British Columbia Pacific Salmon Forum. 2007. http://www.pacificsalmonforum.ca/. Accessed August 17, 2007.

Burke E. 1963. Letter to a Noble Lord on the Attacks Made Upon Mr. Burke and His Pension, in the House of Lords, by the Duke of Bedford and the Earl of Lauderdale, Early in the Present Session of Parliament 1795. In P.J. Stanlis, ed. *Edmund Burke: Selected Writings and Speeches*. Garden City, NY: Anchor Books, p. 561.

Canadian Broadcasting Corporation (CBC) News. Environmentalist defamed B.C. fish farm company: judge. http://www.cbc.ca/news/story/2007/01/17/bc-lawsuit.html. Accessed August 17, 2007.

Coastal Alliance for Aquaculture Reform. 2007. http://www.farmedanddangerous.org/. Accessed August 17, 2007.

Costa-Piece, B. 2002. *Ecological Aquaculture: The Evolution of the Blue Revolution*. Second Edition. Oxford: Blackwell Science.

Department of Fisheries and Oceans Canada (DFO). 2001. Report of the Task Force on Fostering a Sustainable Salmon Farming Industry for Atlantic Canada. http://www.mar.dfo-mpo.gc.ca/communications/maritimes/taskforce-e.html#fnt3. Accessed August 17, 2007.

Department of Fisheries and Oceans Canada (DFO). 2005. Overview: Qualitative Research Exploring Canadians' Perceptions, Attitudes and Concerns Towards Aquaculture. http://www.dfo-mpo.gc.ca/misc/focus-aquaculture_e.htm. Accessed September 16, 2007.

Department of Fisheries and Oceans Canada (DFO). 2006a. 2005–2006 Department Performance Reports. http://www.tbs-sct.gc.ca/dpr-rmr/0506/FO-PO/fo-po03_e.asp. Accessed August 17, 2007.

Department of Fisheries and Oceans Canada (DFO). 2006b. Roles and Responsibilities. http://www.dfo-mpo.gc.ca/aquaculture/governance/role_e.htm. Accessed August 17, 2007.

Department of Fisheries and Oceans Canada (DFO). 2007a. Fisheries and Oceans Canada Fast Facts. http://www.dfo-mpo.gc.ca/communic/facts-info/facts-info_e.htm. Accessed August 17, 2007.

Department of Fisheries and Oceans Canada (DFO). 2007b. Centres of Expertise. http://www.dfo-mpo.gc.ca/science/coe/index_e.htm. Accessed September 16, 2006

Department of Fisheries and Oceans Canada (DFO). 2007c. Aquaculture Collaborative Research and Development Program (ACRDP) – Program Information. http://www.dfo-mpo.gc.ca/science/aquaculture/acrdp_e.htm. Accessed August 17, 2007.

Food and Agriculture Association of the United Nations (FAO). 2007. *The State of World Fisheries and Aquaculture*. Rome: FAO.

Food and Agriculture Association of the United Nations (FAO). 2007. Glossary of aquaculture. http://www.fao.org/fi/glossary/aquaculture/default.asp?lang=en. Accessed August 17, 2007.

Genome Canada. 2004. *Genome Canada Annual Report 2003–2004*. http://www.genomecanada.ca/xcorporate/publications/0304_AR_English.pdf. Accessed August 17, 2007.

Government of Canada. 2007. Mobilizing Science and Technology to Canada's Advantage. http://www.ic.gc.ca/cmb/welcomeic.nsf/vRTF/PublicationST/$file/S&Tstrategy.pdf. Accessed August 17, 2007.

Hampshire S. 2001. *Justice Is Conflict*. Princeton: Princeton University Press.

Kitasoo/Xaixais First Nation. 2007. Fisheries. http://www.kitasoo.org/fisheries/index-fisheries.html. Accessed August 17, 2007.

Meyers B. and R.A.Worm. 2003. Rapid worldwide depletion of predatory fish communities. *Nature*. 423:280–3.

Office of the Prime Minister. 2007. Prime Minister releases national science and technology strategy to strengthen Canada's economy. http://www.pm.gc.ca/eng/media.asp?id=1656. Accessed August 17, 2007.

Organization for Economic Cooperation and Development. 2001. *The Well-being of Nations: The Role of Human and Social Capital*. Paris: OECD Publishing Centre for Educational Research and Innovation.

Organization for Economic Cooperation and Development. 2007. Glossary of Statistical Terms. http://stats.oecd.org/glossary/detail.asp?ID=6865. Accessed August 17, 2007.

Rittel H. and M.Webber. 1973. *Dilemmas in a General Theory of Planning. Policy Sciences 4*. Amsterdam: Elsevier Scientific Publishing. pp. 155–159.

Ruse, M and D. Castle. 2002. *Genetically Modified Food: Debating Biotechnology*. Amherst, NJ: Prometheus Press.

The Economist. 2003. The promise of a blue revolution. August 7, 2003.

The Tyee. 2007. Economy vs. ecology in fish farms debate. http://thetyee.ca/News/2007/05/24/FishFarm2/. Accessed August 17, 2005.

Turtle Island Native Network. 2006. UBCIC Chiefs Council Passes UBCIC Resolution 2006–42 Calling For Fish Farm Moratorium. http://www.turtleisland.org/discussion/viewtopic.php?p=7804#7804. Accessed August 17, 2007.s

World Bank. 2007. Governance. http://go.worldbank.org/MKOGR258V0. Accessed August 17, 2007.

World Fish Center. 2007. www.worldfishcenter.org. Accessed August 17, 2007.

Worm B., Barbier E.B., Beaumont N., Duffy J.E., Folke C., Halpern B.S., Jackson J.B.C., Lotze H.K., Micheli F., Palumbi S.R., Sala E., Selkoe K.A., Stachowicz J.J., and R. Watson. 2006. Impacts of biodiversity loss on ocean ecosystem services. *Science*. 314:787–90.

Part I
Animal Welfare in Aquaculture

Introduction

There is growing scientific evidence that many of the most common finfish species farmed throughout the world have the neuro-physiology capacity to feel pain, a capacity which leads some to hypothesize that fish suffer (Chandroo et al. 2004). What this implies for the aquaculture industry is uncertain, however. Sport fishers, and more recently aquaculturists, have faced pressure from animal welfare activists to halt, or at least drastically modifiy, their practices with respect to finfish. It is true that many aquaculture finfish species compete aggressively, kill and eat each other, have diseases and parasites, are sometimes predated upon, and are ultimately slaughtered for human consumption. But until very recently, aquaculture finfish species were below a moral water line for welfare consideration because the best arguments from analogy with terrestrial farmed animals like chickens and pigs were not very compelling in the case of fish. Now, with better scientific evidence suggesting that fish have the capacity to suffer, animal welfare activists are stepping up their efforts to draw media attention to welfare in aquaculture, and to persuade governments to regulate the industry.

This section begins with Felicity Huntingford's evaluation of the scientific evidence supporting claims that finfish have the capacity to suffer. Shellfish such as mussels seem to be acknowledged as too simple to warrant discussion of pain and suffering. Although there have been doubts raised (Rose 2002), teleost fish including salmon, char and trout, appear to have pain receptors called nociceptors that allow them to avoid noxious stimuli (Sneddon et al. 2003). While fish may not seem radically distinct from other vertebrates in this respect, they are otherwise physiologically and behaviourally distinct. Novel experimental methodologies have thus been devised to understand how these physiological capacities bear on the actual experience of suffering on the part of fish. These studies are producing a composite picture of fish welfare, Huntingford suggests, in which evidence from physiological, behavioural and hormonal changes in fish supports a general hypothesis that fish can suffer. On the basis of the evidence that she gathers, Huntingford suggests ways that the aquaculture industry adapts its management practices to lessen the potential for fish suffering.

Rudimentary capacity for the sensation of pain and physiological or behavioural tendancies are enough for some commentators to suggest a particular ethical response: that fish be given the benefit of the any doubts about their capacity for suffering (Lund et al. 2007). Gilly Giffin responds to Huntingford by noting that ethical consideration of animal welfare predates the scientific evidence. With new scientific evidence to make a case that fish suffer, what has really changed is the potential for implementation of governance mechanisms for the aquaculture industry. Griffin describes many of these initiatives, including the precautionary approach adopted by the Canadian Council on Animal Welfare, frequently illustrating the historical backdrop to welfare in aquaculture.

Media attention paid to aquaculture husbandry, Griffin rightly observes, is typically tied to food safety concerns, and to a much lesser extent fish welfare itself. The connection between the treatment of aquaculture species and human health might include aquaculture species other than finfish species (Clarke 2001). This situation, as it turns out, works well for the aquaculture industry. As Sunil Kadri points out in his response to Huntingford, the aquaculture industry has already taken steps to meet her recommendations for science-based husbandry. Aquaculturists now choose species that flourish in aquaculture and breed selectively for this trait. Second, the industry adopts innovations that arguably improve welfare by reducing stress on fish through passive size grading, and through better slaughter techniques. Finally, and noteworthy from the standpoint of thinking about how the connection between pain, sentience, suffering and welfare is now thought to have a basis in science, Kadri argues that the aquaculture industry supports and uses research to develop on-farm indices of welfare. Why might industry do all of this? Because physically healthy fish make for economically viable businesses with happy customers.

Perhaps, in light of what Huntingford, Griffin and Kadri say, animal welfare issues pose less difficulty for the aquaculture industry than has been generally thought. After all, the industry is science and technology-based, so where evidence of potential productivity and quality improvements coincide with welfare considerations, industry compliance is likely to be high. In fact, one can expect that under many circumstances industry will self-regulate according to guidelines it develops from its own learning. These approaches will continue to evolve alongside other standards for research (for example, The Times Online 2007) and for industry (for example, the U.K. RSPCA 2007). Where attitudes about, and evidence for, fish welfare do not align with industry practices, direct regulation may be necessary, in which case the aquaculture will have to adapt to its social conditions. One thing that favours the industry as a whole is timing. People for the Ethical Treatment of Animals (PETA 2007) has driven many food manufacturers to change their industries after their production and value chains became firmly entrenched. The aquaculture industry might be able to avoid public relations catastrophes by being a late-comer to the debate.

References

Chandroo K.P., Duncan I.J.H. and R.D. Moccia. 2004. Can fish suffer?: Perspectives on sentience, pain, fear and stress. *Applied Animal Behaviour Science*. 86:225–250.

Clarke, T. 2001. Stressed oysters sicken. *News@Nature*. May 11, 2001. http://www.nature.com/news/2001/010511/full/news010517-1.html. Accessed July 10, 2008.

Lund L., Mejdell C.M., Röcklinsberg H., Anthony R., Håstein, T. 2007. Expanding the moral circle: farmed fish as objects of moral concern. *Diseases of Aquatic Organims*. 75:109–118.

People for the Ethical Treatment of Animals (PETA). 2007. Fishing hurts. http://fishinghurts.com/. Accessed August 15, 2007.

Rose J.D. 2002. The neurobehavioral nature of fishes and the question of awareness and pain. *Reviews in Fisheries Science*. 10:1–38.

Royal Society for the Prevention of Cruelty to Animals (RSPCA). 2002. RSPCA Welfare Standards – Salmon. http://www.rspca.org.uk/servlet/Satellite?pagename=RSPCA/RSPCARedirect&pg=welfarestandards&marker=1&articleId=1125906255589. Accessed August 15, 2007.

Sneddon L.U., Braithwaite V.A. and M.J. Gentle. 2003. Do fishes have nociceptors? Evidence for the evolution of a vertebrate sensory system. *Proceedings of the Royal Society of London B*. 270:15–1120.

The Times Online. 2007. 'Stressed' fish get bill of rights. July 1, 2007. http://www.timesonline.co.uk/tol/news/uk/ science/article2010193.ece. Accessed August 15, 2007.

Chapter 1
Animal Welfare in Aquaculture

Felicity Huntingford

Introduction

The welfare of farmed fish has become an increasingly important issue for the general public (both those who eat farmed fish and those who do not) and for governments and regulators. In response to such concerns, the aquaculture industry is taking steps to ensure good welfare in farmed fish, most particularly by setting up Codes of Practice for responsible aquaculture that take fish welfare into account. An example is provided by the Federation of European Aquaculture Producers, which includes "respect the considerations for welfare that apply to the species being raised" as one of their guiding principles for responsible aquaculture (FEAP). Clearly it is important that any action taken should be based on reliable information on and proper understanding of fish welfare, otherwise it is unlikely to be successful. This is a very complex subject with many areas of uncertainty and the aim of this chapter is to examine some of these, including: how welfare is defined, whether fish have the capacity for suffering, how we can identify and measure fish welfare, what scientific study can tell us about the impact of production systems on fish welfare and possible strategies for improving the welfare of farmed fish. For clarity, it should be noted that my perspective is that of a biologist, though biology is only one of a number of academic disciplines with a contribution to make in this context. In addition, I concentrate on finfish, though the welfare of shellfish (particularly decapod crustaceans) is also a matter of concern.

Definitions of Animal Welfare

The concept of animal welfare is complex and hard to define and the relationship between welfare and suffering even more so (Dawkins 2006; FSBI 2002; Duncan 2006; Fraser 1999). It might therefore be argued that it is not an appropriate

F. Huntingford
Fish Biology Group, Division of Environmental & Evolutionary Biology,
Institute of Biomedical & Life Sciences, University of Glasgow, UK
e-mail: f.huntingford@bio.gla.ac.uk

K. Culver, D. Castle (eds.), *Aquaculture, Innovation and Social Transformation*,

© Springer Science+Business Media B.V. 2008

subject for scientific research or an appropriate principle to use for driving policy with respect to agriculture. In many areas of science and science-based policy, however, words in everyday speech are used to describe complex things that are actually hard to define precisely. Aggression research is a case in point; acres of text have been devoted to arguing about exactly what aggression means, but repeated attempts to replace it with new, unloaded terms have failed because the word encapsulates something that, though complex, is real, important and amenable to scientific study (Huntingford & Turner 1986). The same is true for animal welfare. The important point is to define exactly how such words are being used in any given case and, when disagreements occur, to check that these are not the result of using different definitions before assuming that people are getting genuinely different results. As far as definitions of welfare are concerned, most fall into one of three broad categories (FSBI 2002). It is important to note that none of these is right or wrong; they simply capture different aspects of what the word means.

Feelings-based definitions are couched in terms of subjective mental states, good welfare requiring that the animal is free from negative experiences (e.g. pain or fear) and has positive experiences (e.g. companionship). This definition depends on the controversial point that animals have conscious feelings, which will be discussed briefly later in this chapter.

Function-based definitions depend on the animal's being able to adapt to its environment and keep in good health, with all its biological systems functioning appropriately. This definition rests on the assumption that an animal whose physiological systems are functioning poorly is suffering, and the converse.

Nature-based definitions are centred on the idea that good welfare requires the animal to be able to lead a natural life and express its natural behaviour. Put another way, animals suffer if they cannot express the full repertoire of behaviour that they show in the wild. This view is based partly on the general principle that what is natural is good and partly on the more specific assumption that animals have "behavioural needs" that must be met if they are not to suffer. This may or may not be true, depending on exactly how each specific aspect of behaviour is organised and controlled. For example, if wild Atlantic salmon migrate long distances at sea because they are tracking food, there is no reason to believe that they suffer when they are held in cages, provided they have plenty of food. If they have some sort of intrinsic drive to swim long distances regardless of food supply, however, then confinement may cause the fish to suffer. Knowledge on this point can only be gained by empirical research; one cannot extrapolate across species, across life history stages or across behavioural systems.

Welfare and the Five Freedoms

Given the difficulty of defining animal welfare, an alternative approach to the problem of protecting animals from suffering is to identify conditions that must to be fulfilled if they are to have a good quality of life (and death). These famous "Five Freedoms" (FAWC 1996) or "domains of welfare" (Mellor & Stafford 2001) have been expressed in various ways, but one influential framework, derived mainly for mammals and birds, is as follows:

- *Domain 1. Water and food deprivation, malnutrition.* Animals should have ready access to fresh water and an appropriate diet in sufficient quantities and with a composition that maintains full health and vigour.
- *Domain 2. Environmental challenge.* Animals should have a suitable environment, including shelter and a comfortable resting area, whether outdoors or indoors.
- *Domain 3. Disease, injury and functional impairment.* Disease should be prevented or rapidly diagnosed and treated.
- *Domain 4. Behavioural/interactive restriction.* Animals should have sufficient space, proper facilities and where appropriate, the company of the animal's own kind.
- *Domain 5. Mental and physical suffering.* Conditions that produce unacceptable levels of anxiety, fear, distress, boredom, sickness, pain, thirst, hunger and so on should be minimised.

This conceptual framework is useful in flagging areas of potential concern, but making a direct link with welfare is not necessarily simple. It seems fairly obvious that an animal with a gaping wound (and therefore failing in domain 3) experiences poor welfare, but identifying the conditions in which animals feel (for example) fear and distress (and therefore fail in domain 5) is much more tricky. Nor is it simple to determine whether an animal that is denied the opportunity to fight (for example, as is often the case for male Siamese fighting fish, *Betta splendens* kept for ornamental purposes, relating to domain 4) is experiencing good or poor welfare. In addition, in the case of fishes, the framework needs to be adapted to allow for special features of this group. For example, fish do not maintain a constant body temperature and in nature often experience long periods without food with no adverse consequences, so domain 1 is perhaps not so critical as it is for birds and mammals. On the other hand, since fish are in close contact with their environment through the huge area of their gills, water quality (which would come under domain 2) is of central importance to their welfare.

The Capacity of Fish to Experience Suffering

In a recent review, Rose (2002) argues that fish cannot experience pain and suffering because they lack the behavioural capacity and the necessary brain structures for conscious experience. The issues are complex and the argument continues (Braithwaite & Huntingford 2004; Chandroo et al. 2004), but briefly Rose's position is that fish have limited behavioural capacities, are not capable of complex learning and mental processes and tend to behave in a reflex-like way. Therefore, if a fish jumps when it gets an electric shock this is no different from when it strikes in response to food or shows courtship in response to a receptive mate. Additionally, fish brains are very different from those of mammals, and in particular do include the specialised brain structure required for conscious perception of pain in humans (the frontal lobe of the neocortex). This position has been hotly debated and is certainly not resolved, but from recent research we know that:

(1) Fish (rainbow trout) certainly have pain receptors (strictly speaking these should be called nociceptors, so as not to pre-judge the issue of whether fish experience pain) that look and react like the ones humans have and that respond to a range of harmful stimuli. Stimulating such receptors with noxious chemicals increases breathing rate (indicative of stress) and makes the fish show abnormal, rubbing behaviour (commonly seen in mammals when in pain) and suppresses a learned response to food (Sneddon et al. 2003). Since these effects are reversed by analgesia, they clearly perceive noxious stimuli as painful and aversive (Sneddon 2004).
(2) Although the brain of fishes develops and is organised differently from that of other vertebrates, it is still possible to identify many of the same structures (albeit looking different and in slightly different places), doing many of the same jobs. For example, detailed neuroanatomical studies indicate that lateral pallium (part of the forebrain in fish) is homologous to the hippocampus of mammals, even though it looks different and is in a different place. Recent behavioural studies have shown that the lateral pallium is essential for processing spatial information and forming maps of the environment, in much the same way as is the mammalian hippocampus (Broglio et al. 2003). Even if this were not the case, there are many examples of convergent evolution in which animals of very different kinds do the same job with different structures (birds fly with their arms and bats fly with their fingers). So not having a frontal cortex does not necessarily mean that fish do not have brain mechanisms that can generate the experience of suffering, otherwise we would have to conclude that only humans and great apes can suffer.
(3) Fish behaviour is in fact complex and flexible (Laland et al. 2003). For example, fish can recognise companions individually and treat them differently depending on what happened in previous encounters (Griffiths 2003) and learn to avoid places where they had bad experiences (Kelley & Magurran 2003). They are capable of learning complex tasks (for example they can form mental maps of the area in which they live (Braithwaite & de Perera 2006) and of remembering over long periods.

Therefore, fish are clearly not feeling-less robots, but probably do have the capacity to experience suffering, albeit in a different form from human suffering. On these grounds, fish suffering and welfare have real meaning and are things about which we should be concerned.

Natural Responses to Adverse Conditions and How Scientists Measure Fish Welfare

Even if we can agree what is meant by animal welfare and that fish have the capacity for suffering, measuring such a complex phenomenon accurately and objectively (which we must do for effective research and monitoring) is a challenging problem. However, by studying the natural responses of animals to adverse conditions, particularly changes in their physiology, health and behaviour, we can develop tools for assessing their welfare status. Physiological stress responses (release of the hormones adrenaline and cortisol, with resulting changes in a number of body systems that prepare for activity) are part of an animal's natural reaction to challenging conditions (Barton 1997; Wendelaar Bonga 1997; FSBI 2002). Rapid physiological changes are therefore important indices of impaired welfare, but the link between physiological stress responses and welfare is complex. For example, fish are able to anticipate the arrival of food (Lagud & Reebs 2000) and such food anticipation is known to generate short term physiological activation in anticipation of feeding in mammals (Meynard et al. 2005); it is not clear that this should be regarded as an indication of poor welfare.

Behavioural responses such as freezing, avoidance or escape form an early part of the natural response that animals show to adverse conditions, so behavioural traits can be used as direct indicators of impaired welfare (FSBI 2002). Additionally, a number of ingenious behavioural tests have been developed to allow animals to express their natural preferences and, though relatively little used in fish welfare research, such tests have given us valuable insights into the circumstances that promote/detract from animal welfare (Dawkins 2006).

In addition to, and in some cases as a result of, such immediate physiological and behavioural responses, adverse conditions have longer-term consequences (Barton 1997; Wendelaar Bonga 1997; FSBI 2002). These include reduced growth, suppressed reproduction and impaired immune function and can be seen as natural adaptations. Immediate survival in unfavourable conditions are more important for fitness that future growth and reproduction and maintaining health (a dead animal cannot do any of these useful things). Patterns of growth and reproduction and signs of disease are therefore also used as indicators of welfare, though here again the links are complex (Dawkins 2006). For example, if an animal has overt symptoms of disease, it seems reasonable to infer that it is in a poor state of welfare and, since chronic stress is known to impair the body's defences, the presence of disease may be an indirect indicator of poor conditions. The converse is not necessarily true, however, since the welfare of a healthy animal may be compromised in other ways, for example if a social animal is held in isolation.

Such complexities mean that none of these ways of assessing welfare has all the answers; all are to a greater or lesser extent influenced by factors unrelated to welfare. For example, drastic falls in energy reserves may be caused by stress from unsuitable housing conditions, but may also reflect life history stage (some juvenile salmon undergo natural anorexia during the winter, refuse food that is presented to them and lose weight as a consequence (Metcalfe et al. 1988)) or reproductive status (during migration and spawning, Atlantic salmon lose 60–70% of their energy reserves, (Jonsson et al. 1997)). Such dramatic falls in energy reserves are part of the natural life cycle of the species concerned and there is no reason to think that their welfare would be improved by preventing the events that give rise to them.

For this sort of reason, many of the commonly used indices of welfare are noisy variables and welfare researchers often collect data on several different systems, for example looking at the physiology, health and body condition in the same subjects. This gives a suitably broad picture, but raises the problem of how to combine information from different biological systems. Statisticians have developed tools for examining and condensing such multivariate data sets, and these can and have been used to integrate different indices of welfare. Figure 1.1a shows how Principal Component Analysis can be used to integrate four such indices (fin and body condition and plasma glucose and cortisol levels) into a single welfare score for Atlantic salmon in a set of experimental production cages. Figure 1.1b shows that fish identified independently by experienced fish farmers as having poor welfare had

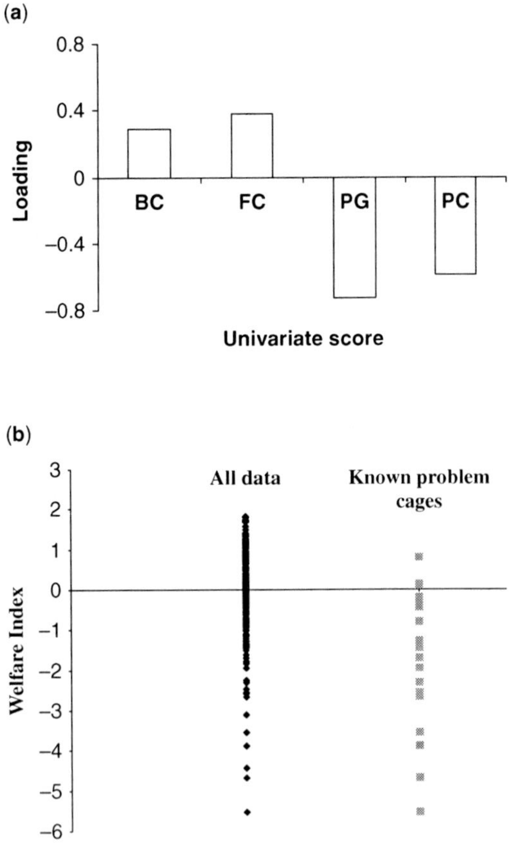

Fig. 1.1 Using principal components analysis to integrate several independent indices of welfare. (**a**) Loadings for (or weightings given to) the 4 univariate measures on the first Principal Component, used as a measure of welfare. BC=condition factor; FC = fin condition; PG = plasma glucose concentration; PC = plasma cortisol concentration (after Turnbull et al. 2005). (**b**) Welfare scores for all fish in the study, compared to those for fish independently identified by the farm manager as causing concern

low welfare scores using this integrated index; this provides indirect validation of this index of welfare for these fish in these conditions (Turnbull et al. 2005).

The Impact of Intensive Farming on Fish Welfare

In principle, one might argue that farmed fish experience better conditions than do wild fish, in that for much of the time, they get a plentiful supply of high quality food, are protected from predators and inoculated against or treated for many diseases. They are often disturbed (and perhaps frightened) by human activities and ultimately they are killed, but wild fish too may be disturbed, frightened and eaten by predators. Against this, however, farmed fish are confined in tanks or cages (which arguably might impair welfare by preventing expression of an instinct for long-distance swimming), in a restricted space and at much higher densities than those usually experienced by wild fish. To be constructive, debate about which of these two points of view comes nearest to the truth must be informed by evidence rather than by assumptions. So what evidence is there that farming practices compromise the welfare of farmed fish?

Much research has been aimed at identifying whether particular aspects of fish husbandry compromise the welfare of farmed fish, using all or some of the measures of welfare mentioned above. Space precludes a complete review of this work in the present chapter; an overview can be found in the Fisheries Society of the British Isles Briefing Paper (FSBI 2002). To give just a few of many possible examples, transportation can induce physiological stress responses in various species (Barton 2000; Sandodden et al. 2001). Physical confinement in otherwise favourable conditions elicits a physiological stress response and impairs disease resistance (e.g. Barnett & Pankhurst 1998). Studies of physiological and behavioural stress responses tell us that all slaughter methods are stressful, but that some are less so than others (Robb et al. 2000).

Numerous studies have examined the effects of holding fish at high densities, and these illustrate just how complicated the factors influencing fish welfare can be. In the first place, effects depend on species and life history stage; for example. high stocking density may impair welfare in some species (e.g. trout, Ewing & Ewing 1995), but enhance it in others (e.g. Arctic charr, Jorgensen et al. 1993). Even where poor welfare has been found in fish held at high density, the causes may be indirect. For example, in rainbow trout, adverse effects of density are largely driven by poor water quality (Ellis et al. 2002), high densities being compatible with good welfare if water quality can be maintained. And even where an effect of density *per se* has been found, this may be small relative to other factors, and/or may interact with other factors in complex ways. In Atlantic salmon, fish held in production cages at densities above a threshold (c23 kg per m^3) show relatively low welfare scores (Fig. 1.2); below this threshold density has no effect. At all densities, other factors such as levels of disturbance are more important determinants of welfare than density (Turnbull et al. 2005).

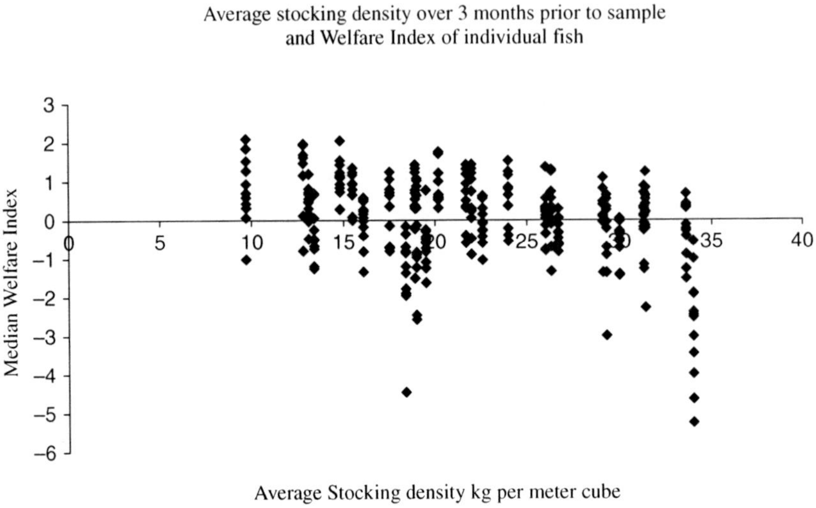

Fig. 1.2 Relationship between an integrated welfare score for farmed Atlantic salmon and mean stocking density over the preceding 3 months (from Turnbull et al. 2005)

Strategies for Improving Welfare in Aquaculture

With the continuing decline of wild fish stocks, aquaculture is likely to play an increasingly important role in providing high quality, healthy food for human consumption. It is also an important component of the economy of many countries, especially in isolated areas. What steps can be taken to protect the welfare of farmed fish while at the same time ensuring that this important industry continues to operate effectively? Because of the difficulties in mapping welfare directly onto holding conditions, touched on in the previous section, it is not possible simply to define husbandry conditions that ensure the welfare of species of farmed fish. This leaves three broad strategies, which are not mutually exclusive and which together could go a long way towards ensuring welfare-friendly aquaculture:

1. *Aquaculture could/should concentrate on species, strains or individuals that, for whatever reason, flourish in the farm environment.* Some species or strains are naturally better suited to intensive husbandry than are others and, arguably, these are the species and strains that should be used in aquaculture. For example, in several lakes, Arctic charr exist in two forms, a benthic form that feeds on bottom-dwelling macro-invertebrates and a pelagic form that feeds on zooplankton. The benthic form tends to be territorial, while the pelagic form is relatively unaggressive and, depending on circumstances, often forms shoals (Mikheev et al. 1996). The pelagic form would seem to be the more appropriate candidate for charr aquaculture. In species such as brown trout and Atlantic salmon that have been farmed intensively for a fair number of generations, there is now

strong evidence that fish from domesticated strains show weaker behavioural and physiological responses to disturbance than do their wild counterparts (Huntingford 2004). Where fish are being reared for food rather than for restocking purposes, and setting aside the problem of escapees and natural gene pools, from a welfare perspective, perhaps these are the fish we should be farming. Within species and strains, there is inherited variation in how individual fish respond to novelty and risk (Huntingford & Adams 2005), often linked to differences in stress physiology (e.g. rainbow trout, Pottinger & Carrick 2001). Where domesticated strains are not available, one possible approach might be to avoid using risk-averse fish by pre-screening fish at the start of the production cycle.

2. *The industry should develop husbandry systems that minimise adverse effects on welfare.* Trying to engineer the behavioural structure of farmed fish populations is complicated and requires a better understanding of the dynamics of fish social interactions that we currently have. An alternative approach is to use what we already know about the biology of fish to develop husbandry systems that minimise threats to welfare. Obviously, it is critical to ensure that fish get sufficient food with appropriate nutrients (based on our understanding of species-specific dietary needs) and live in good quality water (based on our understanding of species-specific physiology). Some less obvious possibilities are based on an understanding of fish behaviour. For example, based on studies of the response of Atlantic salmon in production cages to artificial underwater lights, Juell & Fosseidengen (2004) suggest that such lights could be used to promote even use of the available space within production cages, thereby reducing crowding in farmed fish. Like many other animals, fish adjust their aggressive responses to the costs of fighting (e.g. Grant 1997) and relatively simple manipulations that increase the costs of fighting among farmed fish, such as making fish swim against a current (Arctic charr; Jorgensen et al. 1993; Adams et al. 1995) or introducing older companions (Adams et al. 2000) can reduce aggression and promote even growth. The manner in which food is delivered also has a strong effect on the social environment experienced by farmed fish. In *Tilapia rendalli*, for example, aggressive, dominant fish can monopolise food when this is clumped, but not when it is well distributed (McCarthy et al. 1999). Using interactive feeders that deliver food whenever fish are hungry rather than at predefined, predictable times of day creates a farm environment in which there is much less competition for food (Fig. 1.3a; Andrew et al. 2002) and fish that are poor competitors are able to feed and grow well. As a final example, serotonin is a substance that is produced naturally in the central nervous system and, among other functions, serves to suppress aggression; serotonin activity is higher in the brains of subordinate fish compared to their dominant companions. This is part of a natural system that prevents fish from getting repeatedly involved in hopeless fights. Over a period of 7 days, juvenile rainbow trout, which are a highly aggressive species, fed a diet in which levels of tryptophan were supplemented by as little as 0.15% became significantly less aggressive (Winberg et al. 2001; Fig. 1.3b).

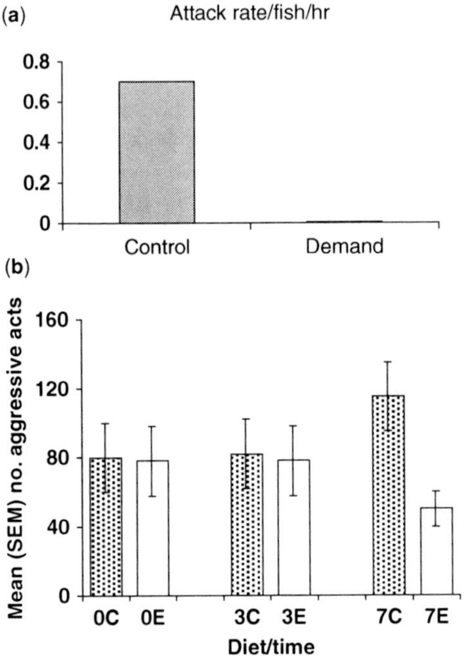

Fig. 1.3 Case studies of the application of behavioural theory to the welfare of farmed fish: (**a**) frequency of aggressive acts in farmed Atlantic salmon post-smolts fed by automatic feeder (control) or on demand (after Andrews et al. 2002). (**b**) Number of aggressive acts performed by juvenile rainbow trout in an intruder test after 0, 3 and 7 days on a diet supplemented with 0.15% tryptophan (after Winberg et al. 2001)

3. *The industry should develop, for each species and life history stage, simple but accurate indices of welfare that can be used on working farms as "early warning systems" for emerging welfare problems.* Based on what we know of the short- and medium-term responses of fish to adverse conditions, there are a number of such indices that could potentially be monitored in fish in production conditions without any need for disturbing them (except perhaps by placing cameras in cages). These include changes in colour, changes in ventilation rate, changes in patterns of swimming, reduced food intake, loss of body condition, damaged fins and slow growth.

Conclusion

It has been argued that research into the welfare of farmed fish is occupational therapy for biologists and is of no interest to fish farmers. It is certainly true that research into fish welfare poses a number of intriguing biological questions, but the answer to these questions can be relevant even to hard-pressed fish farmers. Of course, responsible farmers are concerned for the well-being of their stock as a matter of professional integrity. Additionally, over a wide range of conditions, because of the negative effects of poor conditions on growth, good welfare means good production. There may be a few areas where the two pull in different directions (for example, the best stocking density for maximising profit may be a bit above that for maximising welfare), but in general good farmers should be able to use the results of welfare research to improve production efficiency. Over and above this, there is increasing, legitimate, public concern for the welfare of farmed fish and regulators are responding to this concern. This means that, sooner or later, all fish farmers will have to pay explicit attention of the welfare of their fish; this is particularly the case because welfare status is becoming an additional aspect of product quality from which responsible aquaculture should benefit.

References

Adams, C.E., Huntingford, F.A., Krpal, J., Jobling, M. and Burnett, S.J. 1995. Exercise, agonistic behaviour and food acquisition in Arctic charr, *Salvelinus alpinus*. *Environmental Biology of Fishes* 43: 213–218.

Adams, C.E., Huntingford, F.A., Tumbull, J.F., Arnott, S.A. and Bell, A. 2000. Size heterogeneity can reduce aggression and promote growth in Atlantic Salmon farm, *Aquaculture International* 8: 543–549.

Andrew, J.E., Noble, C., Kadri, S., Jewell, H. and Huntingford, F.A. 2002. The effect of demand feeding on swimming speed and feeding responses in Atlantic salmon *Salmo salar L.* in sea cages. *Aquaculture Research* 33(7): 501–507.

Barnett, C.W. and Pankhurst, N.W. 1998. The effects of common laboratory and husbandry practices on the stress response of greenback flounder (*Rhombosolea tapirinia*, Günther 1862). *Aquaculture* 162: 113–329.

Barton, B. 1997. Stress in fish: past, present and future – a historical perspective. In: *Fish Stress and Health in Aquaculture, Society for Experiment Biology, Seminar Series 62*, Iwama, G.K., Pickering, A.D., Sumper, J.P. and Schreck, C.B., Eds. – 1–31, Cambridge: Cambridge University Press.

Barton, B.A. 2000. Salmonid fishes differ in their cortisol and glucose responses to handling and transport stress. *North American Journal of Aquaculture* 62(1): 12–18.

Braithwaite, V.A. and de Perera, T.B. 2006. Short range orientation in fish: how fish map space. *Marine and Freshwater Behaviour amd Physicology* 39: 37–47.

Braithwaite, V.A. and Huntingford, F.A. 2004. Fish and welfare: do fish have the capacity for pain perception and suffering. *Animal Welfare* 13: 587–592.

Broglio, C., Rodrigues, F. and Salas, C. 2003. Spatial cognition and its neural basis in teolost fishes. *Fish and Fisheries* 4: 247–255.

Chandroo, K.P., Yue, S. and Moccia, M.D. 2004. An evaluation of current perspectives on consciousness and pain in fishes. *Fish and Fisheries* 5: 281–295.

Dawkins, M.S. 2006. Through animal eyes: what behaviour tells us. *Applied Animal Behavior Science* 100: 4–10.

Duncan, I.J.J. 2006. The changing concept of animal sentience. *Applied Animal Behaviour Science* 100: 11–19.

Ellis, T., North, B., Scott, A.P., Bromage, N.R., Porter M. and Gadd, D. 2002. The relationships between density and welfare in farmed rainbow trout. *Journal of Fish Biology* 61: 493–531.

Ewing, R.D. and Ewing, S.K. 1995. Review of the effects of rearing density on the survival to adulthood for Pacific salmon. *Progressive Fish-Culturist* 57: 1–25.

Farm Animal Welfare Council. 1996. Report on the welfare of farmed fish. *FAWC*, Surbiton, Surrey.

Federation of European Aquaculture Producers (FEAP). www.feap.info; accessed: September 5, 2007.

Fraser, D. 1999. Animal ethics and animal welfare science: Bridging the two cultures. *Applied Animal Behaviuoural Sciences* 65: 171–189.

FSBI. 2002. Fish Welfare Briefing Paper 2, Fisheries Society of the British Isles, Granta Information Systems, 82A High Street, Sawston, Cambridge CB2 4H.

Grant, J.W.A. 1997. Territoriality. In *Behavioural Ecology of Teleost Fishes*, J-G. J. Godin, Eds. – 81–98, Oxford, UK: Oxford University Press.

Griffiths, S.W. 2003. Learned recognition or conspecifics by fishes. *Fish and Fisheries* 4(3): 256–268.

Huntingford, F.A. and Turner, A.K. 1986. Animal conflict. London: Chapman & Hall.

Huntingford, F.A. and Adams C.E. 2005. Behavioural syndromes in farmed fish: implications for production and welfare. *Behaviour* 142, Pt 9–10: 1207–1221.

Huntingford, F.A. 2004. Implications of domestication and rearing conditions for the behaviour of cultivated fishes. *Journal of Fish Biology* 65, Suppl A: 122–142.

Jonnson, N., Johnsson, B. and Hansen, L.P. 1997. Changes in proximate composition and estimates of energetic costs during upstream migration and spawning in Atlantic salmon. *Journal of Animal Ecology* 66: 425–436.

Jorgensen, E.H., Christiansen, J.S. and Jobling, M. 1993. Effects of stocking density on food intake, growth performance and oxygen consumption in Artic charr (*Salvilinus alpinus*). *Aquaculture* 110: 191–204.

Juell, J.E. and Fosseidengen, J.E. 2004. Use of artificial light to control swimming depth and fish density of Atlantic salmon (*Salmo salar*) in production cages. *Aquaculture* 233: 269–282.

Kelley, J.L. and Magurran, A.E. 2003. Learned predator recognition and anti-predator responses in fishes. *Fish and Fisheries* 4(3): 216–226.

Lagud, M. and Reebs, S.F. 2000. Food-anticipatory activity of groups of golden shiners during both day and night. *Canadian Journal of Zoology-Revue Canadienne de Zoologie* 78(5): 886–889.

Laland, K.N., Brown, C. and Krause, J. 2003. Learning in fishes: from three-second memory to culture. *Fish and Fisheries* 4(3): 199–202.

McCarthy, I.D., Gair, D.J. and Houlihan, D.F. 1999. Feeding rank and dominance in Tilapia rendalli under defensible and indefensible patterns of food distribution. *Journal of Fish Biology* 55(4): 854–867.

Mellor, D.J. and Stafford, K.J. 2001. Integrating practical regulatory and ethical strategies for enhancing farm animal welfare. *Australian Veterinary Journal* 79: 762–768.

Metcalfe, N.B., Huntingford, F.A. and Thorpe, J.E. 1988. Feeding intensity, growth rates, and the establishment of life-history patterns in juvenile Atlantic salmon *Salmo salar*. *Journal of Animal Ecology* 57: 463–474.

Meynard, M.M., Valdes, J.L., Recabarren, M., Serron-Ferre, M. and Torrealba, F. 2005. Specific activation of histaminergic neurons during daily feeding anticipatory behavior in rats. *Brain Behavioural Research* 158: 311–319.

Mikheev, V.N., Adams, C.E., Huntingford, F.A. and Thrope, J.E. 1996. Behavioural responses of benthic and pelagic Arctic charr to substratum heterogeneity. *Journal of Fish Biology* 49(3): 494–500.

Pottinger T.G. and Carrick T.R. 2001. Stress responsiveness affects dominant-subordinate relationships in rainbow trout. *Hormones & Behavior* 41(3): 419–427.

Robb, D., Kestin, S. and Lines, J. 2000. Progress with humane slaughter. *Fish Farmer* (2000) November/December, 41.

Rose, J.D. 2002. The neuro-behavioural nature of fishes and the question of awareness. *Environmental Ethics* 6: 44–50.

Sandodden, R., Finstad, B. and Iversen, M. 2001. Transport stress in Atlantic salmon (*Salmo salar L.*): anaesthesia and recovery. *Aquaculture Research* 32(2): 87–90.

Sneddon, L.U. 2004. Evolution of nociception in vertebrates: comparative analysis of lower vertebrates. *Brain Research Reviews* 46(2): 123–130.

Sneddon, L.U., Braithwaite, V.A. and Gentle, M.J. 2003. Do fish have nociceptors? Evidence for the evolution of a vertebrate sensory system. *Proceedings of the Royal Society London B* 270: 1115–1121.

Tanck. M.W.T., Vermeulen, K.-J., Bovenhuis, H. and Komen, H. 2001. Heredity of stress-related cortisol response in androgenetic common carp (*Dyprinus carpio* L.). *Aquaculture* 199: 283–294.

Turnbull, J.F., Bell, A., Adams, C.E., Bron, J. and Huntingford, F.A. 2005. Stocking density and welfare of cage farmed Atlantic salmon: application of a multivariate analysis. *Aquaculture* 243: 121–132.

Wendelaar Bonga, S.E. 1997. The stress response in fish. *Physiological Reviews* 77: 591–625.

Winberg, S., Overli, O. and Lepage, O. 2001. Suppression of aggression in rainbow trout (Oncorhynchus mykiss) by dietary L-tryptophan. *Journal of Experimental Biology* 204(22): 3867–3876.

Chapter 2
Science and Governance Issues in Aquaculture Animal Welfare

Gilly Griffin

World aquaculture has grown tremendously during the last 50 years from a production of less than a million tonnes in the early 1950s to 59.4 million tonnes in 2004 (FAO 2004). In Canada, aquaculture production has trebled in the past decade from 66 to 155 metric tonnes (DFO 2005a). According to the Network of Centres of Excellence for Aquaculture (Aquanet 2004), Canada has a significant potential to increase its aquaculture production and expand into the culture of new species, as anticipated growth is expected to continue in step with population growth and dietary trends. Demand for seafood in North America alone is expected to increase by up to 40 per cent by 2010 (Gilbert 2002). As the industry continues to grow, it will have to address public concerns about the quality of the food supply. Recent food related concerns about contaminants, diseases and overall wholesomeness of food (e.g., BSE, avian influenza, levels of mercury, and genetically modified foods) have consumers seeking assurances regarding the source and methods of production of their food. The 2005 strategic plan from the Department of Fisheries and Oceans Canada acknowledges that it will need to work with the aquaculture industry and the Canadian Food Inspection Agency to build credible systems for assuring the quality and safety of farmed seafood products (DFO 2005b). These systems will include quality assurance and control, traceability and on-farm food safety, as well as the environmental sustainability of aquaculture operations. If this mirrors the systems for assuring quality and safety of terrestrial farmed animal products, then assuring quality will include assurance that food has been raised according to certain standards of animal welfare. For example, according to Agriculture and Agri-Food Canada, in addition to complying with legislation and being guided by the Codes of Practice, Canadian producers advance farm animal welfare through continuous learning and adaptation, and via on-farm quality assurance programs that include farm animal welfare components.

Felicity Huntingford has examined the "role" of animal welfare in aquaculture from her standpoint as a biologist in the fish biology group at the University of Glasgow. Dr Huntingford was one of the authors of the Fisheries Society of the

G. Griffin
Canadian Council on Animal Care, Ottawa, Ontario, Canada
e-mail: ggriffin@ccac.ca

K. Culver, D. Castle (eds.), *Aquaculture, Innovation and Social Transformation*,
© Springer Science+Business Media B.V. 2008

British Isles Briefing Paper on Fish Welfare published in 2002, which set out to provide information on fish welfare to assist readers in making informed decisions on a number of related issues (FSBI 2002). It could be argued that historically, Europe, and the UK in particular, has had a greater focus on animal welfare than North America. For example, the first *Cruelty to Animals Act* was enacted in the UK in 1835. In 1964, journalist Ruth Harrison drew attention to the increasing confinement of animals used in food production, resulting in a UK commission of inquiry, and ultimately producing the "Brambell Report" and the "Five Freedoms", designed to improve the manner in which livestock were being raised (Harrison 1964; Brambell 1965). The European focus on animal welfare continues to the present day; in the past year, Welfare Quality® (http://www.welfarequality.net/everyone), a European inter-disciplinary research project, has been established to address numerous issues related to safeguarding animal welfare in the food chain.

Recently, however, rapid changes have occurred in North America. In the United States, the retail sector has been a driving force behind the introduction of standards for animal welfare practices. The Humane Society of the US (HSUS) and the People for the Ethical Treatment of Animals (PETA) have been lobbying fast-food companies to source meat and dairy products from agricultural businesses that are more considerate of the welfare of the animals being raised for food. As well, surveys carried out for the food industry have pointed to consumers becoming more socially responsible and aware, and hence more critical of the processes used in food production (Bob Goldin, executive vice president of Technomic, a food industry research and consulting firm, http://www.technomic.com/facts/index.html cited by HSUS).

It seems likely that public attention is increasingly being drawn to farm animal welfare because of the links being made with environmental and food safety concerns (Bayvel, Abdul Rahman & Gavinelli 2005). As a result of consumer pressure, McDonald's Restaurants (U.S.), Burger King, Wendy's and some grocery chains have each developed separate, although similar, programs for assessment of animal welfare. Sensitive to public concern, each of these fast-food retailers has a statement on its website describing its policies in regard to animal welfare. In particular, Wendy's has a detailed list of requirements that its suppliers must follow (http://www.wendys.com/community/animal_welfare.jsp).

There is also a rapid move toward international harmonization of animal welfare standards, driven by international organizations, transnational companies, and pressure from world trade. In 1924, the OIE (Office International des Épizooties) was established and made contributions to the improvement of animal welfare through its involvement in epizootic disease control. Over the years, and as a result of a 1994 agreement on sanitary and phytosanitary measures to avoid the use of health problems as trade barriers, the standards, guidelines and recommendations of OIE became accepted as international references in relation to animal disease and zoonoses, most notably, the Terrestrial Animal Health Code (OIE 2006) and the Aquatic Animal Health Code (OIE 2007). In 2002, the 167 OIE member countries agreed that OIE should also become the leading international organization in the area of animal welfare. It has been speculated that certificates issued by the official veterinary services of an exporting country to provide evidence that the requirements of the importing country have been met, could also be used as a vehicle to convey information concerning animal welfare (Vallat 2005).

Definitions of Animal Welfare

The concept of animal welfare is hard to define (Chapter 1). It has been described as a term that has arisen in society to express ethical concerns about the quality of life experienced by animals, particularly animals that are used by human beings in production agriculture (Duncan & Fraser 1997). The term animal welfare does not express a scientific concept but points to three different approaches developed by animal welfare scientists in trying to ascertain the quality of life (or well-being) of an animal. These approaches are described as function-based, nature-based and feelings-based.

Function-based approaches to welfare relate mostly to the health of animals, typically the focus of the veterinary community. Arguably, the first focus of aquaculture should be raising healthy fish. This presents a number of challenges with respect to determining and implementing appropriate conditions. For example, there are currently few drugs registered for use in the aquaculture setting. While it is unethical, not to say illegal, to permit food to enter the market place unless drugs used to treat the fish have been found to be safe, and the withdrawal times have been adhered to, animal welfare considerations require treatment of diseased fish in a captive aquaculture situation, whether or not those fish might experience similar disease states in the wild.

Nature-based definitions centre on the importance of a "natural life" or the ability and capacity to express natural behaviors. Huntingford argues that knowledge of the intrinsic drives of fish, such as to swim long distances, regardless of food supply, is essential to understanding the impact of confinement, and that this can only be obtained by empirical research. However, Clubb & Mason (2003) looked at behavioural problems across a wide variety of mammals kept in zoos, and concluded that animals with large ranges in the wild did particularly poorly in captivity. It may therefore be wise to consider the natural habits of the fish when choosing particular species for adaptation to aquaculture.

The feelings-based school faces an especially daunting challenge when attempting to ascertain the quality of life of fish. Fish are phylogenetically distant, and hence physiologically very distinct, from humans; therefore, most people have difficulty in interpreting fish behaviours that might be useful in determining if, or what, a fish "feels." Marion Dawkins has argued for the importance of understanding welfare from the animal's point of view (Dawkins 1980). Emphasizing that what matters to the animal is the animal's experience, a considerable body of work has emerged based on behavioural indicators and preference testing, in particular for farm animals (see, for example, Duncan 2004).

Remembering that animal welfare arose from a genuine societal concern for the quality of life experienced by animals, particularly as agriculture became more industrialized, it is perhaps not surprising that so far there has been less concern expressed about fish welfare and aquaculture. Whereas we have some access to the experience of a cow being forced to hurry along by a stick or cattle prod, we have very little concept of the experience of a fish removed from water for 30 seconds to be vaccinated. Fish are not as visible to the public, and fish behaviour is poorly understood.

The Problem of Pain

There has been considerable debate over the past decade about whether or not the welfare of fish should be given any consideration. In particular, it has been argued that if fish are to be given welfare consideration, there must be a reasonable demonstration of cognitive capacities similar to other sentient beings (i.e. mammals). In 2002, Rose published his somewhat selective and controversial review of the scientific literature concerning awareness and pain in fish (Rose 2002). Rose argues that there is evidence that no animals phylogenetically lower than non-human primates have the ability to perceive pain. His argument hinges on the difference in neuroanatomical structures between species, in particular the presence or lack of a neocortex. During that same period, other authors provided reviews of the literature to examine the current state of knowledge of pain perception in fish, notably Chandroo et al. (2004) and Braithwaite & Huntingford (2004). These authors came to a different conclusion concerning the possibility for fish to perceive pain. In summary, Chandroo et al. (2004) stated that it is becoming clear that fish have the capacity to detect, conceptualize and subsequently respond to a nociceptive stimulus, and concluded that from a neuroanatomical perspective, it appears that modern hypotheses pertaining to the neural basis of consciousness include rather than exclude fish species. In addition, other animals display key behaviours and responses that have long been associated with consciousness in humans (Van Swinderen 2005).

The problem of pain is not simply about whether fish have the capacity to experience pain, but is instead a matter of whether that experience matters to the fish. In other words, do they have a level of conscious awareness that means that they "suffer" as a result? It could be argued that if fish do not actually have an awareness of the experience, then there may be no necessity to concern ourselves with the impact of invasive procedures on the fish.

These are some of the questions that have been of concern to the Canadian Council on Animal Care (CCAC, the national organization responsible for setting and maintaining animal care and use standards in Canadian science), in particular when the CCAC subcommittee on fish was developing *guidelines on: the care and use of fish in research teaching and testing* (CCAC 2005). During the same period, a subcommittee of the American Fisheries Society was preparing US *Guidelines for the Use of Fishes in Research* (Nickum et al. 2004). The CCAC subcommittee made recommendations within the CCAC guidelines based on the potential for fish to experience pain and distress, whereas the US guidelines place a great amount of emphasis on stress, rather than pain. In particular, the US guidelines state that "characteristic of all high-quality guidelines for fish care and health management is emphasis on the importance of avoiding stress, or at least minimizing it" (Nickum et al. 2004), arguably achieving the same ends in terms of protection of the research fish by focusing on minimizing stress as achieved by the CCAC guidelines through addressing the potential for pain.

Legislation and Pain

As long ago as 1789, Jeremy Bentham, the English philosopher and founder of the ethical theory of Utilitarianism, who is widely known for his proposal to include all sentient beings in moral choice-making, stated in regards to animals: "The question is, Not can they reason? Nor can they talk? But can they suffer?" Pain matters to people and therefore, by anthropomorphic analogy, pain in animals matters to people too. Animals have been considered to be worthy of moral concern, primarily because they have the capacity to experience pain (see, for example, Ryder 2001). Polls concerning the use of animals in research in general show that 70–80 per cent of the public is supportive of the use of animals in research, but that support is contingent on strict oversight and minimization of pain and distress (Ipsos MORI 2005; Canadian Public Health Association 2001).

Legislated regulations and guidelines on the use of animals in science have tried to address the question of pain. In the United Kingdom, the *UK Animals (Scientific Procedures) Act*, defines "a regulated procedure" as ". . .. any experimental or other scientific procedure applied to a protected animal which may have the effect of causing that animal pain, suffering, distress or lasting harm" (HMSO 1986). The United States Department of Agriculture Animal and Plant Health Information Service Policy on Painful Procedures defines a *painful procedure* as "any procedure that would reasonably be expected to cause more than slight or momentary pain and/or distress in a human being to which that procedure is applied" (USDA 1997); and within its fundamental *policy statement on: ethics of animal investigation*, the CCAC requires that "Animals must not be subjected to unnecessary pain or distress. The experimental design must offer them every practicable safeguard, whether in research, in teaching or in testing procedures; cost and convenience must not take precedence over the animal's physical and mental well-being" (CCAC 1989).

The definition of pain currently in use by the CCAC is based on the Federation of European Laboratory Animal Science Association (FELASA)'s definition (Baumans et al. 1994), which in turn is based on the definition of the International Association for the Study of Pain: "an unpleasant sensory and emotional experience associated with actual or potential damage or described in terms of such damage" (IASP 1979). This definition points to pain as being purely a conscious experience, with a sensory component and a component of emotional feeling. This definition arose as a description of human pain, and was probably never intended to extend to animal pain, as the assessment of an animal's emotional experience is impossible. For this reason, Bateson (1992) and others have argued that emotion should not feature in the definition of pain in animals. This poses a problem in determining whether animals have an emotional component to the experience of pain; in particular, because a blunt interpretation of animal welfare according to the "feelings" school requires the experience of a painful stimulus to "matter" to the animal, in order for the animal to be worthy of our moral concern.

Bill C-17 (introduced into the Canadian Parliament in 1999, and superseded by Bill C-15 and Bill C-50, which was not passed before the dissolution of the last Parliament), sought to expand the scope of animal protection. While the Bill and the current *Cruelty to Animals Act* are blunt instruments of law in that they can only bring prosecution after an incident of cruelty has occurred, early drafts of the proposed Bill would have made the capacity to feel pain the criterion for protection under Canadian law. During the deliberations of the Senate Committee, the CCAC provided two expert witnesses to answer questions concerning the capacity of animals to perceive pain. As a result of their interventions, the definition of animal under the *Act* was changed from "vertebrate, other than a human being, and any other animal that has the capacity to feel pain" to "vertebrate, other than a human being" (Bill C-50 2005), removing the ability to experience pain as the criteria for protection under the proposed *Act*.

Consciousness

The decision of Canadian legislators to change the definition of animal to exclude the capacity to feel pain is perhaps reflective of the difficulties inherent in determining the conscious experience of pain. Just as there are inherent difficulties in providing a definition of pain that does not rely on anthropomorphic parameters, there are also inherent difficulties in defining consciousness and in describing the conscious experience of an animal. Whether or not one favours the arguments laid out by Nagel in his well known essay on the question first posed by Thomas Sprigge "What is it like to be a bat?", it is perhaps true that at the current time we do not have the tools to answer the question, be it for a bat or a fish. Similarly we have even less understanding of "What is it like to be a bat in pain" (Hanna, unknown date), or the experience of a fish undergoing what would constitute a painful procedure for a human.

The study of consciousness bridges a number of disciplines, including neuroscience, psychology, philosophy and ethology. To date there is no universally accepted definition of consciousness, although there is a generally agreed recognition of two levels of consciousness. Primary consciousness is usually ascribed to animals with the ability to generate a mental picture in which diverse pieces of information can be integrated to direct behaviour (such as use of pre-determined escape route to beat a hasty retreat from a predator). Secondary consciousness is usually described as being of a higher order, involving linguistic capabilities, self-consciousness and self-knowledge. Carruthers (2000), among others, has tried to relate theories of consciousness to animal abilities. He argues that both states of consciousness imply sentience on behalf of the subject, that is, the possibility to conceptualize pain. The question therefore remains whether, and to what extent, consciousness, or some level of consciousness, is important to discussion of how we treat animals. There are those who argue that possession of primary consciousness is sufficient to warrant concern and protection (for example Beckoff 2006), whereas others, including Hanna (unknown date), argue that only animals which have the capacity to suffer, by virtue of their possession of secondary consciousness have intrinsic moral significance and deserve protection. He argues that protection of animals that are at best merely sentient may be for a different reason than prevention of suffering of the animal itself (e.g. we should not harm or kill animals without good reason, since ill-treatment of animals may result in the development of callous behaviour by humans to other humans –Kant's duty ethics, cited by Lund et al. (2007)).

Chandroo et al. (2004) believe that the question of whether or not fish are capable of conscious states must be addressed in order to be able to evaluate their welfare status. However, it seems to me that the assessments of the welfare status of terrestrial animals already occur based on criteria that are not fundamentally related to the level of consciousness of the animal, and further, that this existing approach is morally sufficient. This view is elaborated below.

Welfare Assessment

At the CCAC we are sometimes asked whether animals such as non-human primates used for biomedical research purposes should be given equal or more attention than animals such as fish. The CCAC system focuses on meeting the physical and behavioral needs of the animals, given the constraints of the research environment. Clearly, the needs of a non-human primate are very different from those of a fish, and different again from those of a rat or a rabbit. At this point in time, it is probably true to say that we know much more about the physical and psychological needs of non-human primates or even rats and mice than we do for fish. For example, we know that rats are social animals and cluster in groups when given the opportunity (see, for example, http://www.ratlife.org/Home/0Main-frameset/Mainframeset.htm, accessed May 14, 2007). We know that non-human primates can become bored very easily and need challenges such as a stimulating environment and interaction with humans and other primates (see, for example, Prescott M., *Primate sensory capabilities and communication signals: implications for care and use in the laboratory*, available at http://www.nc3rs.org.uk, accessed May 14, 2007). We know that mice readily build nests when provided with nesting materials (Van de Weerd et al. 1998). However, for fish we understand little about their preferences for housing, what behaviours are important to them, and how those behaviours can be expressed in a research or an aquaculture environment. Perhaps most importantly, we do not yet fully understand the "language" of fish (Tinbergen 1988; cited by Lund et al. 2007).

Dawkins (2004), in an attempt to simplify the measurement of animal welfare, has argued that assessment of animal welfare really comes down to asking two fundamental questions: "Are the animals healthy?" and "Do they have what they want?" The first question "Are the animals healthy?" provides the foundation of all good welfare (Fraser & Broom 1990), and clearly equates with the function definition of animal welfare described by Huntingford. "Do the animals have what they want?" is a more complex question, and requires asking questions of the animals themselves to try to infer, usually from behavioural measures, some indication of what matters to the animal. The challenge is to determine whether a behavioural response to a stimulus is a simple reflex, or whether more complex cognition is involved (Braithwaite & Boulcott 2007). Trying to determine the motivation of an animal to avoid or seek out a particular stimulus has been argued to provide insights into the mental state of the animal (Paul et al. 2005). This approach has been used extensively to gain insight into animal (particularly livestock) preferences, resulting in changes to husbandry practices (Dawkins 2004). In the case of fish, relatively little work has been done to date to look at behaviour and learning. Work carried out by Dunlop et al. (2006) and Yue et al. (2004) evaluated avoidance learning behaviour in trout and goldfish, which is thought to require considerable coordination of information and hence require some form of higher order cognition (Braithwaite & Boulcott 2007). However, as Rose (2007) has recently emphasized, it is difficult to determine whether this type of learning is "associative procedural learning", thought to occur unconsciously, or whether it is "explicit", requiring mediation by conscious awareness.

During the aquaculture boom of the last few decades, concerns regarding animal welfare in aquaculture have largely resulted in increased focus on the health of aquaculture animals. However, with the work of Huntingford, Braithwaite, Sneddon, Yue and others, some of the questions concerning what matters to the fish are beginning to be asked, and are therefore likely to play a greater role in determining appropriate conditions for aquaculture in the coming years.

Governance

Aquaculturists, like their counterparts in terrestrial agriculture, may have to address animal welfare issues if public demand for aquatic products mirrors the expectations associated with other animal commodities. However, to date, public opinion concerning aquaculture appears to have focused on food safety and environmental safety, which are viewed as highly inter-related, along with other aquaculture issues, such as farming operations, escaped fish, diseased fish, pollution, genetic contamination, and capacity (DFO 2005c). There has been relatively little attention to concerns regarding individual animals or groups of animals, in the same way that concerns about beef cattle extend to conditions of transport, and slaughter, public concern for the quality of life experienced by agricultural animals grew considerably as a result of exposure to the confinement of animals by Ruth Harrison (1964), but also because of the ability of the public to identify with those species. In addition, the ability to care for livestock animals is based on centuries of experience, as well as considerable research, and thus knowledge about their anatomy, physiology and veterinary care (Iwama 2007). Although the media has highlighted concerns about aquaculture, so far these have been related to food safety (e.g. levels of PCBs in farmed salmon, Hites et al. 2004) or environmental concerns (e.g. pollution of local environments from excrement collecting on the sea bed beneath salmon cages). As few people have direct experience in production of fish for food, whether farmed or captured in the wild, they appear to have less concern about the welfare of fish in general. In addition, the unfamiliar body language of fish does not encourage human empathy with individual fish (Tinbergen 1988). Hence, so far, it appears as though the public has not extrapolated its concerns for livestock welfare to fish welfare.

The timing of the Rose (2002) review was interesting as it came during the period when a subcommittee of the American Fisheries Society, of which Rose was a member, was revising its *Guidelines for the Use of Fishes in Research* (Nickum et al. 2004). The review article was used as a basis for the guidelines, and within the document, institutional animal care and use committees were urged to read Rose (2002) prior to developing any institutional policies concerning pain. In particular, the guidelines stress that "imposition of regulations based on personal beliefs and values may compromise the purpose, design and scientific value of the research" and that "assumptions and perceptions based on experiences with mammals, especially primates, must not be extrapolated to fish." Nonetheless, the guidelines emphasized the importance of humane treatment of fishes and the authors' "deep respect for life processes and the myriad forms of life. (Nickum et al. 2004)." By contrast, within the same time frame, the Council of Europe was concurrently revising Appendix A of the European Convention for the Protection of Vertebrate Animals Used for Experimental and Other Scientific Purposes (ETS 123) (http://conventions.coe.int/Treaty/en/Treaties/Html/123.htm, accessed May 31, 2007) to include species-specific provisions for fish. These species-specific provisions make no reference to the capacity of fishes to experience pain as this is implicit through article 5 of the Convention, which states "The member states of the Council of Europe have decided that it is their aim to protect live animals used for experimental and other scientific purposes to ensure that any possible pain, suffering, distress or lasting harm inflicted upon them, as a consequence of procedures being conducted on them, shall be kept at a minimum."

We have argued elsewhere (Griffin & Gauthier 2005) that in the face of inconclusive scientific evidence the CCAC has a duty to act on behalf of the Canadian public by taking a precautionary approach. Therefore, the CCAC subcommittee, using the scientific evidence available, developed the guidelines statement that "Fish have the potential to experience pain, and manipulations that provoke stress or avoidance/escape behaviour may be causes of distress", and that "researchers have an obligation to mitigate or minimize pain and distress whenever feasible and consistent with good scientific practice" (CCAC 2005). This perspective was subsequently included in the CCAC *guidelines on: the care and use of fish in research, teaching and testing* (CCAC 2005). Although differences emerged in the approach used by various jurisdictions, there also emerged general principles which could form the basis for efforts towards international harmonization of standards. The most important of these general principles being that fish should not be subjected to stress because of significant physiological and behavioral consequences for the animal (Barton & Iwama 1991).

Governance of Aquaculture

In developing the *CCAC guidelines on: the care and use of fish in research, teaching and testing* (2005), we encountered a genuine concern from scientists that if fish were publicly regarded as animals capable of perceiving pain, a prohibition on fish use in angling, research and aquaculture might follow. This conclusion does not appear to be very realistic. It is more generally accepted that other types of animals such as cattle, pigs, sheep, deer, and moose have the capacity to experience pain; nevertheless these animals are raised for food or hunted. Legislated regulations and guidelines of practice provide criteria for minimizing pain and distress likely to be experienced by the animal; they do not prohibit use of the animals.

As discussed above, consideration of the governance of the aquaculture industry to date has mainly addressed the questions of whether farmed fish are healthy, and how fish farms impact the environment, rather than trying to determine whether the aquaculture environment meets the needs of the fish. The CCAC *guidelines on: the care and use of fish in research, teaching and testing* (2005) provided a list of potential indicators that could be used as behavioural indicators of the well-being of fish; however, much more research is needed to provide measures that can be used by the industry for a sound welfare assessment of fish. Irrespective of whether or not fish have the capacity to perceive pain, it is clear that the public and the industry need a transparent system of governance that provides assurance that fish are raised according to acceptable standards. According to Kadri (Chapter 3), just as good farmers are able to determine the welfare of their animals by visual examination, so fish farmers are already able to make good assessments of welfare in their own stock without use of scientific knowledge. However, a means by which to objectively demonstrate the results of any welfare assessments to stakeholders requires tools produced in collaboration with scientists (Turnbull & Kadri 2007). As more is understood about the things that matter to a fish, there may need to be adjustment to the manner in which fish are housed and cared for, including the manner in which they are killed.

Very poor welfare (e.g. disease, poor growth and mortalities) is not cost-effective for the farmer; therefore, improvements in welfare up to a certain level are likely to improve productivity. A few studies considering factors affecting fish welfare, such as slaughter methods, have indicated strong links between welfare and quality of the final product (Robb et al. 2000 cited by Lymbery 2002). However, other factors of potentially greater importance to the life-time experience of the fish may not have such visible impacts on the final quality of the product, and if not directly beneficial to the farmer may not be easily implemented.

At the international level, the OIE member countries recognized that, as animal welfare is a complex, multi-faceted public policy issue which includes important scientific, ethical, economic and political dimensions, the OIE needed to develop a detailed vision and strategy, incorporating and balancing these dimensions. Even though animal welfare is not covered by the Agreement on the Application of Sanitary and Phytosanitary Measures of the World Trade Organisation (http://www.wto.org/english/tratop_e/sps_e/decisions06_e.htm; accessed July 20, 2007), member countries wished to have guidelines and recommendations to assist them in international negotiations. The OIE International Committee decided that the OIE would give priority to the welfare of animals used in agriculture and aquaculture, and that, within that group, the topics of transportation, humane slaughter and killing for disease control purposes would be addressed first, followed by housing and management. Other topics, such as research animals and wildlife, would be addressed as resources permitted.

The OIE has already submitted the following set of Guiding Principles for Aquatic Animal Welfare, for consideration by its member countries. These principles are almost identical to those laid out for agricultural animal welfare:

1. That there is a critical relationship between aquatic animal health and aquatic animal welfare.
2. That the internationally recognized five freedoms as they apply to aquatic animals (freedom to express normal patterns; freedom from pain, injury and disease; freedom from fear and distress; freedom from physical and thermal discomfort; freedom from hunger, thirst and malnutrition) provide valuable guidance in aquatic animal welfare.
3. That the internationally recognized "Three Rs" (reduction in the numbers of aquatic animals, refinement of experimental methods and replacement of aquatic animals with non-animal techniques) provide valuable guidance for the use of aquatic animals in science.
4. That the scientific assessment of aquatic animal welfare involves diverse elements which need to be considered together, and that selecting and weighing these elements often involves value-based assumptions which should be made as explicit as possible.
5. That the use of aquatic animals in aquaculture, harvest or capture fisheries, research and for recreation (e.g. ornamentals in aquaria), makes a major contribution to the well-being of people.
6. That the use of aquatic animals carries with it an ethical duty to ensure the welfare of such animals to the greatest extent practical.
7. That the improvements in aquatic animal welfare can often improve productivity and food safety and hence lead to economic benefits.
8. That equivalent outcome (performance criteria), rather than identical systems (design criteria), be the basis for comparison of aquatic animal welfare standards and guidelines (http://www.oie.int/downld/SC/2006/A_AAC_MARCH2006.pdf #page=175; accessed July 20, 2007).

OIE guidelines for the transport of fish by boat; guidelines for the land transport of fish; guidelines for the slaughter of farmed fish for human consumption; and guidelines for the humane killing of fish for disease purposes are also under development (http://www.oie.int/eng/bien_etre/en_introduction.htm; accessed July 20, 2007).

Conclusion

The increase in growth of the aquaculture industry will bring it increasingly under public scrutiny. At present that scrutiny extends to the potential contaminants within fish and fish products, and to the impact of fish farms on the environment. While contamination of animal products and the impact of agricultural operations on the environment have also been concerns for the agricultural industry, public concern has also extended to the welfare of animals used for food production. In large part, the concern for animal welfare hinges on minimizing any pain and/or distress likely to be experienced by the animal. Welfare assessment schemes have focused on identification of critical indicators (such as vocalization of cattle in a slaughter house), that can be interpreted as potential sources of pain and/or distress for the animals. In the case of the aquaculture industry, to date, public concern does not appear to have extended to the welfare of fish used for food. This may be due to a lack of identification with the species and only limited knowledge about parameters that can be used as indicators of fish welfare. Recent studies are beginning to give weight to the need to provide protection for research fish by focusing on minimizing stress rather than addressing the more contentious matter of whether fish are capable of experiencing pain. This approach could usefully be employed in the governance of the aquaculture industry as a proactive means of improving fish welfare and ensuring that aquaculture production practices can withstand public scrutiny.

References

Agriculture and Agri-Food Canada, http://www.agr.gc.ca/index_e.php

Aquanet Fact Sheet. 2004. Volume 1 Aquanet Research supporting new marine finfish aquaculture in Canada. http://www.aquanet.ca/English/media/factsheet/factsheet-v1.pdf; accessed May 1, 2007.

Barton, B.A. and Iwama, G.K. 1991. Physiological changes in fish from stress in aquaculture with emphasis on the response and effects of corticosteroids. In: *Annual Review of Fish Diseases*, 3–26. Pergammon Press.

Bateson, P. 1992. Assessment of pain in animals. *Animal Behaviour*. 42: 827–839.

Baumans, V., Brain, P.F., Brugere, H., et al. 1994. Pain and distress in laboratory rodents and lagomorphs. Report of the FELASA Working Group on Pain and Distress. *Laboratory Animals*. 28: 97–112.

Bayvel, D., Abdul Rahman, S., and Gavinelli, A. 2005. Introduction In: Animal welfare: global issues, trends and challenges. *Revue scientifique et technique*. 24(2): 475–482.

Beckoff, M. 2006. Animal passions and beastly virtues: cognitive ethology as the unifying science for understanding the subjective, emotional, empathic and moral lives of animals. *Zygon*. 41:71–104.

Bentham, J. 1789. Principles of morals and legislation. In: *The Collected Works of Jeremy Bentham*, eds. J.H. Burns and H.L.A. Hart 2: 11–12. London: Athlone Press.

Bill C-50: An Act to amend the Criminal Code in respect of cruelty to animals. 2005. House of Commons of Canada, First Session, Thirty-eight Parliament, 53–54 Elizabeth II, http://www.parl.gc.ca

Braithwaite, V. and Huntingford, F.A. 2004. Fish and welfare: do fish have the capacity for pain perception and suffering. *Animal Welfare*. 13: S87–S92.

Braithwaite, V. and Boulcott, P. 2007. Pain perception, aversion and fear in fish. *Diseases of Aquatic Organisms*. 5: 131–138.

Brambell, F.W.R. 1965. Report of the Technical Committee to enquire into the welfare of animals kept under intensive livestock husbandry systems. HMSO: London, UK.

Canadian Council on Animal Care (CCAC). 1989. CCAC policy statement on: ethics of animal investigation. CCAC: Ottawa Canada. http://www.ccac.ca/en/CCAC_Programs/Guidelines_Policies/POLICIES/ETHICS.HTM; accessed May 4, 2007.

Canadian Council on Animal Care (CCAC). 2005. CCAC guidelines on: the care and use of fish in research, teaching and testing. 87. CCAC: Ottawa Canada.

Canadian Public Health Association. 2001. Animal to human transplantation: Should Canada proceed? Public consultation on xenotransplantation final report. 51. Canadian Public Health Association: Ottawa Canada. http://www.xeno.cpha.ca/english/finalrep/reporte.pdf; accessed May 4, 2007.

Carruthers, P. 2000. *Phenomenal Consciousness – A Naturalistic Theory*. United Kingdom: Cambridge University Press.

Chandroo, K.P., Duncan, I.J.H. and Moccia, R.D. 2004. Can fish suffer? Perspectives on sentience, pain, fear and stress. *Applied Animal Behaviour Science*. 86: 225–250.

Clubb, R. and Mason, G. 2003. Captivity effects on wide-ranging carnivores. *Nature*. 425: 473–474.

Dawkins, M.S. 1980. *Animal Suffering: the Science of Animal Welfare*. London UK: Chapman and Hall.

Dawkins, M.S. 2004. Using behaviour to assess animal welfare. *Animal Welfare*. 13: S3–S7.

DFO-Fisheries and Oceans Canada. 2005a. Aquaculture: Canadian Aquaculture Production Statistics. http://www.dfo-mpo.gc.ca/communic/statistics/aqua/aqua05_e.htm; accessed July 14, 2007.

DFO-Fisheries and Oceans Canada. 2005b. Our Waters Our Future 2005–2010 Strategic Plan. http://www.dfo-mpo.gc.ca/dfo-mpo/plan_e.htm#5a4; accessed May 1, 2007.

DFO-Fisheries and Oceans Canada. 2005c. Overview: Qualitative Research Exploring Canadians' Perceptions, Attitudes and Concerns Towards Aquaculture. http://www.dfo-mpo.gc.ca/misc/focus-aquaculture_e.htm; accessed May 31, 2007.

Duncan, I.J.H. 2004. A concept of welfare based on feelings. In: *The Well-Being of Farm Animals, Challenges and Solutions*, eds. G.J. Benson and B.E. Rollins, 85–101. Oxford UK: Blackwell Publishing.

Duncan, I.J.H. and Fraser, D. 1997. Understanding animal welfare. In: *Animal Welfare*, eds. M.C. Appleby and B.O. Hughes, 19–31. Wallingford: CAB International.

Dunlop, R., Millsopp, S. and Laming, P. 2006. Avoidance learning in goldfish (*Carassius auratus*) and trout (*Onchorynchus mykiss*) and implications for pain perception. *Applied Animal Behaviour Science*. 97: 255–271.

Fisheries Society of the British Isles (FSBI). 2002. Fish Welfare. Briefing Paper 2, Fisheries Society of the British Isles. Granta Information Systems: Cambridge UK. http://www.fsbi.org.uk/docs/brief-welfare-refs.pdf; accessed May 2, 2007.

Food and Agriculture Organization (FAO) of the United Nations. (2004) The State of World Aquaculture. http://www.fao.org/fi/website/FIRetrieveAction.do?dom=topic&fid=13540; accessed May 1, 2007.

Fraser, A.F. and Broom, D.M. 1990. *Farm Animal Behaviour and Welfare*, 437. Wallingford: CAB International.

Gilbert, E. 2002. Study No. 7 – The International Context for Aquaculture Development: Growth in Production and Demand, Case Studies and Long-Term Outlook. Office of the Commissioner for Aquaculture Development, Fisheries and Oceans Canada. http://www.dfmpo.gc.ca/aquaculture/ref/Study7_e.htm; accessed July 14, 2007.

Griffin, G. and Gauthier, C. 2005. An approach toward international harmonization: the care and use of fish. In: *Internationalisation and Harmonisation of Laboratory Animal Care and Use Issues*. Proceedings of the 9th FELASA Symposium 4–17 June 2004, Nantes, France, eds. M.R. Gamble and S. Millington. 27–34 Laboratory Animals: UK.

Hanna, R. (unknown date) What is it like to be a bat in pain? http://spot.colorado.edu/~rhanna/bat_in_pain_dft1.htm; accessed: August 21, 2007.

Harrison, R. 1964. *Animal Machines*. London, UK: Vincent Stuart.

Hites, R.A., Foran, J.A., Carpenter, D.O., Hamilton, C., Knuth, B.A. and Schwager, S.J. 2004. Global Assessment of organic contaminants in farmed Salmon. *Science*. 303(5655): 226–229.

Home Office. 1986. *UK Animals (Scientific Procedures) Act*. HMSO: London, UK. http://www.archive.official-documents.co.uk/document/hoc/321/321-xa.htm; accessed May 4, 2007.

International Association for the Study of Pain (IASP). 1979. Pain terms: a list with definitions and notes on usage. *Pain*. 6:249–252.

Ipsos MORI Poll for UK Coalition for Medical Progress. 2005. http://www.ipsos-mori.com/polls/2002/cmp.shtml; accessed May 4, 2007.

Iwama, G.K. 2007. The welfare of fish. *Diseases of Aquatic Organisms*. 75: 155–158.

Lund, V., Mejdell, C.M., Röcklinsberg, H., Anthony, R. and Håstein, T. 2007. Expanding the moral circle: farmed fish as objects of moral concern. *Diseases of Aquatic Organisms*. 75: 109–118.

Lymbery, P. 2002. In too deep – the welfare of intensively farmed fish. Compassion in World Farming, Petersfield. http://www.ciwf.org.uk/publications/reports/in_too_deep_2001.pdf; accessed May 4, 2007.

Nickum, J., Bart, H.L. Jr., Bowser, P.R. et al. 2004. Guidelines for the use of fishes in research. American Fisheries Society: Bethesda MD, USA. http://www.fisheries.org/afs/ publicpolicy/guidelines2004.pdf; accessed May 3, 2007.

Office International des Épizooties (OIE). 2006.Terrestrial Animal Health Code. http://www.oie.int/eng/normes/Mcode/en_sommaire.htm; accessed July 20, 2007.

Office International des Épizooties (OIE). 2007. Aquatic Animal Health Code. http://www.oie.int/eng/normes/fcode/en_sommaire.htm, accessed July 20, 2007.

Paul, E.S., Harding, E.J. and Mendel, M. 2005. Measuring emotional processes in animals; the utility of a cognitive approach. *Neuroscience and Biobehavioral Reviews*. 29: 469–491.

Prescott, M. 2006. Primate sensory capabilities and communication signals: implications for care and use in the laboratory. National Centre for the Replacement, Refinement and Reduction of Animals in Research (NC3Rs). http://www.nc3rs.org.uk/news.asp?id=187; accessed: May 14, 2007.

Rose, J.D. 2002. The neurobehavioral nature of fishes and the question of awareness and pain. *Reviews in Fisheries Science*. 10(1): 1–38.

Rose, J.D. 2007. Anthropomorphism and 'mental welfare' of fishes. *Diseases of Aquatic Organisms*. 75: 139–154.

Ryder, R. 2001. *Painism: A Modern Morality*. UK: Opengate press.

Tinbergen, N. 1988. The study of instinct. 2nd Edn. Oxford: Clarendon Press.

Turnbull, J.F. and Kadri, S. 2007. Safegurding the many guises of farmed fish welfare. *Diseases of Aquatic Organisms*. 75: 173–182.

USDA. 1997. United States Department of Agriculture Animal and Plant Health Information Service Animal Care Resource Guide, Policy 11: Painful Procedures, 11.1–11.2. http://www.aphis.usda.gov/ac/policy/policy11.pdf; accessed May 4, 2007.

Vallat, B. 2005. Preface. In: Animal welfare: global issues, trends and challenges. *Revue scientifique et technique*. 24(2): 467–474.

Van de Weerd, H.A., Van Loo, P.L.P., Van Zutphen, L.F.M., Koolhaas, J.M. and Baumans, V. 1998. Strength of preference for nesting material as environmental enrichment for laboratory mice. *Applied Animal Behavioural Science*. 55369–382. http://igitur-archive.library.uu.nl/dissertations/01801846/c5.pdf; accessed May 14, 2007.

Van Swinderen, B. 2005. The remote roots of consciousness in fruit fly selective attention? *BioEssays*. 27: 321–330.

Yue, S., Moccia, R.D. and Duncan, I.J.H. 2004. Investigating fear in domestic rainbow trout Onchorynchus mykis, using an avoidance learning task. *Applied Animal Behaviour Science*. 87: 343–354.

Chapter 3
Welfare and Aquaculture Industry Practice

Sunil Kadri

Intensive finfish aquaculture has undergone exponential growth over the past two decades, particularly in Europe, Canada and Chile. This growth has been accompanied by many supporting technological advances and innovations, some of which allow fish to be grown at higher stocking densities (e.g. aeration systems), and others which provide for growing of fish in previously prohibitive conditions (e.g. offshore cages). The "industrialisation" of fish culture has brought with it public concerns with respect to sustainability. Such concerns are usually focussed upon issues such as environmental impact, impact upon wild fish and food safety. Fish welfare also comes under the "sustainability umbrella" and has received increasing attention from policy makers over the past decade or so in Europe (Kadri et al. 2005). While this attention stems from public concern, the "public" in this case appears not to be the "general public" mentioned in Huntingford's introductory sentence, but rather that of activists. For fish consumers on the whole, fish welfare in aquaculture appears to remain "below the radar" (Aarset et al. 2004; Kadri et al. 2005; North et al. 2008). Hence, the industry finds itself walking a tight-rope: developing and publicising codes of practice in order to foster public relations, whilst not wishing to make too much of an issue which is not yet at the forefront of consumers' minds.

Science and the Question of Whether Fish Suffer

Huntingford has presented a good case for the argument that fish suffer, while Rose (2002) argues that they do not, again presenting "scientific evidence" to support his contention. These convincing but opposing arguments from reputable scientists both use analogies and "circumstantial evidence" in order to make their point, as science by itself cannot currently give us a definitive answer as to whether or not fish are capable of suffering. Science is based upon observation and objective measurement, however it is not possible to measure attributes such as self-awareness or

S. Kadri
Aquaculture Innovation & Fish Biology Group, Division of Environmental & Evolutionary Biology, Institute of Biomedical & Life Sciences University of Glasgow
e-mail: sunil@aquainnovation.co.uk

© Springer Science+Business Media B.V. 2008

conciousness in fish: hence the approach using analogous evidence. For this reason the issue should not be left to scientists alone, but rather engage a multi-disciplinary forum, to include philosophers, economists and scientists together. Economists will bring the financial and other cost-benefits of aquaculture into persepective for the purpose of discussing aquaculture and fish welfare from a utilitarian perspective, while ethicists can provide a vital contribution to the debate by establishing and maintaining the frameworks for discussion. This will enable not only a dialogue between disciplines, but also with other stakeholders (e.g. consumers and NGOs). This will allow us to move forward with a proper understanding of the issues at hand and whether or not the aquaculture industry and society should be paying attention to them.

Potential Industry Approaches to Fish Welfare in Aquaculture

Leaving aside the question of whether or not science can definitively demonstrate that fish suffer, what approaches might the industry take with regard to fish welfare and what to do about it? Some possibilities include, for example, taking a marketing approach in which the issue of fish welfare is promoted as a marketable trait. If fish retailers claim their farmed fish have been raised with good welfare, were killed humanely and should be bought for this reason, they may gain a premium price for it. However, the implication is that wild-caught produce sold by the same retailers should not be bought as these animals have "suffered" a slow, "painful" death.

A second approach to fish welfare is to focus on the value to production of considering fish welfare. In a production approach, if good welfare improves production and or quality in aquaculture, then fish welfare will be taken care of in standard husbandry and need not be an issue for concern. Bell (2003) compared a multi-variate scientific approach to welfare assessment (Turnbull et al. 2005) with a welfare questionnaire for fish farmers. This study demonstrated that fish farmers are already able to make good assessments of welfare in their own stock without use of scientific knowledge. However, a means by which to objectively demonstrate the results of any welfare assessments to stakeholders would certainly require some sort of tools produced in collaboration with scientists. On the other hand, Huntingford has suggested that while improved welfare will generally result in good production, conditions for "best welfare" may be different from those for best production. This possibility is discussed as an economic reality by Peddie & Stott (2003), building on experience from terrestrial farming systems. More targeted investigations will be required in order that this issue can be properly evaluated and addressed.

A third approach is explicitly ethical in its focus, and asks the most basic question of all: "Should we care?" As mentioned above, ethical concerns regarding fish welfare in aquaculture have arisen from activism but do not seem to have engaged consumers en masse. The experience of the fish farming industry is that the type of activism we are faced with is not likely to be satisfied, no matter what the industry does to address stated concerns. This does not however, provide an "escape clause" for aquaculture, but rather the industry and stakeholders need to find a moral balance between fish interests and human interests. In this regard, it must be realised that man's "progress" is inevitably at the expense of other species, and that in finding a moral position with respect to sustainability, there are some questions to be answered, with important implications for the future of our societies. These include issues such as whether domestication is acceptable; whether we should cease farming (of all types) if an absence of animal suffering cannot be guaranteed and whether we should ban commercial and sport fisheries if fish do indeed suffer.

The ethical stance of today's consumer appears to be one where human interests come first and foremost. This was demonstrated by farmed salmon sales having dropped significantly following the PCB/Dioxin scare of January 2004. In this case, the consumers were no doubt concerned that farmed salmon would have negative effects upon their own health. Other issues surrounding salmon farming, including fish welfare, have also been raised in the media in a negative manner, but these had no effect upon sales. Many consumers currently seem to believe that wild salmon is "better", as indicated by the tendency for wild salmon to fetch higher prices, even when sold frozen or pre-frozen and defrosted. Why? Could it be that they support annihilation of fish stocks and consumption of endangered species? Or that they perhaps prefer to eat animals that have suffered a slow painful death? Presumably not, but the present trend favouring wild salmon consumption could not exist in a climate where the consumer was greatly concerned for the welfare of fishes.

Considering all of the above, it seems inevitable that we ask the question of whether we are protecting fish welfare in general. Presumably scientists, consumers and other stakeholders' pursuit of fish welfare is made easier through a focus on aquaculture without paying much attention to the welfare of wild fish, that is, there is an implicit assumption that we have a greater responsibility for the animals we rear than those we take from the wild, and so we are required to provide good welfare for farmed fish, regardless of whether it can be proved that they suffer or feel pain.

So What are We Doing and What More Should We Do?

The aquaculture industry has not only developed technologies which provide for more fish to be grown in more places, but also those which can improve fish welfare, for example through provision of advanced feeding, passive grading, and humane slaughter (see for example www.aq1systems.com, www.gradingsystems.com & www.seafoodinnovations.com.au respectively). The capture industry on the other hand, has not developed in this manner, although R&D initiatives aimed at improving the welfare the situation in wild fisheries are beginning to appear.

The innovations in aquaculture, referred to above, have not come about solely because they improve welfare, but because welfare improvements such as feeding fish according to appetite improves production, while reducing handling stress at slaughter improves flesh quality. Such innovations will continue, given the industry's recognition that maintaining good fish welfare benefits both production and public relations. This area of technological development falls within Huntingford's second point on strategies for improving aquaculture, i.e. husbandry systems should be devised to minimise the adverse impacts of fish farming upon welfare. It should be pointed out here that most of the improvements made to the lives of fish in aquaculture come from industrial innovation which, while harnessing biological knowledge, has not often resulted from collaboration with academic scientists (Kadri et al. 2005 unpubl. data). However, research collaborations with both academics and regulators are ongoing and provide the means by which to progress further than would be possible by the industry alone, both in terms of public credibility and scientific capability, particularly with respect to development of assessment tools for welfare in aquaculture (see below).

Overall then, the industry does not dispute the need for good welfare among farmed fish and has been pro-active in meeting this challenge. Aside from the technological advances mentioned above, codes of good practice have been developed; e.g. the Scottish Salmon Producers Code of Good Practice (2007), developed in consensus with industry players and other stakeholders; the increasing production under the RSPCA (2007) "Freedom Food" standards (an animal welfare based quality label), and other niche producers such as Loch Duart Sustainable Salmon, which have set high welfare standards for themselves, even to the extent that production volume may be compromised (e.g. through limitations on stocking density and times of year at which fish can be transferred to sea).

Effort in working with researchers and regulators with respect to development of operational welfare indicators (OWIs) has become particularly pressing as the industry comes under increasing pressure from large retailers to meet vague or potentially misguided welfare standards. Thus several such research programmes have been aimed at ensuring that welfare requirements are based upon a sound knowledge of fish and fish farming environments and are being carried out at the national, trans-European and trans-Atlantic levels. The research questions being asked in these projects become more and more applied as time passes, with the most recent European Commission funded project in this area being aimed at cost-benefit analysis and modelling of the implementation of OWIs in aquaculture (European Commission 2007).

Research to date has shown that we cannot apply simple rules (e.g. a maximum stocking density) in order to guarantee welfare of farmed fish (e.g. Turnbull et al. 2004), but rather several parameters must be looked at together in order to provide a full picture regarding welfare status. Thus practical tools for on-farm assessment of welfare will need to be developed by scientists and industry working together, as is currently taking place. Future regulations should be developed in consultation with industry and other stakeholders and based on a sound knowledge of the species and environments concerned. Such regulations would possibly involve regular self-monitoring of stock, overseen by the regulatory bodies concerned.

In conclusion, the aquaculture industry is already making efforts with respect to the three broad strategies put forward by Huntingford: Choice of species and selective breeding have and will continue to favour animals which flourish in an aquaculture environment; the industry has developed, and continues to develop technologies and husbandry systems which improve the welfare of the fish and the need to develop repeatable, objective on-farm indices of welfare for monitoring and regulation purposes is supported through involvement in current research programmes. There can be no doubt, however, that there is always room for further improvement, and wider involvement of the industry; with welfare appearing as a buzzword in aquaculture these days, more of the industry is paying attention and will have to make appropriate efforts and investments in ensuring continued maintenance and improvement of the welfare of fish in aquaculture.

References

Aarset, B., Beckmann, S., Bigne, J., Beveridge, M.C.M., Bjorndal, T., Bunting, M.J., McDonagh, P., Mariojouls, C., Muir, J.F., Prothero, A., Reisch, L.A., Smith, A.P., Tveteras, R., Young, J.A. 2004. The European consumers' understanding and perceptions of the organic food regime: The case of agriculture. *British Food Journal* 106: 93–105.

Bell, A. 2003. The effects of some management practices on the behaviour and welfare of farmed Atlantic salmon (*Salmo salar*) in marine cages. PhD Thesis, University of Stirling, Scotland, UK.

European Commission. 2007. BENEFISH. www.benefish.eu. Accessed September 2007.

Kadri, S., Sandøe, P. and Huntingford, F.A. 2005. Fish welfare – catering for the needs of the fish and the expectations of the consumer. In: Aquaculture Europe 2005, "Lessons from the past to optomise the future", 5–9 August, 2005, Trondheim, Norway–Oostende: European Aquaculture Society, (EAS Special Publication 35)-p. 320–322.

North, B.P., Ellis, T., Bran, J., Knowles, T.G. and Turnbull, J.F. (2008). The use of Stakeholder Focus Groups to identify Indicators for the On-form Assessment of Trout Welfare. In: Branson E.J. (ed.) Fish Welfare pp. 243–266, Blackwell.

Peddie, S. and Stott, A. 2003. An economic perspective on fish welfare. *Fish Farmer*. 26(1): 38–39.

Rose, J.D. 2002. The neurobehavioural nature of fishes and the question of awareness and pain. *Reviews in Fishheries Science* 10: 1–38.

Royal Society for the Prevention of Cruelty of Animals (RSPCA). 2007. http://www.rspca.org.uk/servlet/Satellite?pagename=RSPCA/FreedomFood/FreedomFoodHomepage. Accessed September 25, 2007.

Scottish Salmon Producers. 2007. Code of good practice. www.scottishsalmon.co.uk/aboutus/codes.asp. Accessed September 25, 2007.

Turnbull J.F., Bell A., Adams C.E., Bron J.E. and Huntingford F.A. 2005. Stocking density and welfare of cage farmed Atlantic salmon: application of a multivariate analysis. *Aquaculture* 243: 121–132.

Part II
Knowledge Management and Intellectual Property Issues in Aquaculture

Introduction

Proponents of aquaculture sometimes jokingly contrast images of 'fish farmers' with the 'fish killers' of the capture fisheries, insinuating that aquaculture is more orderly and civilized than traditional fishing. One particular point of pride among fish farmers is the extent to which their enterprise is scientifically and technologically knowledge-intensive. This is not, many will hasten to say, to deny the knowledge of fishers, but to to identify its proper place as the local or traditional knowledge of the hunter/gatherer who follows and understands a process, yet falls short of using knowledge to manipulate an entire process in the way fish farmers do. Aquaculture's reliance on a scientific and technological knowledge-base is underscored by the interactions fish farmers have with had with research councils and university laboratories, from joint work to understand the physiology of aquaculture species to collaboration in understanding and breeding for desirable traits. Given aquaculture's reliance on a range of scientific and technological practices which change constantly as competition drives innovation, aquaculture might be expected to rely on associated commerical practices including use of intellectual property rights to protect the knowledge which enables intensive aquaculture production. Here we arrive at this section's motivating question: Does success in aquaculture involve extensive use of intellectual property rights?

Performing a search for 'aquaculture' in all fields for patent applications and patents granted on the United States Patent and Trademark Office (USPTO) website is suggestive about the importance of patenting in aquaculture. Since 1790, the search yields 1273 patents granted, 1261 of which were granted since 1976. From 2001 until the time of writing, 1205 applications searchable for 'aquaculture' were filed. Searching with terms arguably restricted to the context of aquaculture is also possible. 'Finfish' produced 59 hits for patents granted since 1976, and 63 applications since 2001; 'netpen' 13 and 15 hits respectively, 'sea lice' 18 and 14 hits respectively. To put these numbers in perspective, from 1976 to 2006, an average of 113,827 patents were granted by the USPTO each year, an average which swells in the period 2001–2006 to 181,798 patents granted per year (USPTO 2007). As for patent applications, from 1976 to 2005, 202,245 applications were filed with the USPTO, a number which again swells from 2001 to 2005 to 373,583.

Quick patent searches are not methodologically robust, and only severely limit conclusions can be reasonably drawn from the results. Nevertheless, aquaculture does not appear to be a patent-intensive field. Compare, for example searching for 'canola,' which generates 4338 hits for patents granted since 1976 and 5190 hits for 2001–2005 patent applications. The patent gap between aquaculture and agriculture is even more striking when considered in light of aquaculture's tremendous growth. According to the Food and Agriculture Organization of the United Nations, aquaculture of food fish and aquatic plants increased in production 6.9% in quantity and 7.7% in value from 2002 to 2004, with a total value of US $70.3 billion in 2004 (FAO 2006). Truly impressive worldwide growth of a technology-intensive field is associated with very little patenting activity. What explains this surprising situation?

The context for this question is the broadly held view that patents and innovative activity go hand-in-hand. The idea is an old one. Patents provide limited-term, monopoly-like rights to inventors, thereby rewarding them for their efforts and risk-taking behaviour while enabling public disclosure of the invention for further use and development. The incentive-access paradigm, as this is known, is thought to fuel innovative activity because the patent system becomes a *de facto* incentive system. There are well-known problems with this view. Patents are sometimes used defensively to block innovation, and some believe they create resource anti-commons (Heller and Eisenberg 1998). Patents might just be outputs, not inputs that stimulate innovation. In this respect they would be similar to publications or commercializable products – tradeable, but not in themselves equivalent to money. Given, then, the incentive-access paradigm with its *caveats*, what can be said of aquaculture if patenting activity is generally low, but there is innovation-fueled growth worldwide?

The answer is 'intangible assets,' not patents, according to Keith Culver. Patents have long been over-used and abused as direct measures of innovation. States have become accustomed to relying on few measures other than patents counts to evaluate the return on investment from research and development (R&D) expenditure. Activity-based program evaluation by progam funders does not get at the real goal of aquaculture R&D, which is to generate a competitive and sustainable industry. The innovative activity is necessary to achieve this goal sometimes shows up as patents, but they are like the tip of a firm's iceberg of knowledge. The vast majority of the knowledge assets – the intangible assets held by a firm in contrast with its material assets – are unseen. They are not publicly disclosed, they are often not recorded in any formal way, and if recorded or 'codified' they are often protected as trade secrets, whose value is notoriously difficult to express.

Today, intangible assets can easily make up 70% or more of a firm's total value, a shift from an average of 20% in the 1970s (Blair and Roe 1999). Accordingly, to really measure innovation, Culver argues that knowledge management techniques supporting long-term, competitively sustainable industries must be developed and deployed. With this, Brad Hicks agrees, although from his perspective as an industry insider his starting point is somewhat more abrupt. According to Hicks, there is very little in the aquaculture industry that can be patented. Important innovations are generally about husbandry know-how, which is not generally patentable subject matter. Even if it were, Hicks contends, small firms would lose control of the value they create because they have limited resources to defend patents. Perhaps this will change if firms become larger through consolidation, and develop knowledge portfolios worth patenting. Tom Sephton is largely sympathetic with what Culver has to say about the need to have more meaningful measures of innovation and longer-term aspirations for research networks that support industry growth. Yet there is an undercurrent of tension between their views, since Sephton emphasizes the importance of attracting significant contributions to research and attracting international partners in industry. How and whether this will happen in a results-based management framework with intangible assets remains to be seen.

References

Blair, M.M. and M.J. Roe, editors. 1999. *Employees and Corporate Governance*. Washington D.C.: Brookings Institution Press.

Food and Agriculture Organization of the United Nations (FAO). 2006. *State of World Fisheries and Aquaculture*. ftp://ftp.fao.org/docrep/fao/009/a0699e/a0699e.pdf. Accesed August 15, 2007.

Heller, M.A. and R.S. Eisenberg. 1998. Can patents deter innovation? The anticommons in biomedical research. *Science*. 280:698–701.

United State Patent and Trademark Office (USPTO). 2007. U.S. Patents Statistics Chart: Calendar Years 1963–2006. http://www.uspto.gov/web/offices/ac/ido/oeip/taf/us_stat.htm. Accessed August 16, 2007.

Chapter 4
The Mark of Innovation in Aquaculture: The Role of Intangible Assets

Keith Culver

Introduction

Aquaculture faces a complex future in developed states. Few doubt that aquaculture brings at least some social and economic benefits to coastal and rural communities, yet there are many reasonable worries about the social and environmental sustainability of aquaculture, and many obstacles to implementation of research-based solutions to those worries. Governments have nonetheless continued to subsidise R&D efforts in aquaculture, with special preference given to research efforts which bridge the gap between university researchers' experimental results and solution of industry issues. Prominent organizations include the recently discontinued Canadian research network AquaNet which operated from 1999 to 2006 (www.aquanet.ca for legacy webpages) and the European consortium Seafoodplus (www.seafoodplus.org) which represent two especially sophisticated attempts to develop a new mode of R&D. These government-subsidised research bodies aim to provide researchers and end-users with incentives to collaborate in problem-focussed research generating practically relevant, near-term results for end-users, in turn accelerating and increasing production of social and economic wealth. These attempts are closely watched by governments seeking return on national investment in aquaculture R&D (OECD 1996, 1999). Here our trouble begins: what, exactly, is the mark of successful bridging of the gap, sufficient to warrant continued confidence in the ability of aquaculture to innovate in response to emerging challenges? And aside from this question regarding the conditions for successful aquaculture R&D, what about the success of aquaculture R&D marks it as especially deserving of continued national investment, as just one innovative sector amongst many seeking national investment within national and supra-national systems of innovation?

I offer a two-part argument, largely in the context of the Canadian national system of innovation, but with direct relevance to similarly structured systems present in many other OECD states engaged in aquaculture. The first part argues that it

K. Culver
Department of Philosophy and Centre for Social Innovation Research, Centre for Social Innovation Research, University of New Brunswick Fredericton, New Brunswick E3B 5A3, Canada
e-mail: kculver@unb.ca

© Springer Science+Business Media B.V. 2008

is a mistake to treat production of intellectual property, and in particular patents, as a reliable indication that aquaculture R&D has crossed the research-to-industry gap in national systems of innovation. This misplaced reliance results in misleading measurements of national return on investment in aquaculture R&D, and hampers the efficiency of interactions between research, governance, and commercial elements of R&D in national systems of innovation. In the second part I argue that reformulation of the typical goal of aquaculture R&D – a sustainable aquaculture sector – can lead to more useful indicators. From the motivating idea of *competitive sustainability* in aquaculture I argue that measurement of *knowledge management* and *innovation management* ought to absorb and supplant measurement of intellectual property production in a comprehensive picture of the intangible assets created by knowledge-intensive R&D for knowledge-intensive aquaculture production.

This thesis is not an ordinary piece of science or social science, so it is perhaps worth saying a little more about the kind of thesis I take it to be, and the kind of evidence needed to support it. This is a thesis about sound public policy, seeking to improve evaluation of complex national investments by establishing a kind of reflective equilibrium amongst competing priorities in evaluation of the merit of the results of particular national investments in innovation. The evidence offered for this thesis is accordingly an interdisciplinary blend of conceptual analysis, and empirical case study of the particular case of aquaculture in the Canadian innovation system. Whether I am absolutely right will be difficult to assess; whether the option I present as an alternative to current methods of evaluation of aquaculture R&D is more coherent, and less internally contradictory will be rather easier to assess.

Intangibles in National Systems of Innovation

National systems of innovation are heavily implicated in development of the so-called new economy (Roberts 1998). The nature of this economy, and the importance of intangible assets in it, is neatly summarized by Stanford economic growth theorist Paul Romer:

> How can it be that we are wealthier today than people were 100 years ago?. . . This question is puzzling because, if you add up all the things we own, it is clear that the underlying quantity of raw materials has not changed over time. To put the point in extreme form, the total physical mass here on earth is the same as it has ever been, and now we have to divide this up among a much larger group of people. So how could it be that we have more total wealth per person than we ever did before?
>
> When you think about this in the very broad sweep of things, there's only one explanation for this increase in wealth. We took this raw material that was available to us and rearranged it in ways that made it more valuable. We took stuff that was not very valuable and made it much more valuable. So it is not the raw material, or the mass of things living here on earth, that really lies behind economic success and high standards of living; it is this process of rearrangement. And what lies underneath this process of rearrangement are instructions, formulas, recipes, methods of doing things – the things that accountants classify as *intangible* assets if they recognize them at all. They tell us how to take something that is not very valuable and rearrange it into a new configuration that is more valuable (Romer 1998).

Romer marks both the core role of intangible assets in innovation and the difficulty in recognizing, classifying, and evaluating their effects. This difficulty is largely attributable to the fact that we are quite unfamiliar with intangible assets, as their rise in importance has been rapid and recent. Leonard Nakamura famously drew attention to the sudden rise in the importance of intangible assets in his provocatively titled article "A Trillion Dollars a Year in Intangible Investment and the New Economy," observing that "... in the long sweep of US data since 1953, we see that R&D expenditures have more than doubled as a proportion of corporate gross domestic product (GDP) [table and figure omitted]. By contrast we see that tangible investment in plant and equipment (as a proportion of GDP) was no higher in the 1990s than in the 1950s and 1960s [table omitted] (Nakamura 2003)." What has this investment in R&D produced? Nakamura explains:

> These investment expenditures give rise to economically valuable, legally recognized *intangible assets*, including copyrights ("Harry Potter" and Windows 2000) and patents (Viagara and Mach 3) that give the investing firms the exclusive right for a certain period to sell the newly developed products. These assets can be extremely valuable. Pfizer sold over $1 billion worth of Viagara in 1999 after its introduction in April of the previous year.... (Nakamura 2003).

It is now conventional wisdom that in the last two decades the emergence of the new economy has changed the nature of businesses. As Al-Ali reports the view, "formerly, business resources comprised 80 percent of tangible and capital resources, with intangible assets making up around 20 percent. Gradually this changed with intangible assets reaching 80 percent of the assets of most organizations by 1999 (Al-Ali 2003)."

The intangibles-heavy new economy presents several challenges to governments promoting systems, clusters, networks, or other configurations of innovators. There are *organizational* difficulties, as government-sponsored innovation is quite different from innovation within a single firm or consortium of firms. National investments typically involve public sector funders, researchers in university and federal labs, and private sector knowledge receptors. Collaboration across institutional cultures is not always smooth, particularly when discovery-oriented university scientists are asked to re-conceive themselves as cogs in the national economic engine (Lee 2000). Yet government cannot simply demand institutional change as a block of shareholders might do, since government is simultaneously an advocate for innovation and a regulator charged with guarding longer term public interests.

In governments' regulatory role we begin to see the difference in *goals* between private firm innovation, and a national system of innovation. Governments' goals are not satisfied by brief moments of flourishing followed by bankruptcy and dissolution, and governments cannot easily merge or acquire in order to adapt to challenges. It is an exaggeration, but only just, to say that governments desire risk-free innovation producing concrete social and economic benefits within the exclusive control of the investing state's citizens (Albagli et al. 1996). Yet the nature of innovation means this desire for control over risk-free innovation is not easily satisfied. As Zucker observes, "Innovations are generally treated in the growth literature as a nonrivalrous good – freely useable by an unlimited number of potential users at a zero marginal cost (Zucker et al. 1998)." In the absence of assertion of intellectual property rights, innovations can easily be used by others, to the competitive disadvantage of the original innovator. National governments are accordingly very keen to promote robust intellectual property rights regimes, and to further promote active use of those regimes by the nationally-approved beneficiaries of national R&D investments, to turn nonrivalrous goods into rivalrous goods for controlled distribution by nationally-approved commercialisers.

Intellectual property rights can also be used as a measurement of the total amount of innovation occurring within the national system of innovation. A massive literature has emerged tracing patenting rates, rates of mutual citation among patent applications, correspondence between patenting and profits, patenting and stock prices, and so on (Caenegem 2007; Kingston and Scally 2007). This literature has arisen for good reasons. In their influential essay "The Determinants of National Innovative Capacity" leading authorities Furman, Porter, and Stern observe that "the pitfalls associated with equating patenting with the level of innovative activity are widely recognized ... As Griliches (1990, p. 169) point out, 'not all inventions are patentable, not all inventions are patented, and the inventions that are patented differ greatly in 'quality', in the magnitude of inventive output associated with them' (Furman et al. 2002)." This caveat aside, patents are treated by these influential authors and national systems of innovation depending on these authors' methods as the *de facto* measure of innovation. Furman et al. say that:

> Ultimately our approach is based on the assessment that patenting rates constitute 'the only observable manifestation of inventive activity with a well-grounded claim for universality.' (Trajtenberg, 1990, p. 183), a judgment reflected in prior work employing international patent data (Evenson, 1984; Dosi et al., 1990; Cockburn and Henderson, 1994; Eaton and Kortum, 1996, 1999; Korum 1997; Vertova, 1999). (Furman et al. 2002)

Much more can be said about patenting, but this much is sufficient to support a general claim about national systems of innovation. In light of the special challenges they face in demonstrating return on investment, intellectual property generally, and patents specifically, are tremendously attractive as the best available indicator, measure, and guardian of nationally-subsidised innovations. Whether this appearance matches reality in practice is another matter, one to which we will turn now in the context of aquaculture innovation in Canada.

Innovation in Canada, Aquaculture, and Commercialisation of R&D

The Canadian innovation system takes intangible assets seriously, and takes especially seriously the commercialization of intangible assets created by nationally subsidized, largely university-based research. In the national innovation strategy launched in 2002 by the then Liberal government, the central document of the strategy, *Achieving Excellence*, identifies as a key goal to "Leverage the commercialization potential of publicly funded academic research (Government of Canada 2002)." The method for protection and exploitation of this commercialization potential is expressed in the context of competition with American institutions whose success sustains the productivity and wealth gap between the US and Canada.

> Canadian investments in university research, both private and public, are also generating economic benefits. In 1999 Canadian universities and research hospitals earned $21 million in royalties, held $55 million in equity, generated 893 invention disclosures, were issued 349 new patents and executed 232 new licences. To date they have spun off as many as 818 companies, posting a strong record relative to the U.S. The Association of University Technology Managers estimates that the commercialization of academic research in Canada resulted in more than $1.6 billion in sales and supported more than 7300 jobs in 1999. The evidence suggests that universities can contribute to economic growth and benefit from industrial funding without compromising their role as key performers of basic research, and without compromising their ability to disseminate results widely through publishing.
>
> A comparison of the 139 U.S. and 20 Canadian universities that report to the Association of University Technology Managers suggests that there is still room for improvement. Although the U.S. universities perform about 14 times as much research as their Canadian counterparts, they receive 49 times as much licensing income — a key indicator of the value of innovations. Recommendations by the Advisory Council on Science and Technology centred on the need for the government to provide financial support to enable universities to increase their level of effort.
>
> In return, universities need to focus on areas of excellence, train greater numbers of highly qualified people in the skills required by the private sector and government, and more aggressively seek out commercial applications for publicly funded research. Key commercialization performance outcomes should at least triple over the next decade (Government of Canada 2002).

Whatever the merits of this bargain between government and researchers, its terms are clear. The Canadian government of the time offered increased funding of research enabling increased academic output in the form of peer-reviewed publications, on the condition that publication is matched by a tripling of commercialization outputs within a decade, using as the preferred, gold standard output the financial consequences of licensing of patents. Since that time a minority Conservative government has taken control, yet interestingly, little has changed but the relative emphasis on the importance of the private sector to commercialization. The Conservative government's May 2007 Science and Technology Strategy uses the same knowledge translation and mobilization rhetoric, working from the title *Mobilizing Science and Technology to Canada's Advantage* to call for further investigation of mechanisms for better commercialization pathways from universities to the private sector, while nonetheless using patents and licensing as a key indicator. It is worth quoting in full a relevant paragraph of the strategy to mark the similarity in policy between successive governments led by opposed parties:

> Businesses in Canada need to do more to improve their productivity. Canada's private-sector R&D investment as a proportion of GDP is below levels in Japan, the United States, Germany, and France. Similarly, the number of patents produced in Canada is low compared with many other OECD countries. Canadian firms also invest less in new machinery and equipment, which embody the latest innovations, than do many of their competitors. Since these investments are a key productivity driver, it is crucial that the private sector increase its investments in S&T and advanced technologies (Industry Canada 2007).

The 2007 Conservative strategy is however, likely an advance on the prior federal strategy to the extent that it recognizes that successes to date in patenting and licensing represent nonetheless what at least appears to be underperformance by university-private sector collaborations.

The strategy explains:

> Efforts to support the transfer of technology from Canadian universities to the private sector are resulting in spin-off companies, technology licensing agreements, and patent filings. More can be done to encourage technology transfer at both ends of that process. A review will be launched to uncover factors that might be inhibiting S&T collaboration between industry and the higher-education sector (universities and colleges). This review will include an assessment of whether a new approach to intellectual property management of university research is warranted. In the meantime, the government will pilot new approaches to university and government laboratory technology transfer; greater involvement by the private sector in the design of these new approaches is needed (Industry Canada 2007).

Even as a small move away from emphasis on patenting is visible, it remains quite remarkable just how firmly successive governments remain committed to a 'push' version of innovation famously articulated by Vannevar Bush in his 1945 policy paper "Science the Endless Frontier" (http://www.nsf.gov/od/lpa/nsf50/vbush1945.htm). On the 'push' version of innovation, discoveries in science and technology are 'pushed' from the scientific explorers of university, government, and corporate labs out to commercialisers, with patents serving a crucial role as the exchangeable currency of innovation.

It might be thought that a patent-intensive, push model of national R&D investment would suit aquaculture very well. Aquaculture depends on a blend of basic and applied biology with engineering and other disciplines, and much of that expertise lies in universities, organizations increasingly keen to see their patented products and processes licensed to private sector commercialisers. Other organizations such as the Network of Centres of Excellence are similarly focused on push innovation using translation devices and measures which prominently include patents – throughout AquaNet's life, for example, reporting requirements specifically listed patents as a distinct reporting category, and patenting and licensing featured prominently in the annual reports of the wider Network of Centres of Excellence program. Yet despite these policy pressures and incentives to patent, Canadian aquaculture is not currently a patent-intensive R&D environment. A search of the Canadian Intellectual Property Office patent database reveals just 91 patents applied for or issued listing 'aquaculture' within the patent's claims, contrasted against over nearly 2800 listing 'crop.' This is a rough measure, to be sure, but what is important is that aquaculture is not a near rival, but a distant also-ran. There are other indicators which, while equally blunt, point in the same direction. For example, no patent applications resulting from AquaNet-funded research have been reported to AquaNet. What might the causes of this situation be? A failure on the part of university researchers to recognize the importance of protecting discoveries so Canadian commercialisers are protected from unscrupulous would-be thieves of Canadian-developed aquaculture technologies? A failure on the part of the industry to invest in patent protection out of some dismal motive like a 'get rich quick' attitude with little regard for the long-term sustainability of the sector? Does this mean Canadian aquaculture is not innovative and so does not deserve national investment? There are several legal, cultural, and strategic reasons to reject this conclusion, inclining instead toward the conclusion that patents are ill-suited to measurement of aquaculture innovation.

Legal Considerations

Patents are certainly one measure of intellectual property production, but patents are not exhaustive of intellectual property relevant to innovation. Copyright is increasingly important as a method of tracking contribution of ideas to scientific debate, and in high-value software used in biotechnology research and in areas such as genetic management of aquaculture fish stocks. Changes in aquaculture animal diets to include plant material may similarly increase the importance of Plant Breeders' Rights, as the Department of Fisheries and Oceans reported in 2005 that "fish farmers help to ensure that the [capture] fishery remains healthy and abundant by reducing their use of wild fish up to nearly 40 percent in the last 5 years. They have done so by using more vegetable-based protein in the diets given to salmon" (http://www.dfo-mpo.gc.ca/media/backgrou/2005/salmon_e.htm). These publicly visible, easily counted assertions of property rights are of course not the only kinds of intellectual property: innovators may rely on trade secrets to protect their ideas. Trade secrets are an inexpensive, easily employed, and legally enforceable way to protect intellectual property, however their role in innovation is notoriously difficult to assess, since examination of them carries the danger of destroying their secrecy.

Cultural Considerations

One response to the shortcoming of patent and licensing rates as indicators of innovation is to simply add other forms of intellectual property to the list of success conditions for national systems of innovation. Yet even this expanded point of view is likely insufficient to capture fully the intangible assets generated by R&D investment. Even if evaluators assess research outputs in light of a fuller understanding of intellectual property, there are reasons to suppose that there are cultural obstacles to full reporting of intellectual property discovery. University-based research is still heavily devoted to discovery and problem-solving, and not to acquisition of property rights. While some researchers clearly have oriented toward themselves toward both publications and patents (Di Gregorio and Shane 2003; Owen-Smith and Powell 2001), researchers have varied incentives to patent. Faculty collective agreements across Canada vary regarding ownership of intellectual property created by employees, just as universities vary in their financial and administrative capacity to allow or encourage faculty to participate in further development of an idea captured in a patent. Even those researchers who are patent-oriented are not following a uniform regional, national or disciplinary standard for detection of intellectual property, and innovation systems researchers are aware that little is known regarding the efficiency of innovation activities prior to patent filing. As Agarawal explains in the context of discussion of differences between firm-focussed innovation studies and university-focussed innovation studies, focus on patenting has left a gap in our knowledge of just where universities as knowledge-generators innovate and communicate innovation. Agarawal writes:

> Much of the work in the area of *university characteristics* has focused on determining the effects of the Bayh-Dole Act. While this is certainly an interesting and important topic, there are many others that have been largely ignored. For example we know very little about the amount and type of technology transfer that occurs outside the formal route of the technology transfer office, but many scholars have suggested that there is reason to believe this could be quite significant (Agrawal 2001).

Strategic Considerations

Funder-mandated incentives to patent may also be overcome by strategic considerations. At a global level, problems of common concern may warrant joint efforts to resolve those problems together with efforts to communicate the solution widely for the health of the entire sector. For rather different reasons, where problems have genuinely local solutions there may be little incentive to patent. Local solutions are often incremental and specialized to local circumstances, as, for example, innovation in Canadian cod farming is focused in New Brunswick and Newfoundland, and there is little danger of losing competitive advantage protected by trade secrets so long as researchers do not provide the international research community with specific recipes regarding competitive advantages gained through Canadian R&D. As Zucker et al. explained in discussion following a passage I cited earlier, while the research results of a group of researchers may be a non-rivalrous good easily taken up and used by others, the intellectual capital of a particular group of researchers is a rivalrous good. Zucker et al. write that

> We believe that some innovations, particularly a breathrough 'invention of a method of inventing' (Griliches, 1957), may be better characterized as creating (rivalrous) human capital – intellectual human capital – characterized by natural excludability as opposed to a set of instructions for combining inputs and outputs which can be protected only by intellectual property rights. This natural excludability arises from the complexity or tacitness of the information required to practice the innovation (see Nelson, 1959; Kenneth J. Arrow, 1962; Nelson and Sidney G. Winter, 1982; and Nathan Rosenberg, 1982). (Zucker et al. 1998)

Groups of researchers may develop a knowledge-generating method not easily conveyed in recipe format, and in the absence of defection of bearers of tacit knowledge about this method, the products of this group have the power of 'natural excludability.' There may, then, be little net value in expending effort and funds on small patents in a small, new industry when there are no immediate competitors who cannot be kept at bay by less costly means. A wide variety of tactics are available: trade secrets, non-disclosure agreements between researchers and industry partners, natural excludability or even simply professional ethics and obligations of confidentiality.

The specific characteristics of aquaculture R&D raise a final worry about the role of patents in measurement of aquaculture innovation. In revising the weight given to patent applications and granted patents in assessment of the innovation capacity of the aquaculture sector, care must be taken in constructing inferences from patenting *rates* to conclusions about aquaculture R&D institutions' contribution to the national innovation system in turn supporting the new economy. Patenting rates are measures of activity over time, and the typical time-frames of aquaculture R&D – typically closely aligned with seasonal and annual cycles – are not conducive to use of patenting as a reliable indicator of innovation. Concerns regarding public accountability mean that research grants are typically limited to research periods of 2–3 years prior to assessment of progress and advice requiring course correction. Even longer-term organizations such as AquaNet, with a 7 year mandate, are evaluated at the close of 3 years, on a mid-term basis. This is not to say that patenting is irrelevant to all research networks – some networks can and do patent effectively, but short evaluation time-frames applied to a seasonal and multi-year aquaculture cycle likely count as an exception to reasonable expectations regarding research network patent activity. The subject matter of particular networks matters as much as the formal organization of the network as an R&D organization. Given that patent applications are themselves not final evidence of eligibility for patent, and the fact that roughly 3 years elapses between application and issuance of a successful patent application, it seems that patents are most useful as evidence of innovation-creation present in longer-term R&D projects, sometimes present in the Canadian innovation system but much more likely to be associated with R&D areas not dependent on seasonal or annual growth and development cycles. Moreover, supposing that patents are indicative of innovation-creation in these longer term projects still leaves unexamined the fact that possession of a patent is not equivalent to possession of a sound business plan, and the further fact that a sound business plan built on patents is only as strong as the patents themselves. (Consider, for example, Canadian technology firm Research In Motion's costly 3-year litigation with a firm claiming patent infringement by the products Research In Motion contended were its inventions. Whether Research In Motion was in the right is of less relevance than the fact that it is time-consuming and organizationally costly to manage these suits.) There may well be some desirable ratio of patent applications to innovations in aquaculture R&D, but that ratio is unlikely to be easily found in near-term evaluation of patenting. The value of a patent to innovation is better assessed once it has been licensed, used, and its importance to Canada's competitive advantage in the world can be demonstrated and the wisdom of the patent owner's choice to invest in patenting (rather than protecting the research result as, e.g., a trade secret) can be assessed. All of this points to a feature of the nature and value of intangible assets which tends to go missing in patent-focussed evaluation of innovation systems: the fact that the point of strategic innovation is not to generate some specific number of innovations. Rather, our larger goal is to create what is sometimes glibly referred to as 'continuous innovation,' more precisely expressed as an infinitely repeatable 'method of inventing' creating a power of natural excludability for those outside the controlled community of government-subsidised innovators. In evaluating whether researchers are keeping their end of the bargain with government with respect to commercialization of research results, our goal is to know whether they are doing the kinds of things which later result in innovations sometimes justifiably protected by patents in relatively late stages of the transfer of research results to commercialisers. What, then, might indicate more reliably the mark of intangibles-intensive innovation in Canadian aquaculture R&D?

One Step Back, Then Two Steps Forward

Even with the demise of AquaNet as a network and meeting place for Canadian aquaculture R&D, it remains possible to discern something like a common sector goal, usefully and generally summarized as "an environmentally sustainable aquaculture sector benefiting all Canadians." There is nothing unusual about this goal, and it has the advantage of synergy with the goal of adjacent sectors, such as forestry and mining. Yet it is worth considering whether the core goal of 'sustainability' is so vague as to be unhelpful as a guiding standard which guides tactical judgements regarding funding and direction of research programs, evaluated against ground-level progress measures such as production of intellectual property. Vagueness in goal-setting runs the risk of encouraging vague measures of progress toward that goal, achieving precision without accuracy as we struggle to relate absolute numbers of, e.g., patents issued, to the goal of sustainability. Put in context of *competitive sustainability*, however, the goal of sustainable aquaculture can be judged comparatively against competitors.

In this new context, indicators and absolute measurements can be given more nuanced interpretation. Canadian patenting rates in aquaculture, for example, can be compared against patenting rates in competing innovation systems such as those of the USA, Norway, the UK, and France. That analysis will allow us to ask whether there are geographic or network spill-over effects as high patenting rates in one area or network spill-over into adjacent areas or networks, in the form of patent use, cross-citation in filing of improvement patents, or entirely novel patents in aquaculture. Reporting of Canadian patenting rates can become contextualized in a way which will enable more nuanced management of intellectual-property seeking, examining not just the rate of patenting but the impact of patenting in a new economy area. Have Canadians sought key patents at the foundation of new technologies? Or are Canadian researchers and developers consistently claiming to innovate while in reality their patents are follow-ons to others' innovation? Are Canadians' patents in other jurisdictions referred to by other patent-applicants in those jurisdictions, demonstrating international recognition of our leadership? And where other jurisdictions patent less yet still provide us with stiff competition, what non-IP intangible assets are our competitors using to maintain their competitiveness?

If this proposed emphasis on competitive sustainability is justified, heavily contextualized indicators and measures should be anticipated, and not just for patenting but in evaluating and defending choices *not* to patent. Competitive sustainability can be pursued and the national interest in return on investment can be satisfied so long as those choices offer some other means to transfer knowledge, secured for the benefit of Canadians by other means such as natural excludability – or when we get the transfer from research to end-user right, 'transferrable natural excludability.'

What, then, might these heavily contextualized indicators and measures look like? And what indicators and measures should apply to aquaculture R&D organizations?

Here we may borrow usefully from the private sector's experimentation with knowledge management (Bontis 2001), a set of experiences which have led to useful distinction of three purpose-differentiated phases of knowledge creation and application in the continuum from research to innovation. *Knowledge management* is now widely taken to consist in mobilization of tacit knowledge within organizations to face the problems and goals of those organizations (Bennet 2004). Leading academics tend to be quite good at knowledge management, communicating regularly at conferences, fostering graduate students into the profession through informal pub-based chats mulling through problems and methods, and circulating pre-publication papers for comments. Yet *innovation management* is relatively foreign to academics, consisting in organization- or system-wide selection of knowledge capable of development into a product, process, or service likely to become an innovation in the marketplace when introduced and managed there by a skilled commercialiser. Innovation management involves elements of end-user relationship-building, anal ysis of market opportunities, and early-stage intellectual property management as decisions are made regarding the kind of intellectual property protection appropriate to the proto-innovation. *Intellectual property management* is the final stage, as the marketplace innovation's intellectual property must be protected, perhaps licensed to others, and trademarks and logos which represent the innovation must be sought and protected.

Seen through the lens of knowledge management, innovation management, and intellectual property management, aquaculture R&D organizations such as AquaNet and Seafoodplus appear to be properly regarded as knowledge management organizations whose mission intersects with innovation management. Production of intellectual property should be expected to play a minor role in these organisations' efforts to conduct timely, industry-relevant research delivering near-term results to industry. Reporting of intellectual property production and rates of production is therefore unlikely to be an accurate indication or measure of R&D success. More relevant indicators and measures must account for the particular character of these organizations and the private sector receptors of knowledge created. Particular attention should be given to the way knowledge is transferred between phases of R&D activity and between research and end-user organizations. Our rather lengthy organizing question, then, must be:

What are indicators and measures of increased competitive sustainability in aquaculture resulting from:

- investment in a rivalrous 'method of inventing' whose use of tacit and explicit organizational knowledge provides
  - *rapid* response to
  - *industry-relevant* R&D questions and delivers
  - *near-term* results secured for Canadians' benefit through
    - intellectual property protection, or
    - rivalrous 'know-how' whose dissemination can be controlled?

This question can be unpacked and indicators found across knowledge management, innovation management, and intellectual property management phases of R&D-driven innovation in the context of Canadian aquaculture's competitive sustainability.

Analysis of the effectiveness of an organisation's development of a 'method of inventing' is analogous to examination of maintenance of the physical plant. In the case of aquaculture R&D typically spread across networks of researchers, the key competitive feature of the method of inventing is its capacity to make the network of investigators somehow more than the sum of their parts. In order to reach beyond misleading and incomplete indicators in the form of counts of intellectual property produced and counts of researchers involved, indicators must capture the amount and quality of communication through the organisation. One indicator and accompanying measure might be drawn out of the following three questions: "Is there an organization communications plan and mechanism for systematic sharing of information? How frequently can it be used and how frequently is it used?" We might then contextualize this information by asking "How does the communications mechanism compare to that of competitors?" Similarly, we might ask as an indicator "Is there provision for exchange of researchers and technicians amongst research institutions?" and follow up asking "How many researchers or technicians are actually exchanged" and "How does this exchange mechanism and rate compare against competitors' practices?"

The operational effectiveness of the organization can then be measured by indicators located at the interstices of a matrix composed of the intersection of x-axis categories of knowledge management, innovation management, and intellectual property management, and y-axis categories of rapidity of response, industry-relevance, and achievement of near-term results. In the knowledge management phase of R&D, for example, indicators for rapidity of response or 'timeliness' of engagement of industry relevant questions might include questions such as "What formal and informal mechanisms exist for contact with industry?" and measures might include provision of logs of numbers of telephone or email contacts, use of a fund for non-fieldwork visits to industry members, or use of a fund for preliminary exploration of emerging questions. Once a project or larger, integrated research initiative achieves milestones near research goals with commercial application, 'timeliness' as rapid-response remains relevant. Indicators might include questions asking "What mechanism exists, e.g., short-term secondment, to rapidly focus organizational expertise on troubleshooting of a proof of concept?" and measures might include frequency of use of the available mechanisms and comparison of the measures and their use against competitors' analogous activities. Further indicators and measures can be developed through the matrix in a similar manner, always looking to examine where knowledge has been transferred in a way which disseminates knowledge in a targeted fashion creating Canadian rivalrous human capital who 'know how' to do what competitors do not, and non-rivalrous knowledge protected by intellectual property provisions prior to publication or other public disclosure.

In all of this I have left untouched the questions of who ought to set the indicators and measures, and what they might accept as an incentive for doing so. Let me simply offer without supporting argument the suggestion that it is in the best interest of accountability-driven public servants *and* publicly-funded aquaculture researchers to *collaborate* in development of novel indicators and measures to gain a better picture of the merits of aquaculture as a new economy activity. Contrary to the dark muttering sometimes heard at conferences regarding the tendency of disciplines to hold onto funds whether they need them or not, aquaculture researchers seem to me, in my experience as Chair of the Research Management Committee of AquaNet, to be quite convinced of the merit of aquaculture and positively hungry for opportunities to show aquaculture's promise. For these researchers, the prospect of continued funding of open and publicly scrutinized aquaculture science seems incentive enough to adopt new ways of understanding the return on national investment in aquaculture R&D.

Conclusion

I have argued for a course-change away from regarding intellectual property, and particularly patents, as the measure of aquaculture R&D organisations' contribution to the Canadian national system of innovation. The intangible assets those organizations create include but are not limited to patents and other forms of intellectual property. Organisations operating on short-term funding and evaluation periods ought to emphasise instead the value of the rivalrous non-intellectual property intangible assets they create, in human capital capable of solving emerging problems, and the variety of ideas and know-how which can be employed to benefit Canada without protection as intellectual property. This emphasis requires aquaculture R&D organizations to actively study their own contribution to innovation, and to promote increased internal sharing of knowledge and transfer of that knowledge of innovations near to market to those end users prepared to commercialise those innovations. This is no small task, and it will require unusual depths of collaboration between governance and research elements of the national system of innovation, perhaps to the extent of requiring routine secondments between governance and research elements to create a cadre of science-policy-innovation experts. An existing program can be used in support of this program. See the 'Interchange Canada' program at http://www.psc-cfp.gc.ca/ic-ec/index_e.htm). Anything less than this deliberate blurring of institutional cultures puts at risk a valuable opportunity to move an adversarial model demanding that aquaculture 'show me' its success, to a systemic, collaborative model working to understand 'what we have accomplished' in advancing Canada's competitiveness in industry-led, research-driven, sustainable aquaculture. (Leitner and Warden 2004; Smith 2000, 2003).

References

Agrawal, A. 2001. Research on university-to-industry knowledge transfer: framework of existing literature and unanswered questions. *International Journal of Management Reviews* 3: 285–302.

Al-Ali, N. 2003. *Comprehensive Intellectual Capital Management.* Hoboken, NJ: John Wiley and Sons.

Albagli, A., P. Dawson, and S. Hasnain. 1996. Competitive science and technology intelligence. *International Journal of Technology Management* 12: 320–328.

Bennett, A. 2004. Alive with the fire of shared understanding: implementing knowledge management in the Department of the Navy. In M. Koenig and T. Srikantaiah, eds. *Knowledge Management: Lessons Learned*, 279–292. Medford, NJ: American Society for Information Science and Technology, 2004.

Bontis, N. 2001. Assessing knowledge assets: a review of the models used to measure intellectual capital. *International Journal of Management Reviews* 3: 41–60.

Caenegem, W. 2007. *Intellectual Property Law and Innovation*. Cambridge: Cambridge University Press.

Di Gregorio, D. and S. Shane. 2003. Why do some universities generate more start-ups than others? *Research Policy* 32: 209–227.

Furman, J., Porter, M., and S. Stern. 2002. The determinants of national innovative capacity. *Research Policy* 31: 899–933.

Government of Canada. 2002. *Achieving Excellence: Investing in People, Knowledge and Opportunity,* Appendix A, ISBN 0-662-31225-253564B. Ottawa: Government of Canada.

Industry Canada. 2007. *Mobilizing Science and Technology to Canada's Advantage*. Ottawa: Industry Canada.

Kingston, W. and K. Scally. 2007. *Patents and the Measurement of International Competitiveness: New Data on the Use of Patents By Universities, Small Firms and Individual Inventors*. Cheltenham: Edward Elgar.

Lee, Y. 2000. The sustainability of university-industry research collaboration: an empirical assessment. *Journal of Technology Transfer* 25: 111–133.

Leitner, K. and C. Warden. 2004. Managing and reporting knowledge-based resources and processes in research organizations: specifics, lessons learned and perspectives. *Management Accounting Research* 15: 35–51.

Nakamura, L. A trillion dollars a year in intangible investment and the new economy. In Hand, J and B Lev, eds. 2003. *Intangible Assets: Values, Measures, and Risks*, 19–47. Oxford: Oxford University Press.

OECD. 1996. *Innovations, Patents and Technological Strategies*. Paris: OECD.

OECD. 1999. *Managing National Innovation Systems*. Paris: OECD.

Owen-Smith, J. and W. Powell. 2001. To patent or not: faculty decisions and institutional success at technology transfer. *Journal of Technology Transfer* 26: 99–114.

Roberts, R. 1998. Managing innovation: the pursuit of competitive advantage and the design of innovation intense environments. *Research Policy* 27: 159–175.

Romer, P. 1998. The soft revolution: achieving growth by managing intangibles. Journal of Applied Corporate Finance 11: 8–14. Reprinted in Hand J and B Lev, eds. 2003. *Intangible Assets: Values, Measures, and Risks*, 63–94. Oxford: Oxford University Press.

Smith, J. 2000. From R&D to strategic knowledge management: transitions and challenges for national Laboratories. *R&D Management* 30: 305–311.

Smith, J. 2003. Building an entrepreneurial knowledge culture in a national research laboratory. *R&D Management* 33: 231–237.

Zucker, L., M. Darby and M. Brewer. 1998. Intellectual human capital and the birth of US biotechnology enterprises. *American Economic Review* 88: 290–309. Reprinted in Hand J and B Lev, eds. 2003. *Intangible Assets: Values, Measures, and Risks*, 185–206. Oxford: Oxford University Press.

Chapter 5
New School Fish Production vs Old School Fish Harvesting

Brad Hicks

Introduction

Fish farming has been practiced for about 5000 years. Until the mid 1970s, however the majority of fish farming practices were extensive. By the mid 1970s the knowledge of fish diseases, fish nutrition and fish husbandry had advanced to the state that intensive farming was becoming possible. Early attempts at intensive fish farming were still plagued by significant information and knowledge gaps which resulted in many failures. By the mid 1980s enough progress on the know-how of fish farming had been made that intensive fish farming was becoming a global reality. Between the mid 1980s and the present, intensive fish farming has been making great strides. For example, by 1996 the production of salmon on farms exceeded the production of salmon from wild sources and by 2005 approximately 70% of the world salmon was originating from farms. Such rapid increases in production have been made possible only as a result of the rapid increase in the development of intangible assets.

This chapter has two over-arching goals in addressing Culver's views about the role of intangible assets in aquaculture:

- emerging social issues arising because of innovation in aquaculture; and
- responsible governance models for aquaculture innovation.

An Emerging Industry Embraces Innovations

The practice of aquaculture today is far different than it was just 20 years ago. Rapidly growing modern day aquaculture depends on sophisticated financial, human resource, animal husbandry, marketing and other systems, which in turn rely heavily on intangible assets. The Canadian aquaculture industry has grown and expanded in the "new." Commercial aquaculture is still populated with risk-taking founders: Brian Blanchard of Scotian Halibut, Glen Cooke of Cooke Aquaculture and Larry

B. Hicks
Vice-President, Taplow Feeds, 21222 24th Ave, Langley, BC, V2Z 2AB, Canada
e-mail: bhicks@direct.ca

K. Culver, D. Castle (eds.), *Aquaculture, Innovation and Social Transformation*,

© Springer Science+Business Media B.V. 2008

Ingalls of Ocean Horizons in New Brunswick, Gord Cole of Aquacage Fisheries in Ontario and Tom May of Paradise Trading, Gus Angus of Totem Seafoods, Spencer Evans of Creative Salmon and, Terry Brooks of Canadian Sable Fish. All entered the industry in Canada during the pioneering phase and continue pioneering husbandry techniques such as organic aquaculture and the rearing of novel species such as Halibut and Atlantic cod on the East coast and Sable fish on the West coast. With the continued commitment of these founders of Canadian aquaculture, it is not surprising that aquaculture embraces new ways of doing business and has rapidly adapted to an industrial format sustained by extensive intangible assets. Aquaculture is still in the experimental phase of development with inventions and innovations occurring daily. Research and development are key components of new and emerging industries with the result that the aquaculture industry is also heavily populated with managers with research backgrounds.

Intangible assets having substantial commercial value include: manufacturing systems producing feed, nets and pen systems; hatcheries with complicated recirculating water systems; complicated breeding programs; complex on-farm rearing strategies, sophisticated harvesting methods and processing facilities heavily reliant on automated machinery. Automatic feeding systems, the use of video cameras to monitor feeding, automatic pneumatic stunning devices for harvesting fish, and automated computerized filleting machinery for processing are now common. This sophistication also extends to the administration of modern aquaculture enterprises. Accounting and marketing systems are very reliant on the use of intangible assets such as computer modeling for inventory control, monitoring and forecasting production (http://www.superior.no/index.cfm?id=116840).

The rapid development of intangible assets as production aids and the almost immediate adaptation of these intangibles is a hallmark of modern aquaculture. Research has contributed greatly to all aspects of aquaculture production, from feed production and feed ingredients onward. Early diets relied on frozen horse spleens, while modern diets are based on the nutritional needs of the fish and are based on mainstream agricultural commodities with the inclusion of fish meals and fish oils and single amino acids and modern vitamin and mineral premixes. In hatchery design and rearing techniques, initially all hatcheries were flow-through using either troughs or raceways where design was based on simple methods to hold and handle water and, where very little if any consideration was given to the requirements for the fish. By contrast, modern hatcheries use recycle systems which allow for optimum conditioning of the water, and holding facilities are designed with both engineering and animal husbandry being taken into account. With respect to cage and net design, initial cages were small and frequently made of wood and as a result were fragile and had to be placed in well protected areas where often water quality was poor. Modern steel cages are well engineered and are much large resulting in much better housing (fish thrive in larger cages) and the better engineered cages can be placed in more open waters which are often of better quality. Modern on-farm husbandry with better housing, better feeds and improved handling techniques have decreased time to harvest by one half and improved the feeding efficiency by about 400%. Production efficiencies have been further increased by sophisticated, automated harvesting and processing technologies. Initial harvesting was by hand brailing and early processing was by hand. Now fish are harvested by large vessels equipped with either large mechanical wet brailers or live fish pumps and processing is primarily by sophisticated computer controlled processing machinery. Based on these innovations modern aquaculture had moved from a system of growing fish in an inefficient, old school, agriculture model which relied heavily on husbandry skills and farm labour to a modern production format which rivals the most sophisticated food production systems. This is clearly illustrated by examining on-farm per employee production, which has increased dramatically from approximately 15 tonnes per person in 1990, to 55 tonnes per person in 1995 (EAO 1995) and to in excess of 100 tonnes per person by 2007 (Fisheries and Oceans Canada 2007). This compares well with Scotland where the production per person soared from 17 tonnes per person in 1989 to 62 tonnes per person by 1997 (Fisheries and Oceans Canada 2007).

One result of this marked increase in efficiency has been a significant decrease in the cost of production and concomitantly, an increase in production and a decreased market price in response to increasing supply (Asche 1997 Highlands and Islands Enterprise 1999). The price of salmon declined by 54% to 92% depending on species between 1988 and 2002 (Eagle et al. 2004).

The rise of the aquaculture industry, and the fall of the commercial salmon fishery, have resulted in significant social stress and upheaval. Here we are speaking of Canadian harvest declines on west coast, with next to virtual absence of east coast commercial salmon fishery since salmon was declared an endangered species in Maine. Salmon production in Canada for the last century was based on harvesting wild salmon. The harvesting of wild salmon was regulated and entwined in a complicated web of government regulations and policies which were designed to conserve the resource while at the same time meeting constitutional requirements to supply fish to First Nations groups as well as supplying fish to commercial and sport fisheries (Noakes et al. 2002, Schwindt et al. 2003). There is also an overriding policy that harvesting of these fish must somehow maximize employment. As a result of these competing demands, the wild salmon fishery has become very inefficient with many participants and extensive over-capitalization (Eagle et al. 2004, Schwindt et al. 2003). The end result of the social engineering of the wild fishery is that there has actually been a decrease in the per tonne production per harvester and an increase in the cost of production (Schwindt et al. 2003). The situation of the commercial salmon fishery is much different from the situation of the relatively prosperous, expanding aquaculture sector, and the difference has not gone unnoticed. As the cost of production has increased for wild fish and decreased for farmed fish the commercial fishery has experienced financial difficulties with serious adverse social consequences. The loss of financial viability and the significant loss of social status which has accompanied this insolvency has resulted in angry, frustrated, fishers and their allies lashing out at the aquaculture sector (CBC 2003; Friends of Wild Salmon Pinkerton, 1999). Protecting the wild commercial fishery from further erosion has motivated the remaining wild harvesters to adopt the tactic of demonizing the aquaculture sector as a strategy to influence public policy in ways that will hamper the growth and efficiency of the aquaculture industry. The wild fishery has had some success using this tactic (Price Waterhouse Coopers 2003).

Significant pressure on the design and implementation of public policy has exposed a weak spot in the Canadian government's ability to respond to such pressures. The government does not have what Culver calls "a cadre of science-policy-innovation experts" capable of dealing with this new and innovative industry. This lack of intellectual capacity had resulted in the government developing reactionary policies, which interfere with aquaculture production. By adapting such policies the Canadian government is making it increasingly more difficult for the Canadian aquaculture industry to compete internationally. The rapid growth of aquaculture in Canada has slowed and there is a fear that unless the national government increases its ability to deal with this industry that Canada could become only a minor player in aquaculture production.

Inventions and Innovations are Not Risk Free

Governments by nature are conservative, non-risk-taking-institutions. Invention and innovation are risk-taking activities. And although Culver is right to identify the fact that, "governments desire risk-free innovations producing concrete social and economic benefits," there are no risk-free innovations, where innovations are understood as necessarily involving some kind of change to a product or process. In this context, risk is often misunderstood. Risk is a personal and sometimes group perception, not an impersonal, specific and independent entity. Often group perceptions of risk are associated with shared comprehensive worldviews which may conflict with the worldviews of other groups. In the case of aquaculture the farmer perceives the risks to be the vulnerability of the crop and the vagaries of the market place. The commercial fishers perceive the risks as competition from the farmer resulting in loss of livelihood and loss of status in the community. The environmentalist or preservationist sees the risk as change in the way things are done and these changes may include risks to their perceptions of how the environment functions.

The fishers, environmentalists and preservationists continue lobbying government to develop public policies which stifle innovations. These efforts to influence public policy in Canada are severely hampering the ability of Canadian firms to extract value from their intangible assets (where know-how, at a minimum, would mean the know-how to raise fish) and these intangible assets are then being exported out of the country where their value can be realized. One way of evaluating the loss of intangible assets in Canadian aquaculture is to reflect on the loss of investment in human capital which has been migrating from Canada to other counties such as Chile where invention and innovations are embraced (United Nations 2006) We can identify by name several people who have migrated to Chile to take advantage of the opportunities in Chile and list several Canadian companies which expanded into Chile both on the farming and on the health and equipment supply side. The Chilean aquaculture industry, which is worth about $ 2,000,000,000 (United Nations 2006) is now three times the value of the Canadian industry (Fisheries and Oceans Canada 2006 Statistics Canada 2005) even though the Canadian industry had a 5-year head start and the Canadian industry is closer to the main market for both Canadian and Chilean aquaculture products. The failure of the Canadian government to recognize the value of intangible assets (human capital), their role in modern aquaculture, and the mobility of intangible assets is crippling the competitive advantages of the Canadian aquaculture industry.

The aquaculture sector was very quick to adapt to the new economy. It had the financial resources and the intellectual capacity to make the switch. The commercial sector and the government were not prepared to adapt. Support to develop experts within intellectual capacity to understand this new industry is essential if Canada is going to participate in the fisheries of the future. Sadly, this support has been in short supply to date.

Where Does Aquaculture Research Fit in the Race for Government Funding?

Aquaculture is a relatively small industry in Canada. There are limited funds available for aquaculture research. Compared with other food production systems, aquaculture is very poorly funded. Part of the reason is that the majority of food producing systems are led at the federal level in Canada by the Department of Agriculture and Agri-Food, while aquaculture is led by Department of Fisheries and Oceans (DFO). The implications of this division are worth noting precisely. Agriculture and Agri-Food Canada has a mandate to help food producers:

> **Our Mandate:** Agriculture and Agri-Food Canada provides information, research and technology, and policies and programs to achieve security of the food system, health of the environment and innovation for growth (AAFC 2007).

Yet the Department of Fisheries and Oceans has a much broader mandate, leaving aquaculture one priority among many, including the department's longstanding experience and cultural comfort with conservation and protection of wild fisheries. As the Department mandate explains:

> The Department of Fisheries and Oceans (DFO), on behalf of the Government of Canada, is responsible for: policies and programs in support of Canada's economic, ecological and scientific interests in the oceans and freshwater fish habitat; the conservation and sustainable use of Canada's fisheries resources in marine and inland waters; and safe, effective and environmentally sound marine services responsive to the needs of Canadians in a global economy.

As outlined above, the department's mandate is extremely broad. It covers:

- management and protection of the marine and fisheries resources inside the 200-mile exclusive economic zone;
- management and protection of freshwater fisheries resources;
- marine safety along the world's longest coastline;
- facilitation of marine transportation;
- protection of the marine environment;

- support to other federal government institutions and objectives, as the government's civilian marine service; and
- research to support government priorities such as climate change and biodiversity (Industry Canada 2005).

Conflicting elements in the Department of Fisheries and Oceans' mandate have meant that much of the money being spent on aquaculture research is spent investigating perceived threats to the wild fishery from aquaculture rather than research into ways to increase productivity. The DFO website has indications of various projects in sealice detection, indicating if not a bias in favour of its older "conservation" mandate, at least a rather reactive approach to development of aquaculture as a new industry. More generally, DFO says this in its "Science" section indicating the "mission" of aquaculture-focussed research, a mission consistent with the assertion above, that perceived threats to the wild fishery are a key concern: "The mission of Aquaculture science is to conduct research and to transfer the latest technologies to the rapidly growing aquaculture industry. DFO scientists monitor wild and cultured stocks of finfish and shellfish for diseases and carry out research on the interactions between wild and cultures stocks. They are also responsible for implementing regulations and carrying out inspections to ensure the health of Canada's wild and cultured fisheries resources (Fisheries and Oceans Canada 2007b)."

A review of the Report on Plans and Priorities: 2007–2008 Estimates of DFO (Fisheries and Oceans Canada 2007) combined with the values of the wild fishery and the value of the aquaculture sector (Fisheries and Oceans Canada 2005) is revealing. Although approximately 30% of the value of fish produced in Canada comes from the fish farming sector (Fisheries and Oceans Canada 2006) the DFO has only targeted $3.4 million of a total fisheries budget of $899.9 million for aquaculture or 0.4% compared to the 30% it should be spending on aquaculture. This means that the targeted spending and support for aquaculture is only about 1/60th of what is should be. For me this is the strongest and clearest evidence for the systemic lack of support for aquaculture in Canada. Canada needs an institution which can allocate funding based on production needs rather than political needs. AquaNet was the one research-funding source that had the ability to do research on productivity increase and environmental and social issues. Because AquaNet was at arm's length from the government and it has a rigorous evaluation process to allocate research funding, AquaNet was in an excellent position to help Canada with the allocation of the research funds available for aquaculture so that these funds were not distracted by political agendas.

Are the Only Measures of Success Peer-Reviewed Publications and Patents?

Aquaculture is an internationally competitive food production system. Success is measured by the ability of the Canadian industry to compete globally. Research that aids and abets the ability of the Canadian industry to compete will be research with a real lasting contribution to Canada. Research, which results in patents, is not a measure of successful application of research dollars if it produces little or no benefit toward aiding the competitiveness of the aquaculture industry. Culver focuses on an alternative measure, clearly more relevant to aquaculture's blend of tangible and intangible knowledge assets. Culver writes in this volume: "Trade secrets are an inexpensive, easily employed, and legally enforceable way to protect intellectual property, however their role in innovation is notoriously difficult to assess, since examination of them carries the danger of destroying their secrecy."

One of the best ways to capitalize on an innovation is to keep it secret. One of the quickest ways to lose the intangible value of an invention is to patent it. This is especially true for small companies. By keeping an innovation or invention a secret the smaller enterprise can simply exploit the value of the innovation – both its patentable content and intangible knowledge regarding its use. If a small enterprise patents an innovation or invention a competitor will often be able to steal it without serious repercussions. Protecting an invention by patenting simultaneously requires the ability to access expensive and time consuming legal services. For many small enterprises the most prudent way to benefit from their innovation is to simply keep it a secret. Most inventions and innovation "trade secrets" in aquaculture have not been patented because patenting would be greatly diminish the value of the innovation to the innovator.

Cultural Considerations

The aquaculture research community and the aquaculture industry in Canada are still populated by the founders. The founders are driven by passion to succeed in this pioneering industry. They are typically focused on production gains and not the building of assets – tangible or intangible. As both the research community and the industry mature there will be an increased interest on asset accumulation. As the industry consolidates and the companies involved become larger firms with greater legal resources for protecting patents, patents will become more relevant. As described in Culver's discussion of cultural considerations in understanding the place of intangible assets in aquaculture innovation, there is not a lot of motivation for many university faculty members to pursue patents. Patenting is time consuming, can be costly and the benefit is often elusive. Perhaps the most rapid way to motivate faculty to patent their discoveries is to make patents a prominent part of the promotion and tenure process, and to have university technology transfer offices across Canada give a level-playing field of support for patenting researchers' inventions. This kind of change would give me more confidence in what should otherwise be viewed as Culver's too-rosy picture of the current attitudes of aquaculture researchers, as he suggests that "aquaculture researchers seem to me to be quite convinced of the merit of aquaculture and positively hungry for opportunities to show aquaculture's promise. For these researchers the prospect of continued funding of open and publicly scrutinized aquaculture science seem incentive enough to adopt new ways of understanding the return on national investment in aquaculture R&D".

Conclusions

The major steps forward in the aquaculture industry – better feeds, better stocks, better vaccines, better cages, the development of algae exclusion barriers, the development of methods to rear larval marine species and improved methods for inventory control – are very remarkable and have greatly contributed to the viability of the Canadian industry. However, there are to my knowledge *no* patents with respect to any of these major innovations developed from collaborations amongst university and government researchers and industry partners.

The counting of patents is an inaccurate and incomplete measurement of innovation in aquaculture for several reasons. First, patenting by smaller firms or individuals destroys the value of the patent to that firm or individual when they do not have the resources to defend the patent. Second, many of the innovations have to do with how to do things and how to mange biological entities and these are generally issues of husbandry, which usually, cannot be patented. Third the private sector aquaculture industry in Canada is very young, less than 25 years old. During the first 10 years the majority of the focus was on growth. The following 10 years focused on marketing and business management to stabilize an industry which is in a very competitive international environment. The last 5 years have been spent consolidating the industry so that the value of economies of scale can be exploited. In this context, only a few larger firms have the resources required for protecting and managing patents, and have to date seen little reason to patent. Once consolidation produces larger players with the resources and motivation to protect and manage patents, and the motivation to protect inventions in the event their staff leave the firm, there will be an increase in the patenting rates. Looking now, however, for patenting, is premature and misleading as an indication of innovation in the industry and in research and development partnerships.

Aquaculture has permanently changed fishery in Canada and globally. Aquaculture is a substantially different type of fishery than the capture fishery and requires a completely different knowledge base. The Canadian government is only now waking up to the fact that aquaculture is a different way of doing things and that the government needs people with new knowledge and different intellectual capacities to make rational policies for this new industry. For example, it is only since December of 1998 that aquaculture has been given any internal prominence within the Department of Fisheries and Oceans, as the Office of the Commissioner for Aquaculture Development was established, becoming in 2001 the Aquaculture Management Directorate. As Culver rightly points out, "This is no small task, and it will require unusual depths of collaboration between governance and research elements of the national system of innovation, perhaps to the extent of requiring routine secondments between governance and research elements to create a cadre of science-policy-innovation experts."

Behind the issues Culver and I have raised, a larger question remains: is Canada ready for aquaculture? Is the national government prepared to invest in its own ability to deal with this new and innovative industry? Will the national government invest enough in the development of its own human capital resources so that it can deal with this new industry rationally or will it continue to deal with the industry in a reactionary and draconian fashion? For Canada to compete in the new fishery, the national government must invest heavily in developing Canadian know-how, an intangible asset with real value which must be measured using the right metrics, not the most convenient metrics.

References

Agriculture and Agri-Food Canada. 2007. Our Mandate. http://www.agr.gc.ca/index_e.php?s1=info&page=mandat; accessed September 9, 2007.

Asche, F. 1997. Trade Disputes and Productivity Gains: The Curse of Farmed Salmon Production? *Marine Resource Economics* 12(1): 67–73.

Canadian Broadcasting Corporation (CBC). 2003. Disclosure: Fish Farm Flap. http://www.cbc.ca/disclosure/archives/030204_salmon/main.html; accessed September 9, 2007.

Eagle, J., Naylor, R. and Smith, W. 2004. Why farm salmon out compete fishery salmon. *Marine Policy* 28(3): 259–270.

Environmental Assessment Office (EAO). 1995. Salmon Aquaculture Review Report- Socio-Economic Impacts and Related Technical Papers, British Columbia Government, volume 4, pp. 259–270. http://www.eao.gov.bc.ca/epic/output/documents/p20/1036108359021_8f7f46b7f5d34a4880f886282a95d52f.pdf; accessed September 8, 2007.

Fisheries and Oceans Canada. 2005. Commercial: 2005 Value of Atlantic & Pacific Coasts Commercial Landings, by Province. http://www.dfo-mpo.gc.ca/communic/ statistics/commercial/landings/seafisheries/s2005pv_e.htm; accessed September 9, 2007.

Fisheries and Oceans Canada. 2006. Report on Plans and Priorities: 2006–2007 Estimates. http://www.tbs-sct.gc.ca/rpp/0607/fo-po/fo-po_e.pdf; accessed September 9, 2007.

Fisheries and Oceans Canada. 2007a. Report on Plans and Priorities: 2007–2008 Estimates http://www.tbs-sct.gc.ca/rpp/0708/fo-po/fo-po_e.pdf; accessed September 8, 2007.

Fisheries and Oceans Canada. 2007b. Mission. http://www.dfo-mpo.gc.ca/science/mission/mission_e.htm; accessed September 9, 2007.

Friends of Wild Salmon. Who Cares About the Skeena? – Commercial Fishermen Care. http://www.friendsofwildsalmon.ca/commercial; accessed September 9, 2007.

Highlands and Islands Enterprise. 1999. Economic Impact of Salmon Farming/July 1999 Highlands and Islands Enterprise, Bridge House, 20 Bridge Street, Inverness, IV1 1QR Scotland.

Industry Canada. 2005. Research, Technology, Innovation: Department of Fisheries and Oceans. http://strategis.ic.gc.ca/epic/site/rti-rti.nsf/en/te01453e.html; accessed September 9, 2007.

Noakes, D.J., Beamish, R.J. and Gregory, R. 2002. British Columbia's Commercial Salmon Industry. (NPAFC Doc. No. 642), p. 13 Dept. of Fisheries and Oceans, Sciences Branch – Pacific Region, Pacific Biological Station, Nanaimo, British Columbia, Canada V9T 6N7.

Pinkerton, E. 1999. Factors in Overcoming Barriers to Implementing Co-Management in British Columbia Salmon Fisheries. *Conservation Ecology* 3(2): 2. http://www.consecol.org/vol3/iss2/art2/; accessed September 9, 2007.

Price Waterhouse Coopers. 2003. A Competitiveness Survey of the British Columbia Salmon Farming Industry. Government of British Columbia. http://www.agf.gov.bc.ca/ fisheries/reports/Competitiveness_Survey.pdf; accessed September 9, 2007.

Schwindt, R., Vining, A.R. and Wejman, D. 2003. A Policy Analysis of the BC Salmon Fishery. *Canadian Public Policy – Analyse de Politique* 29(1): 73–94.

Statistics Canada. 2005. Aquaculture Statistics. http://www.statcan.ca/bsolc/english/bsolc?catno=23-222-XWE#formatdisp; accessed September 9, 2007.

United Nations. 2006. A Case Study of the Salmon Industry in Chile. United Nations Publication UNCTAD/ITE/IIT/2005/12, ISSN Number: 1817–3225. http://www.unctad.org/en/docs/iteiit200512_en.pdf; accessed September 9, 2007.

Chapter 6
Return on Investment or How Not to Pay Commercial Licenses for Your Own Technology

Thomas W. Sephton

Innovation and Its Measurement

Culver presents an interesting perspective on performance measurement and the evaluation of intellectual property and research in general. It deals with the common issue of getting the best knowledge that science and social science can produce into the public forum for informed, enlightened debate when dealing with the issues of the day, and for use in promoting the social and economic well-being of Canadians. A practical example of this would be the transformation of primary research on fish nutrition and diet experiments conducted under the auspices of the Department of Fisheries and Oceans Canada, Aquaculture Collaborative Research and Development Program (a public good program) and the need for industry confidentiality to secure a market for the new product. At a pre-determined point in time, the novel results obtained from this collaborative research project will be put into the public domain. Another would be the collection and translation of multi-beam bathymetric data by the DFO Canadian Hydrographic Service into not only electronic charts for public marine traffic safety but also to claim Canadian sovereignty over the continental shelf under the auspices of the 1982 *United Nations Convention on the Law of the Sea (UNCLOS)*. This Convention provides coastal States with exclusive sovereign rights to explore, exploit, conserve and manage fisheries within 200 nautical miles (370.4 km) of their shores. The social and economic spin-offs of a claim of this magnitude are enormous and far-reaching, even for development of off-shore aquaculture. Many times, however, the actual field data are collected under collaborative agreements with the private sector (fisheries and natural resource entrepreneurs) who negotiate its exclusive use for a pre-determined period of time, thereby complicating the release of the data for public use.

Culver's premise is that the success and relevance of aquaculture research should not be measured only by the fashionable yardstick of patents and other intellectual property rights. He argues that science and technology networks like the Networks of Centres of Excellence for Aquaculture (AquaNet) were geared towards the rapid

T.W. Sephton
Department of Fisheries & Oceans, Bedford Institute of Oceanography, Dartmouth, NS, B2Y 4A2
e-mail: sephtonT@mar.dfo-mpo.gc.ca

K. Culver, D. Castle (eds.), *Aquaculture, Innovation and Social Transformation*,

© Springer Science+Business Media B.V. 2008

translation of research into development. AquaNet was a knowledge management organisation whose mission is to transfer 'rivalrous' information from a normally competitive group of researchers to end users in industry. By 'rivalrous' knowledge, Culver means that knowledge which is finite and excludable, for example, the invention of a recirculation pump whose use of significantly less energy than other comparable pumps gives it a competitive advantage. The technology of the pump can be patented, and 'rivals' for the knowledge behind the pump can be excluded from using the pump without compensation to its inventors.

Unfortunately, what is profitable is not always patentable or otherwise easily captured by conventional measures of rivalrous knowledge. Innovation in aquaculture that will drive the industry as a whole forward, and will give countries or companies comparative advantage, may not always be patentable subject matter. Troubleshooting approaches, and even good laboratory habits can contribute to competitive advantage as downtime and errors are reduced, yet it is difficult to capture the value of this 'non-rivalrous' or 'intangible' knowledge. Equally significant is the fact that a collaborative model of research dissemination is being encouraged by agencies funding R&D networks. There is a push towards the increased sharing of knowledge among researchers as well as the transfer of knowledge to end-users to quickly commercialize those innovations. There are complex tensions at play as normally competing researchers are asked to collaborate, and as they do so in ways which produce non-rivalrous knowledge which does not appear in patent statistics, they are also asked to transfer knowledge to industry more quickly than ever before – a process that is not well suited to measurement by the relatively drawn out process of seeking a patent. Fish feed is the most expensive part of any aquaculture operation and its most efficient use is of prime concern to farm-site managers. The incorporation of modern remote sensing and control technology (high speed underwater video surveillance, satellite uplinks, robotic feeders) with fish feeding behaviour under local environmental regimes into modern husbandry practices is slowly making its way throughout the salmon aquaculture industry. Interestingly, the application of this technology is not only company specific (and their independent ability to invest in new technology) but also site specific. Ultimately, the implementation of the new technology depends on the experience and local knowledge of the site managers who will 'tweak' the new technology to suit their local operations.

The use of patents as a yardstick for measuring research network innovation raises broad issues about the use of performance frameworks for measuring the performance of outputs and outcomes of R&D networks, and we have just scratched the surface in mentioning above the difficulties found in fostering collaboration among researchers who normally think of themselves as competitors. In addition to fostering cultural change, research networks must struggle to develop formal performance measurement criteria, often without pre-existing, industry relevant standards to guide them. Instead, research networks are asked to participate in complex new developments in governments' methods of managing public expenditures. In Canada, the federal government moved away from evaluating research investments against input criteria to output and outcome criteria. This is well documented in the *Canada School for Public Service* paper "Modernizing Government Accountability: A Framework for Reform", which states that

> Results-based reporting to Parliament is an integral element of a reform effort, now almost a decade old, to improve reporting by departments to Parliament in order to improve accountability not only to Parliament, but also to ministers and the public. It is also a crucial part of a larger effort that seeks to improve management through the improved measurement of results (Aucoin and Jarvis 2005).

More recently, the Treasury Board Secretariat has adopted and posted principles of results based management which characterizes the results-based management as a 'lifecycle' approach to planning, implementation and monitoring, and reporting. The discussion of Reports on Plans and Priorities (RPPs) and Departmental Performance Reports (DPRs) makes explicit links between effective public reporting and accountability to Parliament (TBS 2006).

Outputs and Outcomes

The basic design of results-based (or performance) reporting has now become common across Western democracies. Results include performance in the production of outputs and the achievement of desired outcomes. Outputs are the goods and services provided by government in pursuit of its policy objectives; desired outcomes are the desired effects that the government intends from these outputs. The shift to output and outcome measures is clearly the better approach as opposed to the old, input-based system which often declared success once an allocated budget was spent. There are, however, some remaining issues for the new approach to face. Measures of outputs and outcomes measures must be well-characterized as easily applied tests, and, to the fullest extent possible, the values that support judgments about whether one has succeeded and done something well must be as transparent as possible. Design and application of these tests is far more difficult in practice than in theory, as practice involves communities of people who need to learn accountability systems, and for practical reasons need those systems to be consistent with one another.

To give a sense of the complexity of the measurement problem, it is worth considering the different facets of R&D network activity that can be rewarded. For example, it is possible to identify and measure those qualities that establish scientific credibility and reputation, for example membership of researchers in editorial boards of reputable journals, awards won, or impact of publications. Alternatively, one can try to evaluate what gives a researcher or their team that competitive edge over others. So how do funding organisations actually know if they are spending their funds on the right people, teams and ideas, and rewarding their collaborations and work output? The general problem raised by these different criteria is that there is no universally accepted indicator of overall science and technology performance for a country or a sector. This explains why there is a plethora of tailor-made performance measurement frameworks, not all of which are inter-operable. Canada's Network of Centres of Excellence Program, for example, lists five main criteria which will be satisfied and shown by a successful research network. A successful network investigating an approved focal area will demonstrate: (1) "excellence, focus and coherence … [in] the research program"; (2) "development of highly qualified personnel"; (3) effective "networking and partnerships" among a wide range of public, private, and other sectors; (4) "knowledge and technology exchange and exploitation" including "Effective management and protection of intellectual property" resulting from network-funded research and (5) the network must have an effective management structure (NCE 2005). These are the criteria by which aquaculture-relevant research is, and was, evaluated within various Networks of Centres of Excellence including OPEN, the Ocean Production Enhancement Network, and AquaNet, the aquaculture network.

By contrast, Genome Canada, which also funds aquaculture-relevant research, explicitly separated scientific and social research excellence and requirements in its evaluation criteria in its 2004 competition. Genome Canada requires "Scientific excellence of the proposed research as affirmed by peer review," and additionally requires involvement of researchers from the 'GE3LS' community comprised of experts in Genomics, Ethics, Environment, Economics, Legal and Social issues. As the criteria put it, "All applicants must consider the GE3LS aspects of their proposed research and, where appropriate, seek advice from one or more GE3LS experts (as a co-applicant, collaborator, or through membership on an advisory committee) to develop a plan to address those GE3LS issues directly raised by the proposed research." Clearly the researchers within Genome Canada projects are being held to a different standard of productivity – one which includes social benefits in addition to the NCE's aim to achieve "knowledge and technology exploitation." This difference is further emphasized by Genome Canada's spelling out in much greater detail than the Network of Centres of Excellence just what is meant by a social benefit, as Genome Canada research is required to have a plan not just for knowledge transfer, but for economic development resulting from research. Successful proposals in 2004 were required to have a "plan for the transfer, dissemination, use or commercialisation (as appropriate) of the anticipated results of the research proposed. Demonstration of how the research results will contribute to job creation and economic growth in Canada and their impact on society, quality of life, health, and the environment, including the creation of new policies in these areas (Genome Canada 2004)." These are the standards which will be used to judge the progress of currently supported projects in aquaculture salmon and cod.

Further observations can be made regarding the contrast between the two research organizations both funding aquaculture. But the general point is already made. Researchers progressing through their careers may move from research funder to research funder, and with each move a different way of measuring success is imposed on the researchers. Directly and indirectly, their so-called research 'track record' is re-interpreted each time and what was once considered 'cutting edge' may be re-interpreted as 'status quo.' Sometimes, as in the NCE research, it appears to be optional or less important that science-based teams take social impacts into account. In other instances, such as Genome Canada research, it seems that scientists no longer have independent control over the success of their projects, to the extent that they are now interdependent with social scientists, even when these two groups might have little beyond mutual good will to guide them in their collaborations and little more than good will when they end up on interdisciplinary peer-review panels looking at the diverse research outputs of these collaborations. There is no question that some of these changes are resulting in improved accountability to Canadians and even improved research in the sense that more research leaves the lab and the journal page to benefit Canadians. There is still reason to worry, however, that more can and should be done to improve the way we identify what success will look like, and measure our progress toward success. Thinking about this next task can be approached by looking at our evolution away from traditional measures of progress.

The traditional yardsticks of peer-reviewed publications and production of highly qualified personnel are measures still used routinely by social and natural science managers. In an attempt to integrate across the traditional boundaries, some managers are now using 'annual total research expenditures' processed through their accounting departments as the measure of success: although it looks good in a presentation to senior officials, it blurs the reality of the situation. The idea is relatively simple is equivalent to the common economic term: gross domestic product. The annual total research expenditure is a combination of salary (staff), operating (lab and field) and capital (small and large equipment) funds but it sheds no light on the distribution among the three categories and to whether a department is growing, maintaining the status quo or in a state of decline in terms of its research effort and output. A young, state of the art, contract salary research team will have a very different funding throughput (low salary, high operating and high capital) than that of a late career, comfortably tenured one (high salary, low operating, low capital) even though each may have the same total research expenditure.Those interested in exploring this further should consult the federal public service document which motivated this discussion: Moving from the Heroic to the Everyday: Lessons Learned from Leading Horizontal Projects (CCMD 2001).

Despite the several and diverse challenges which stand in the way of finding metrics for evaluating government investment in innovation, perhaps one criterion stands above all else. Whatever tool is used to measure a network's success, the research outputs and outcomes must be understood by, and relevant to, the general public. Society's expectations of targeted R&D investments must be understood and met if the funding base – the taxpayer – is to be satisfied that they got good return on their investment. Securing society's buy-in to targeted research is a never ending process and the so-called public consultation ranges from using Management Boards with 'wise' people selected from the client base (more akin to client consultation) to using professional public opinion polls and surveys (true public consultation). This is where the natural science community can make greater inroads by working collaboratively with the social science community to create an inter-disciplinary working environment.

Government's Role in Enabling and Regulating Innovation

A cross section of the Canadian Federal Government science and technology community met in 2002 in a forum setting (*Transforming Federal Science and*

Technology for the Future, Hull, Québec) and they produced a vision statement that included a statement on the measures of success:

> The benefits to Canada of federal science are realized through its implementation in policy and innovation. Federal scientists will measure their success by the cumulative contribution of their research to public policy and to innovation, by the timeliness and pertinence of their solutions to issues of national priority, and by the success of their partners in innovation and public service. Federal science will continuously assess emerging challenges and opportunities and reorient and refocus its resources accordingly (Government of Canada 2002).

The statement is vague enough to cover many different science-based government departments and agencies (SBDA) and it tries to measure success for the government researcher in terms of both policy development and innovation. Unfortunately, these can be conflicting and competing measures, as stakeholders are eager to point out to federal departments such as Fisheries and Oceans Canada, since those Departments are empowered to be both regulators and developers. A brief history of the aquaculture dilemma facing DFO and provincial government agencies is chronicled in the 2001 interim report by The Standing Senate Committee on Fisheries (TSSCF 2001) who met with a good cross section of competing public interests on both the Atlantic and Pacific coasts. The frustrations of environmentalists, general public, aquaculturists and government agencies are captured clearly in the report. Since that report, DFO has tried to address this issue face-on by incorporating the Aquaculture management portfolio (and the former office of the Aquaculture Commissioner) into Fisheries Management to create a single sector: Fisheries and Aquaculture Management under a single Assistant Deputy Minister. This forces all policy and management discussions of either mandate through the same public policy review and debate for the benefit of all Canadians.

I believe that it is in the best interests of all SBDA and science networks to move off of patents as the sole condition of identifying success to include some other important measures, including public accountability. Culver has suggested in his paper that the wealth generated by an R&D network includes the tangible assets normally focused on, but also includes intangible assets that could be counted if one decided to swing one's approved metrics in that direction. If this is right, and intangibles should count, then we would value an aquaculture network more highly if it generated a diversity of forms of intangible assets in aquaculture R&D, with particular attention to the value of trade secrets, or Sveiby-style 'total knowledge asset' measurement – see www.sveiby.com.

Investment in the Public's Interests

The Canadian taxpayer has an expectation that the goal of most research conducted by SBDA is to resolve 'public good issues' or, at the very least, 'client group issues' with adequate consideration given to how and when knowledge and technology is communicated back into the published literature for public consumption. This counters the role of the private sector research organisations and, perhaps, universities, charged with commercialization and innovation for the creation of wealth for their private sector partners who must stay ahead of their competition.

Aquaculture R&D in Canada is a special blend of public and private sector investment in a multi-disciplinary research activity that can take from 12 to 15 years to see the successful transformation of a laboratory experiment (eureka, it's growing) to a sustainable commercial activity (yes, it grows well and we can make a profit). Throughout this developmental rollercoaster, there are opportunities for different types of research to break through bottleneck issues and raise the level of knowledge and innovation required to be successful. Unfortunately, a developing aquaculture partner cannot succeed on the back of research projects alone: there must be an infusion of capital funds to elevate the pilot scale activity to full commercial operations. Therein lies the major stumbling block for many Canadian aquaculture partners trying to develop and commercialize a new seafood product for the protein supermarket. Private investors and venture capitalists demand a fast turnaround of their capital commitments, complete with a significant return on the original investment. Most marine finfish and shellfish take years to grow to a market size and that in turn drives many investors to turn to other biotechnology fields. Governments no longer provide grants and forgivable contributions to burgeoning industries as this would be viewed as a direct subsidy in the world of globalisation and open to countervailing duties and appeals to arbitrators. The only funding recourse would be low interest re-payable government loans (as are usually available from regional funding agencies like the Atlantic Canada Opportunities Agency) but these only add to an already staggering debt load carried by many aquaculture companies who have ventured into the world of product diversification on their own. There is no easy way to get out from under this Canadian way of doing business in a global economy other than to broaden our current research base across international boundaries and to become a true partner and collaborator with the international research community, such as the European Union.

References

Aucoin, P. and Jarvis, M.D. 2005. Modernizing Government Accountability: A Framework for Reform. Canada School for Public Service (CSPS), http://www.ccmd-ccg.gc.ca/Research/publications/pdfs/p131_e.pdf; accessed September 9, 2007.

Canadian Centre for Management Development (CCMD). 2001. Moving from the Heroic to the Everyday: Lessons Learned from Leading Horizontal Projects. http://www.ccmd-ccg.gc.ca/Research/publications/pdfs/horinz_rt_e.pdf; accessed September 9, 2007.

Genome Canada. 2004. Guidelines and Evaluation Criteria for Competition III, July 2004. http://www.genomecanada.ca/xresearchers/competitions/c3/GuidelinesFinal.pdf; accessed September 9, 2007.

Government of Canada. 2002. The 2002 Report on Federal Science and Technology. http://www.innovationstrategy.gc.ca/gol/innovation/site.nsf/en/in05195.html;accessed September 9, 2007.

Networks of Centres of Excellence (NCE). 2005. Letter of Intent Guide 2005 Competition for New NCE Networks-Program Criteria. http://www.nce.gc.ca/comp/comp2005/loiguide2005_e.htm#criteria;accessed September 9, 2007.

The Standing Senate Committee on Fisheries (TSSCF). 2001. Aquaculture in Canada's Atlantic and Pacific Regions, 37th Parliament, 1st Session. http://www.parl.gc.ca/37/1/parlbus/commbus/senate/com-e/fish-e/rep-e/repintjun01-e.htm#THE%20FEDERAL%20ROLE; accessed September 9, 2007.

Treasury Board of Canada Secretariat (TBS). 2006. Results-based Management. http://www.tbs-sct.gc.ca/rma/rbm-gar_e.asp; accessed September 9, 2007.

Part III
The Environmental Sustainability of Aquaculture

Introduction

A few years ago, in its review of the major fish stocks around the world, the Food and Agriculture Organization of the United Nations reported that half are fully exploited, 18 percent are overfished, 9 percent are depleted, 21 percent are moderately exploited and 1 percent are slowly recovering (FAO 2002). Just 4 percent of fish stocks are under-utilized according to 1998 data, and as the FAO notes, the most "worrying finding of the analysis is that during the last 25 years, the proportion of stocks exploited beyond their maxium sustainable yield (MSY) level has increased steadily (FAO 2003)." Stocks have since either plateaued or declined since, due to the euphemistic "overcapacity" of fishing fleets (FAO 2007). Boris Worm and his colleagues report on a deteriorating situation in which oceanic ecosystem services are so seriously threatened that global fisheries might collapse within 50 years (Worm et al. 2006). The collapse is foreseeable since it is based on a simple, long-established pattern of fisheries: stocks are discovered and then exploited using ever more efficient methods of capturing fish (Roberts 2007). In addition to the pressure on aquatic species populations, there is also evidence that fisheries are powerful selective pressures that drive evolutionary change in targetted species (Palumbi 2001). Despite attempts to harmonize global fisheries management by developing a Code of Conduct for Responsible Fisheries (FAO 1995), the trend toward exhaustion of global fish stocks continues as demand for fish protein increases.

The environmental impact of the capture fishery provides a powerful motivation to find a sustainable alternative in aquaculture. As it stands, capture fishery has plateaued, and the demand for fish, crustaceans and shellfish – nearing 20 kg per person each year – is being met by the doubling of aquaculture output in the last 10 years (FAO 2002). Perhaps the aquaculture industry, seeing that it provides a solution to the fisheries problem, realizes that environmental sustainability is the single most important production criterion from the perspective of concerned citizens, NGOs and government regulators. For example, the Canadian Aquaculture Industry Association claims that the next decade's projected growth of the Canadian industry, from approximately CDN$500 million to nearly CDN$3 billion, will be environmentally sound (CAIA 2007). One of the greatest challenges the industry has faced is developing feeds based on sustainable inputs that improve feed conversion ratios. Since feed accounts for up to 75 percent of finfish aquaculture costs, there is a strong incentive to develop sustainable, efficient feeds from vegetable sources such as soy and canola to reduce reliance on fish meal. Apart from feeds, CAIA and similar organizations around the world have recognized the need to respond to environmental concerns by selecting sites where waste impact on the environment can be managed, monitoring fish health, and sparing use of the few antibiotics approved for aquaculture.

Public scrutiny of aquaculture is high and appears to have significant effects on government and private sector approaches to aquaculture management. In Canada, several NGOs monitor the aquaculture industry, and are characteristically adept with the media. The David Suzuki Foundation devotes considerable resources to fisheries and aquaculture issues, as does the Sierra Club of Canada, each organization taking a Greenpeace-style approach to sparking issues and fanning the flames in the media. Remarkably, Patrick Moore, co-founder and former director of Greenpeace, intervenes between the Suzuki foundation and the aquaculture industry to deliver "positive aquaculture awareness" through his foundation Greenspirit. It appears that regulators are paying close attention to these debates, at least in their written words. For example, in an interview for *Snapshot 2006–2007 Oceans and Sciences Branch Gulf Region*, Dave Dunn, Division Director, Oceans and Habitat Division, responds to the question "How does you section's work impact the average Canadian?" by pointing out the accountability of government officials to protect the environment: "The Habitat Provisions of the *Fisheries Act* are among the strongest environmental legislation in the country and environmental groups hold DFO accountable for its environmental decisions. The *Oceans Act* provides a vehicle to sustain our marine environment for future generations."

Despite global awareness of the need for food protein of the kind aquaculture promises to produce sustainably, the debate over those promises continues, with little end in prospect even as regulators in countries such as Canada appear to be taking seriously the views of a wide range of stakeholders beyond government-employed scientists. This chapter's contributors articulate many of the complexities driving the debate. Kenneth Black discusses the many different kinds of environmental impact aquaculture can have, such as the effects of waste on the seabed, nutrients, medicines and chemicals introduced into water, parasites, escapes, and the sustainability of feed. He describes a "limiting factor" approach to deciding if a farm is unsustainably drawing down on a resource. The physical limits of aquaculture, if known, provide farmers with options to self-regulate or to be regulated by relevant authorites. In the second chapter, Marc Saner inquires into what would count as an appropriate regulatory regime by considering what sustainability means in the context of good industry governance. He articulates a five principle governance framework, and suggests that Black's limiting factor approach meets some, but not all aspects of the framework.

In the third chapter, Fiona Cubitt, Kevin Butterworth and Scott McKinley provide an extensive overview of the environmental issues facing the British Columbia farmed salmon industry. Their review of the evidence for the sustainability of salmon farming challenges some of the generalizations made by Black, which suggests that any discussion of sustainability ought to be based on local conditions.

They conclude that discussions of sustainability are dominated by mixed messages whose use of available scientific evidence is often sporadic and incomplete. They voice hope that a thorough, comprehensive review of the evidence might resolve at least some issues and move the debate on toward remaining issues. This hope is admirable in its desire to find an end to debate which is acceptable to all participants, but some may see within this wish a tendency on the part of scientists to suppose that they are the ones with the evidence which has been underappreciated by stakeholders – and so resolution of issues is a matter of *other* stakeholders making up the deficit in their knowledge of the issues by learning more about the scientific evidence. Yet as may be seen in other contributions to this volume, particularly in reflections on the role of local and traditional ecological knowledge, issues surrounding social acceptance of aquaculture are least as much issues about *what counts* as relevant knowledge as they are about the choices to be made in light of relevant knowledge. It may be worth considering whether the debate over environmental sustainability needs to proceed with *both* more and better scientific evidence, *and* more and better attention to transparent policy development methods which visibly integrate reliance on government and university science, local and traditional ecological knowledge, industry and media perspectives, and broader social concerns.

References

Canadian Aquaculture Industry Association (CAIA). 2007. Canadian Aquaculture Industry Profile. http://www.aquaculture.ca/facts/industryprofile.htm. Accessed September 13, 2007.

Food and Agriculture Organization of the United Nations (FAO). 1995. *Code of Conduct for Responsible Fisheries*. Rome: FAO.

Food and Agriculture Organization of the United Nations (FAO). 2002. *Summary Report: World Agriculture: Towards 2015/2030. An FAO Perspective*. Rome: FAO.

Food and Agriculture Organization of the United Nations (FAO). 2003. *World Agriculture: Towards 2015/2030. An FAO Perspective*. Edited by Jelle Bruinsma. London: Earthscan Publications.

Food and Agriculture Organization of the United Nations (FAO). 2007. *The State of World Fisheries and Aquaculture*. Rome: FAO.

Palumbi S.R. 2001. Humans as the world's greatest evolutionary force. *Science*. 293:1786–90.

Roberts C. 2007. *The Unnatural History of the Sea*. Washinton: Island Press.

Worm B., et al. 2006. Impacts of biodiversity loss on ocean ecosystem services. *Science*. 314:787–90.

Chapter 7
Environmental Aspects of Aquaculture

Kenneth D. Black

Introduction

Scientifically defensible and transparent regulation of aquaculture is necessary to protect the marine environment, ensure the supply of environmental services for all users, and promote the adoption of best environmental practices.

The sustainability of aquaculture is a difficult concept to define and measure (Black 2001). Overall sustainability is the product of the sustainability of a range of issues none of which are static in a rapidly evolving, technology-driven industry. There is an important relationship between internal regulation (i.e. good practices voluntarily adopted by the industry) and external regulation, (i.e. practices imposed on the industry by regulators). Improvements in the sustainability of certain aspects of aquaculture are influenced by both these factors, although the balance varies between countries (and often regionally within countries).

In this chapter, the main effects of marine fish farming on the receiving environment are briefly reviewed and the main issues of each impact type discussed. For each area of aquaculture-ecosystem interaction, regulatory and societal issues are considered. A "limiting factor" approach to aquaculture development is discussed, as are some issues related to development policy. The chapter ends with some general conclusions on sustainability.

Effects of Organic Wastes on the Seabed

Particulate organic wastes from aquaculture arise from feed pellets that are not consumed by the fish, from faecal material produced by the fish, and detached fouling organisms that settle and grow on nets and cage structures. The degree to which pellets are wasted depends on the appropriate matching of the supply of feed to the fish and their appetite. Appetite varies with species according to both long-term

---

K.D. Black
Ecology Department, Scottish Association for Marine Science,
Dunstaffnage Marine Laboratory, Oban, Scotland, UK
e-mail: kenny.black@sams.ac.uk

K. Culver, D. Castle (eds.), *Aquaculture, Innovation and Social Transformation*,
© Springer Science+Business Media B.V. 2008

patterns, e.g. temperature, and short term events e.g. disease and stress. As feed is an expensive input, farmers have developed sophisticated systems for matching supply to demand to minimise waste and, for environmental reasons, the inclusion of such controls on waste may be an important part of the decision process for the granting of a new development, or the continued licensing of existing enterprises. Typically, feed is supplied by an automatic system that provides pellets according to well-established fish-growth models under a regime where the frequency and timing of the meals has been optimised for environmental parameters such as temperature and day-length. Coupled to this, many farmers use camera systems (or pellet traps with sensors) deployed within cages to monitor feeding performance and fine-tune feed input to the feeding behaviour of the fish. It is easy to see that the scope for both financial loss and environmental damage are high if such systems are not employed on large sites. For example, a salmon farm will typically supply 0.7% of biomass as feed per day in the summer months: a farm of 1000 tonnes biomass will receive 7 tonnes of feed per day. If the fish are over-fed by 5% this will result in the loss of 350 kg feed per day.

Early estimates of the proportion of feed lost were of the order of 20% but, with improved training and technology, this is now thought to be much lower – between 0 and 5%. However, the fraction of feed wasted in open cage systems is notoriously hard to measure and there is considerable debate. A recent study (Strain & Hargrave 2005) estimated waste feed using digestibility and energy retention coefficients and derived a value of 15%. A source of wastage which cannot be controlled using camera or other remote sensing systems is that caused by the abrasion of feed pellets as they are blown through feeding pipes. This fine material can be seen as particles on the water surface or as a slippery coating on farm structures. Such small particles may sink to the seabed immediately or they may be carried some distance from the farm on the water surface.

The effect of farm-origin particles on the environment has received a great deal of attention (e.g. Gowen & Bradbury 1987; reviewed by Pearson & Black 2001). Briefly, organic matter (faeces, feed, detached bio-fouling) sinks to the seabed with a spectrum of settling velocities. Typically, this material is initially deposited on the bed near the farm at a distance that is a function of current speed and water depth, and may be subsequently resuspended and transported if the current velocity at the bed exceeds a critical speed. Particles on the bed are degraded by micro-organisms creating an oxygen demand. When the oxygen demand exceeds supply, the oxidising agent will change from dissolved oxygen to sulphate (producing sulphides) and other oxidised species (Holmer & Kristensen 1992). Finally, if sulphate becomes depleted in the sediment, fermentation can take place leading to the production of methane. Benthic species vary greatly in their oxygen requirements and their sulphide tolerance. Faunal succession across an organic enrichment gradient is well known (Pearson & Rosenberg 1978): highly polluted sediments are dominated by small opportunist worms of a small number of species but at very high abundances that are confined to the surface layer. As the input of organic material reduces away from the source, a more diverse community of deeper-burrowing animals can develop. Burrowing animals increase the rate at which organic matter is degraded by a range of mechanisms, not all of which are fully understood (Heilskov & Holmer 2001).

The extent to which anaerobic processes dominate in sediments depends crucially on the rate of input of organic matter and the rate of oxygen supply (Findlay & Watling 1997). Thus for small sites with low inputs in areas experiencing high water flows, changes to benthic communities will be more subtle that at large, quiescent sites. Changes in the sediment geochemistry and faunal community may be profound, but once the organic inputs have ceased e.g. on cessation or fallowing of a site, recovery processes are initially rapid (Pereira et al. 2004; Brooks et al. 2004). Highly degradable components of the wastes are consumed over a few months and oxygen demand is reduced. This initial recovery in the sediment chemistry typically takes a few months and is followed by a recolonisation of the sediments by a more diverse community of animals. The precise timing of this depends on seasonal factors such as temperature, natural organic inputs and the (often seasonal) supply of invertebrate larvae. Sediments may maintain large inventories of less degradable organic material which continue to exert an oxygen demand over several months to years. The processes of chemical and biological recovery are intimately linked as bioturbation (the process of sediment irrigation and mixing by burrowing animals) allows increased supply of oxygenated water and more rapid removal of reduced chemical species: recolonisation is both a sign of recovery and an important contributory factor to it.

In many jurisdictions, the degree and extent of changes to benthic communities form a key part of scaling farm size to the environmental assimilative capacity and in compliance monitoring. Such relationships have been embedded in computer models (Cromey, Nickell & Black 2002a; Cromey et al. 2002b; Silvert & Cromey 2001; Silvert & Sowles 1996) that combine these with predictions of organic matter flux to the sea bed. Regulators are increasingly turning to such modelling approaches to attempt to predict the scale of effects and to improve the transparency of decision-making.

Different approaches to the regulation of benthic pollution are taken in different legislatures. In Scotland, an important concept is that of the Allowable Zone of Effect or AZE. For any polluting discharge some effect is inevitable near the source. If developments with effluent discharges are to be permitted then some reduction in quality with respect to pre-existing or background conditions must be allowed. The Scottish Environment Protection Agency sets an AZE for benthic pollution such that some change to the seabed is allowed in the immediate vicinity of the cages. Two separate sediment quality criteria are set: (1) a high standard imposed outside the AZE where sediment quality should not depart far from reference conditions and (2) a lower standard within the AZE that allows a greater deviation from reference conditions but still sets minimum standards to retain an active bioturbating infaunal community.

In Scotland, the AZE was previously set at a fixed distance from the edge of the cages (25 m). Although simple to understand and convenient for monitoring, this ran counter to the policy of favouring farms in more dispersive locations where increased depth and stronger currents lead to larger waste footprints. This has been superseded by a modelling approach that attempts to match the size of the AZE to the dispersiveness of the site. The environment is protected by maintenance of the within-AZE sediment quality criteria and a precautionary cap on the maximum farm size. Sites in less dispersive areas are still subject to the 25 m AZE rule (www.sepa.org.uk/aquaculture).

Effects of Nutrients on the Water Column

Fish farms are sources of dissolved inorganic and organic nutrients in the marine environment. A plausible nutrient budget for salmon farming is given in Fig. 7.1. As fish metabolise feed, ammonia is excreted from the gills and other nitrogenous (N) and phosphorous (P) by-products are lost in faeces (Hall et al. 1992; Handy & Poxton 1993; Holby & Hall 1991). The degradation of particulate material in the water column and in sediments also releases N and P into the water column. Ammonia is the most reactive nitrogenous species and it is quickly taken up by the pelagic

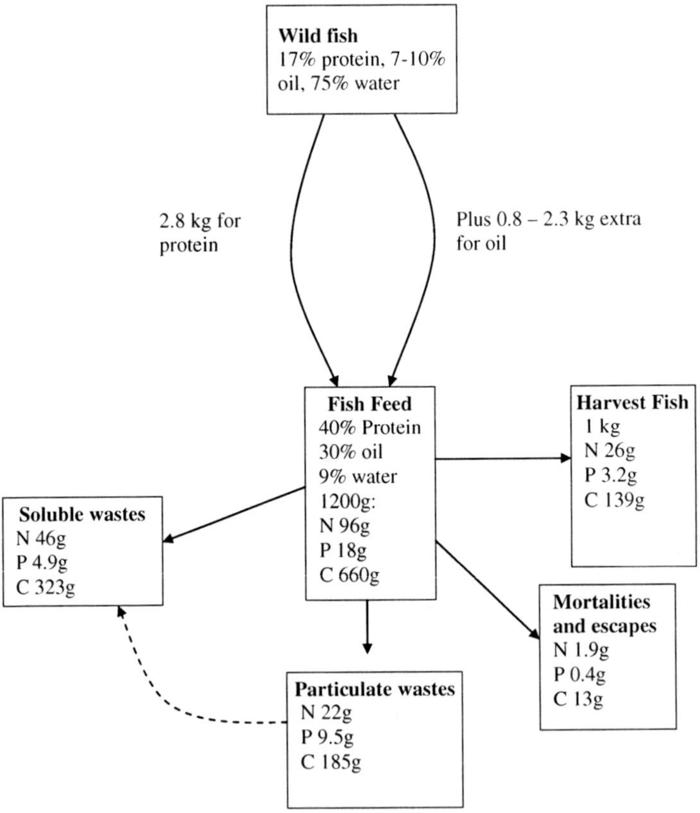

Fig. 7.1 A budget for the flow of nutrients from oceanic wild caught fish to the coastal environment for a harvest of 1 kg of farmed salmon assuming no substitution with vegetable protein or oil and a ration of fish feed to product of 1.2:1

microbial community yielding oxidised nitrogen species such as nitrates. Relatively little research has been undertaken on effects on the microbial heterotrophic community (e.g. bacteria and protozoa): it appears that this community is stimulated by the addition of inorganic and organic nutrients from fish farming but that this is limited to the immediate vicinity of the farm. This effect is a consequence of the rapid generation time of these small organisms (hours) compared to typical water residence times (days). Much more attention has been given to the potential for nutrient additions to enhance autotrophic production (plants: e.g. micro- and macro-algae). The word eutrophication is used to define the undesirable increased production of algae caused by anthropogenic nutrient inputs. The fundamental question concerning nutrients from fish farming is "how much nutrient can a system receive before it becomes perturbed from its natural state?" However, assessment of the "natural state" of a system is complicated by the fact that there are very few areas with sufficiently long time-series of measurements. This is a fundamental obstacle that will only be improved if such time-series measurements are made across a representative set of marine system types. Researchers and regulators largely depend on models of varying complexity when considering this issue (Black et al. 2002).

The relationship between plant production and nutrient inputs in a particular system is a function of the availability of light, the mixing of water in the system and the exchanges between the system and its neighbouring coastline or ocean. It is possible to rank systems on their potential susceptibility to eutrophication. Restricted systems with slow exchange with open water (i.e. long residence times) and where mixing is suppressed by stratification of the water column are likely to be most susceptible to stimulation of algal production by nutrient addition. The greatest risk occurs in late summer when natural nutrient concentrations are low, where temperature stratification is likely and where nutrients outputs from fish farms are at their highest. Where systems exchange with open waters more rapidly than the generation time of phytoplankton (typically 1–3 days), increases in phytoplankton biomass are unlikely to be measurable and the risk of eutrophication is low.

Some potentially undesirable effects from fish farm nutrients in Scottish water have recently been evaluated (Black et al. 2002)

> Three main concerns are raised against the aquaculture industry: a) plant nutrients from fin-fish farms have led to an increased occurrence of algal blooms; b) plant nutrients from fin-fish farms have disturbed the natural ratios of nutrient elements in sea water so favouring the occurrence of toxic species over harmless algae and c) plant nutrients from fin-fish farms have made potentially toxic algae more poisonous. Lack of long term observational data preclude direct comparison of nutrient and phytoplankton levels from pre-fish farm development times and the present day. Modelling studies, however, have shown that only a few sea loch sites are strongly enriched with nutrients to such a level that they might exceed environmental quality standards but, in the main, enrichments are low. In addition, models have shown that the algal production attributable to fish farm nutrients in Scottish coastal areas is small relative to that generated by marine and terrestrial nutrient inputs.
>
> It is concluded that the present level of fish farming is having a small effect on the amount and growth rate of Scottish coastal phytoplankton but that this effect should not be a cause for concern except in a few, heavily-loaded sealochs.
>
> The perturbing effect of fish farm waste on nutrient element ratios in most Scottish cases can be shown to be small. Typical farm waste has a ratio of nitrogen to phosphorus which is

close to natural ratios. However, there is a possibility that because of the absence of silicate in fish foods there may be a danger of exceeding the "safe" N:Si limit of 2.5 locally at heavily enriched sites in summer when background nutrient levels are low and silicate has been drawn down by the Spring Bloom. However, modeling studies suggest that broad area effects should be small. Similarly there is no convincing evidence to suggest that changes in nutrients as a result of fish farm inputs ratios are likely to stress potentially toxic species to cause them to increase their toxicity.

In highly turbid areas, such as the Bay of Fundy on the east coast of Canada, light is a strong limiting factor on phytoplankton growth. Very large tidal flows cause suspension of fine particles that attenuate light. However, in some parts of the Bay fish farming is intense and there is evidence that tidal and subtidal macroalgal communities are stimulated by fish farm-origin nutrients. Being attached to the substrate, macroalgae may be exposed to elevated nutrient levels for long periods of time and considerable research effort is being expended in determining the relationships between fish farming and macroalgal production in the Bay. Increased macroalgal production may have an effect on intertidal and shallow water communities by smothering and, where part of this community is harvested in a fishery, a socio-economic impact is possible. Some macroalgal species are of commercial value and attempts are being made to determine whether this enhancement of macroalgal productivity can be harnessed by culturing seaweeds in fish farming areas. The co-culture of fish and plants may offer economic benefits as well as removing some of the fish farm nutrients from the system (Chopin et al. 2001; Cook et al. 2007; Sanderson et al. 2008).

The recycling of nutrients from fish farming has been examined in a number of locations using suspension feeders such as mussels, omnivores such as sea urchins and autotrophs such as seaweeds separately or together (Troell, Kautsky & Folke 1999a; Troell et al. 1999b). In order for "integrated aquaculture" to have the potential to significantly reduce nutrient wastes from fish farming, the co-cultured species must be grown on a very large scale and this may have aesthetic or environmental dis-benefits. Nevertheless, where profit margins on farmed fish species such as salmon are low, the economic benefits may improve the sustainability of aquaculture. The economics are highly dependent on the capital costs of co-culture and the market prices of the cultured organisms (Whitmarsh, Cook & Black, In press).

This is a complex and evolving area and regulators continue to demand models that offer a robust and defensible estimate of the assimilative capacity of a water body for dissolved nutrients such that water quality does not diminish – a fundamental requirement in Europe owing to the European Union Water Framework Directive (WFD 2000/60/E). The WFD is the most substantial piece of water legislation ever produced by the European Commission, providing the major driver for achieving sustainable management of water in Member States for many years to come. In summary, the WFD requires that all surface waters (lakes, rivers, estuaries and coastal waters) within defined river basin districts must reach at least 'good' status by 2015. 'Good' status must be defined by setting environmental quality objectives (www.euwfd.com).

Effects of Medicines and Chemicals on Marine Ecosystems

A variety of medicines are used to treat infection and parasitic infestation of farmed fish. The range of medicines available to the farmer varies by country. For example, Canadian farmers only have access to one effective medicine for sea lice, Slice (emamectin benzoate), but Scotland's farmers may also use Excis (cypermethrin) and several other products are available in Norway. This is in part a consequence of the differing emphases of environmental regulation and also a result of differing policy support for aquaculture.

Sea lice are crustaceans and there is the potential for medicines used to control infestation on farmed fish to have wider ecosystem effects. Several laboratory studies have shown that these medicines are indeed toxic to a variety of crustacean species. Such toxicity data is used to derive Environmental Quality Standards (EQSs) for these medicines. In general, these products are highly regulated and discharges are only permitted to be at concentrations above the EQS over short temporal and/or spatial scales. A recent 5-year study (Black et al. 2005) into the ecological effects of sea lice medicines in Scotland revealed no clear indication of ecosystem damage from their use and concluded that Scottish regulations offered good protection of the environment (Willis et al., In press).

The most serious chemical threat to the marine environment is from persistent organic pollutants (POPs) e.g. dioxins, PCBs and PAHs. These toxic substances are ubiquitous at low concentrations but cause harm by being concentrated in lipids and passed up food chains where they may cause reproductive failure, disease or mortality to top-predators. The current generation of sea lice medicines do not fall into this category having environmental half-lives in the range of days to months. The regulation of discharges of emamectin benzoate in Scotland limits the frequency as well as the amount of the discharge in order to ensure that cumulative effects are avoided.

In the 1980s and 1990s, antibiotic use was very high in salmon farming owing to the prevalence of the bacterial disease furunculosis. Concern over the potential environmental effects of antibiotic use was high (Capone et al. 1996) but in fact few adverse effects were reported. The advent of effective vaccines in the early 1990s not only greatly reduced antibiotic use (Alderman 2002) but also resulted in much lower mortality and improved fish growth performance.

Recently, researchers in North America and Scotland have considered the potential for copper (mostly from antifoulant coatings but also from fish feed where it is added as a supplement) and zinc (from feed) to have toxic effects in the environment (Brooks & Mahnken 2003; Brooks et al. 2003). These metals can reach quite high concentrations beneath fish farms (Dean et al. 2007) but it is thought that they are largely bound to sulphides produced in anaerobic sediments reducing their bioavailability and toxicity. Certainly sediments beneath fish farms with high copper loadings can still support abundant infaunal communities.

The discharge of toxic chemicals into the marine environment from aquaculture is necessary while there are pathogens and nuisance organisms that affect animal welfare and reduce production efficiency. The risk to the environment is a function of the specific toxicity, the amounts used, the frequency of use and the environmental persistence of the chemical. Improvements can be or have been made to each of these factors through a combination of better animal husbandry, the use of more-targeted products with short environmental half-lives and the continued search for vaccines.

Effects of Sea Lice from Salmon Culture

Sea lice are ectoparasites of fish. There are several species of lice, some of which are highly host specific and others which have a wide range of hosts. Farmed salmon in the Atlantic suffer infestations of two species: the salmonid specific *Lepeophtheirus salmonis* and the non-specific, smaller *Caligus elongatus*. Both of these can pose a serious threat to farmed fish health and, before the recent advent of effective medicines, considerable morbidity and mortality. Patterns of infestation vary between water bodies and countries (Revie et al. 2002).

Sea lice from fish farms represent a threat to wild anadromous salmonids (McKibben & Hay 2004; Penston et al. 2004) and have been blamed by some for declines in wild populations of sea trout and salmon. There is much circumstantial evidence to support the proposition that sea lice of farm origin are a factor contributing to the decline of wild populations in fish farming areas, but they are certainly not the only factor as evidenced from declines reported in areas where there is no fish farming. Other factors may include freshwater habitat damage and changes in oceanic productivity caused by climate changes.

Considerable efforts are being made to control sea lice in farmed fish by both farmer and regulator. These include:

- Regular monitoring of lice numbers
- Coordinated chemical treatments between farms sharing the same water body
- Single generation sites
- Synchronized fallowing of management areas to break lice cycles
- Treatment of lice in the spring when lice numbers are low

These measures certainly have the potential to reduce the lice problem but reductions in lice burdens per fish may be compromised by increased farm production over time (Heuch & Mo 2001). Different legislatures put different emphasis on lice and the medicines to control them. For example, in Norway lice are seen as the major problem both to farmed and wild salmonids and the emphasis is on ensuring low lice levels on farmed stock. In Scotland, the focus has been more on closely regulating medicine use to protect the wider marine environment. Thus the implied environmental quality objectives differ, which has the potential to affect relative economic performance between countries, although there is also evidence that environmental pressures lead to different patterns of lice infection in the two countries (Heuch et al. 2003).

Effects of Escapes on Wild Populations

Fish farmed in sea cages escape. Although great engineering improvements have been made over the last 10 years leading to increased containment security, escapes still occur because of accident or extreme weather. Wild salmon have a high level of genetic diversity both within and between populations. Between-population variability is driven by selection for the particular river (or part of a river) of origin and these differences are maintained by accurate homing during the breeding migration. In contrast, farmed salmon have relatively low levels of genetic diversity and have been selected for traits such as growth and disease resistance. Although inter-breeding between escaped farmed salmon and native wild fish has been studied for many years it has not been clearly shown that such interbreeding damaged wild populations that is, that the genes present in wild populations conferred significant fitness advantages that would be lost if diluted with genes from escapees. A recent seminal study (McGinnity et al. 2003) has shown that local adaptations are highly linked to fitness and that their loss has severe fitness consequences especially in the second generation.

Where populations of wild salmon are small and escapes into that population are large and recurring, significant reductions in survival are likely. Although containment will continue to improve over time through technological advances in net and cage design, for the foreseeable future escapes of farmed salmon are likely to continue. Escapes from salmon farms constitute a major threat to the genetic integrity, and therefore viability, of wild populations. Norway adopted a new regulation in 2003, implemented from January 2006, with a certification scheme, which sets an industry standard for cage construction and mooring systems (NS9415). In Scotland, a Code of Good Practice includes a detailed section on containment (www.scottishsalmon.co.uk/dlDocs/CoGp.pdf).

Sustainability of Feed Supplies

The culture of carnivorous fish species has traditionally depended on the supply of feed based on fish meal and fish oil obtained from industrial pelagic fisheries (Fig. 7.1). There is no doubt that many capture fisheries are in crisis on a global basis (Pauly et al. 2002). This particularly applies to the traditional fisheries catching relatively long-lived species high in the food chain for human consumption, for example cod. Industrial fisheries are generally based on small, short-lived pelagic species that have a high capacity to recover quickly from population crashes caused by over-fishing or environmental variability. The sustainability of these fisheries, and the consequences of population fluctuations for their predator species, is the subject of a great deal of attention. This is too wide a topic to be discussed here. However, even assuming that these fisheries are entirely sustainable at current levels, it is clear that in a few years the supply of fish meal and especially fish oil, will be monopolised by demand from the rapidly growing global aquaculture industry

(Tacon, 2005). This is partly because the diets of herbivorous species such as many carps are being supplemented by feeds containing fish-oils that offer growth performance and fish health benefits to these species.

Two important questions arise from this: (1) will the demands of global aquaculture increase the pressures on wild fisheries rendering them unsustainable (Naylor et al. 2000); and (2) can the requirement for fish meal and oil in the diets of carnivorous species be reduced without reducing the human health benefits derived from eating fish?

The first question is rather difficult to answer as it depends on the behaviour of the aquaculture, fish feed manufacturing and fish meal/oil industries, each of which is subject both to consumer and regulatory pressures. All of the major aquaculture corporations claim to use materials from sustainable sources, and the fish meal industry representative organisations make similar assertions. Ultimately the answer will depend on the ability of the regulators of the world's industrial fisheries to ensure that capture fisheries do not respond to increasing demands with unsustainable increases in catch.

The second question is easier to discuss as a large amount of research has been focussed on determining the precise dietary requirements of farmed fish in terms of proteins and lipids (Montero et al. 2003; Mourente & Dick 2002; Mourente, Good & Bell 2005; Opstvedt et al. 2003). Much progress has been made and already a significant proportion of the protein and oil content of commercial salmon diets is derived from vegetable sources. It appears that salmon can be fed a low fish oil diet until near the end of the grow-out phase followed by a high fish oil diet for the last few months. Fish reared on this diet have lipid profiles that are similar to those fed fish oil throughout the life cycle. The final product has the same human health benefits (high omega-3 fatty acid content) but using only a fraction of fish oil. As the proportion of global fish oil production demanded by the aquaculture industry increases to nearly 100% in the next decade, it is likely that research effort will continue to focus on its efficient use in the growing aquaculture industry.

A Limiting Factor Approach

As discussed above, the main categories of impacts of aquaculture are:

A. Effects on the seabed from particulates
B. Effects on the pelagic ecosystem from dissolved nutrients
C. Effects of chemicals on marine ecosystems
D. Effects on wild populations from the transfer of parasites (or the pathogens) from cultured populations
E. Genetic effects on wild populations from escapes
F. Effects on capture fisheries owing to increased demand for fish meal and fish oil from aquaculture.

Each of these categories is scale-dependent and so is strongly influenced by farm size and density, and also by increased efficiencies of resource use. These are clearly linked by market forces: increased production unit (farm) size leads to economies of scale and lower costs, but increased absolute industry size leads to increased supply and reduced profits with a consequent requirement for cost-reducing efficiencies.

A "limiting factor" approach simply means that the scale of a proposed development is defined by the environmental issue that presents the most risk in a given location. Consider a highly dynamic environment with a very low water residence time: it is likely that dissolved nutrient inputs will not significantly affect water quality (B, in the list of issues above). Strong currents may also mean that impact on the sea bed (A) will be relatively low and perhaps that medicines used (C) will rapidly be diluted below toxic thresholds and, consequently, the farm will be able to use and discharge sufficient medicine to keep lice numbers low (D). However, the hypothetical farm is close to a major salmon river and the dynamic and exposed nature of the site means that it receives considerable mechanical stress from wind and waves, increasing the risk of escapes (E). Ignoring F (feed sustainability) as it is a generic issue and not location specific, the risk of escapes appears to be the limiting factor – the most risky aspect of the development. To mitigate this risk the regulator could require smaller, stronger cage units, a higher than normal mooring specification and stronger nets – as this will be costly, the scale of the farm may be limited by this factor rather than by factors A, B, C and D. Thus the regulatory focus has been on the most risky factor, the limiting factor, and less time and money has been wasted considering factors that have lower risk.

Consider the case where the regulator insists that the risk of escapes is too high even when mitigated by strengthened structures and a site further from the river should be sought. It is worth pursuing the hypothetical farm developer to his alternative new site location to better understand the limiting factor approach. This new location is not near a salmon river, is in a well flushed bay with low residence time but does not experience sufficiently strong currents to resuspend a large proportion of the particulate waste. There is no other farm in the bay and the benthic habitat is unexceptional in that it is common and extensive. At this site, issue C: effect of chemicals on the environment or issue A: benthic impacts, may be the factor limiting the scale of the development. Modelling can be carried out to estimate the scale of farm production that will remain within sediment environmental quality criteria (assuming these have been established). The frequency of medicine treatment required can be estimated (roughly and with a degree of caution) from the experience of farms in similar conditions such that the farmer will have permission to use sufficient medicine to treat his entire stock at peak production biomass. That medicine discharge can be modelled to predict compliance with EQSs and the model repeated iteratively with different biomasses (and hence discharges) until such standards are met. So, from both benthic and medicine perspectives, maximum biomass can be estimated. The factor (A or C) resulting in the lower biomass becomes the factor limiting production and regulatory effort can be targeted at this aspect. There is a direct spin-off for compliance monitoring with this approach as regulators can subsequently focus monitoring on the limiting factor putting less effort into measuring parameters that relate to a non-limiting factor.

Not discussed thus far are the various aesthetic and socio-economic issues, which also have some relationship to scale and are also important in determining the absolute size of the industry that society will accept. For example, areas of high cultural, wilderness, conservation or scenic importance may be so valuable to society that aquaculture development, not seen as a traditional activity, is prohibited or difficult. Similarly, in areas that are already subject to use from fisheries, tourism or some other industry, aquaculture has to compete for a share of the required resource. Planning and development policies help to prioritise this resource use allocation. In many areas the concept of Integrated Coastal Zone Management is being developed precisely to deal with the resource allocation and conflict resolution requirements associated with marine aquaculture development (see the discussion of ICZM in Lane and Chopin, this volume). In the UK the concept of Marine Spatial Planning is currently being enshrined into law (the UK Marine Bill currently (July 2008) in parliamentary committee, and its likely Scottish Equivalent) which will involve a radical change to the way marine activities are planned not least because a single statutory marine management body is proposed replacing the present piecemeal sectoral governance.

In areas of high unemployment aquaculture may be welcomed for its economic benefits. For example, along the coast of Maine there is a gradient of prosperity with the south being relatively prosperous and the north being less so. Consequently the development of fin-fish farming in northeast Maine has been rapid and intense but in the south, where the socio-economic benefits of aquaculture development would contribute only a small fraction to the local economy, proposed developments have met strong local opposition. Although the state government appears to support the expansion of aquaculture across Maine, socio-economic differences across the state might support a socio-economically tailored policy. However, the danger of such a differential approach (at local or national level), is that aquaculture might be permitted at environmentally degrading densities in areas of low prosperity, whereas wealthier communities may enjoy a less developed marine environment. Linking the local socio-economic benefit to planning and development policy might have the consequence of concentrating aquaculture in relatively poor regions. This may be seen as a good thing as it gives employment where it is needed, but it may also be seen as a bad thing if it concentrates environmental problems in relatively less prosperous areas. The key must be to ensure that where aquaculture gives obvious benefits to local communities it should also be well regulated and planned so that it is practiced at a scale that does not cause environmental degradation.

The potential for interactions between marine aquaculture and mammals and birds has received relatively little attention in the literature (Furness 1996; Kaiser 1998; Ross, Lien & Furness 2001; Lepper 2004; Johnston 2002) but they may be the most important issue in areas where there are rare or important populations of these animals that may be disturbed by farm activities.

Sustainability of Cage Fish Farming

Hunting is generally the harvesting of wild animals for human consumption. In many ways commercial fishing is to fish farming as deer-hunting is to farming reindeer. In both land and sea the end product of each activity is approximately the same so farming can be seen as an extension of hunting – ultimately, animals are killed and eaten. The difference between fish farming and domesticated animal production is that most of the intensively farmed fish species (in Western industrialized countries, ignoring for the moment carp in China) are carnivorous and are not fully domesticated. However, farmed fish in the future may be much less dependent on fishery products (i.e. become more vegetarian) and may become more domesticated. The process towards domestication has already begun in salmon with intense selection for traits that increase farm productivity and this is likely to be accelerated with the advent of genetics-based, selective-breeding programmes. Such selection has the objectives of creating faster growth, increased feed conversion efficiency and disease tolerance, and may also in the future select for strains that do better on largely plant-origin diets. Improved domestication may have environmental benefits. For example, improved disease resistance may reduce the need for medicines and improved feed efficiency may reduce the nutrient waste per unit of production.

Farmed fish are typically grown in cages suspended in natural waters. This means that most wastes are immediately transferred to the environment for processing. An alternative system is where animals are cultured in a semi-closed system, usually on land. In such systems it is possible to process the wastes, in the same way as other industries are required to, by means of settlement tanks, filtration and biological digestion. Some commentators promote such technologies as the most environmentally acceptable future for aquaculture (Folke, Kautsky, and Troell 1994, 1997). However, it could equally be argued that such systems have environmental costs, for example energy for pumping large volumes of water, which should be factored into any environmental comparison. It is also worth considering the best use of resources. The main ecosystem service provided to humanity by the shelf seas is that of processing wastes, but emissions to the sea must not exceed its capacity to assimilate them without ecosystem changes or damage to human health. The disposal of human sewage in the sea is becoming more carefully managed in many areas. For example, the dumping of sewage sludge in European waters has now largely ceased. Although much of the motivation for these changes has been to reduce the risks of eutrophication, an additional important benefit is the reduction of the concentration of human pathogens in coastal waters. Not only does this lead to increased safety in bathing waters but it also reduces the risk of disease from consumption of coastal fishery products, notably shellfish. Where there is surplus assimilative capacity in the sea it does not seem unreasonable that fish farming, which does not produce human pathogens, should have access to a appropriate share of it.

Domestication may eventually have additional benefits in that highly domesticated strains of fish may become reproductively incompetent in the wild thus reducing the potential for genetic pollution. In the meantime, the survival of wild salmonids populations in several areas may be intimately linked to the ability of farmers to contain their stock and develop ever better methods of cutting off the flow of infective lice larvae to wild populations.

Benthic impacts are the best studied of the impacts of aquaculture. However, as long as fish farms are not sited over deep, irregularly exchanged fjordic basins, where they may contribute to wide-scale bottom-water hypoxia, and provided that benthic impacts from aquaculture only affect a small proportion of the available seabed habitat in a system, benthic effects are generally restricted to the local scale. Exceptions to this may be in areas of hard substrate where suspended or resuspended particulates may be transported long distances by fast currents, potentially damaging important reef ecosystems (Hall-Spenser et al. 2006).

Global aquaculture is predicted to continue to increase in scale very rapidly over the foreseeable future. It is likely to cause damage to marine ecosystems in areas where short-term economic advantage is prioritised over environmental health. In such areas environments may be come degraded to the disadvantage of many users. However, where such degradation is caused by the overloading of organic matter and dissolved nutrients, the environment is likely to recover quickly when the source of the pollution is brought under control and so mistakes are unlikely to have permanent consequences. Much more worrying is where degradation comes from habitat destruction such as smothering of reefs and clearing of mangroves, or from the use of persistent chemicals, or from the introduction of a disease or alien "weed" organism, as such changes may profoundly affect ecosystem function in an irreversible manner.

In the North Atlantic context, sustainability is likely to be a function of markets. Can production of relatively low-value, mass-produced food products continue to increase in countries that typically have high standards of living and high labour costs? Or will globalisation give overwhelming advantages to economies with fewer regulations and lower labour costs? The sustainability of fish farming in western economies may depend on being able to demonstrate the highest possible environmental standards and consequently develop high-value niche markets among environmentally aware consumers.

Conclusions

Aquaculture is here to stay as a global industry. It is as legitimate an activity as the farming of livestock on land. In the sea, it is as legitimate a user of the coastal resource as fisheries. However, in the private desire for profit and the public desire to provide economic alternatives in rural communities, the development of aquaculture has not always been well planned. The advent of intensive aquaculture has highlighted deficiencies in legislative and regulatory frameworks in many countries – regulators have had to play catch-up as the industry realised rapid expansion. While there is now much better understanding of the environmental issues, and a greater capacity to make rational decisions regarding new developments, fish farm sites remain that are poorly located, or at too high a density, causing real environmental problems.

Faced with growing regulatory and consumer scrutiny in a low-profit market, many aquaculture enterprises and associations have championed industry standards, codes of best practice and formal environmental management systems. These have reduced the environmental "footprint" per unit of production of the industry. Legislators too are attempting to improve the efficiency and transparency of the decision making process, using good science and bespoke computer models to encapsulate understanding of environmental processes, and have attempted or are attempting to formalise their long-term development policy for the industry.

Significant challenges, such as the impact on wild salmonids from escapes and from lice, remain in many areas. The science community must improve its understanding of the factors linked to optimum siting of salmon farms to minimise transmission of lice larvae to wild fish and to better understand the effects of farming on hard-substrate fauna. Better understanding of the assimilative capacity of the environment is a prerequisite. In addition, understanding the environmental interactions of "new" fin-fish species, such as cod, requires further work. Technologies to improve containment are necessary as well as more widespread use of existing and new technologies to minimise feed losses. All of these developments require a profitable industry – low profit remains the most significant challenge to sustainability in many western countries.

Acknowledgments This work was supported by the EU Framework Programme 6 project ECASA, Ecosystem Approach for Sustainable Aquaculture (www.ecasa.org.uk). The author is grateful to Averil Wilson for her comments on the manuscript.

References

AGJ, Tacon, 2005. State of information on salmon aquaculture feed and the environment. Salmon Aquaculture Dialogue, pp. 81. www.worldwildlife.org/what/globalmarkets/aquaculture/WWFBinaryitem8840.pdf

Alderman, D. J. 2002. Trends in therapy and prophylaxis 1991–2001. *Bulletin of the European Association of Fish Pathologists* 22: 117–125.

Black, K. D. 2001. Sustainability of aquaculture. In *Environmental Impacts of Aquaculture*, (ed. K. D. Black), pp. 199–212. Sheffield Academic Press, Sheffield.

Black, K. D., Cook, E. J., Jones, K. J., Kelly, M. S., Leakey, R. J., Nickell, T. D., Sayer, M. D. J., Tett, P. and Willis, K. J. 2002. *Review and Synthesis of the Environmental Impacts of Aquaculture*. Scottish Executive Central Research Unit, Edinburgh. http://www.scotland.gov.uk/cru/kd01/green/reia.pdf.

Black, K. D., Blackstock, J., Cromey, C. J., Duncan, J., Gee, M., Gillibrand, P., Needham, H., Nickell, T. D., Pearson, T. H., Powell, H., Sammes, P., Somerfield, P., Walsham, P., Webster, L. and Willis, K. 2005. The ecological effects of sea lice treatment agents, p. 286. Scottish Association for Marine Science, Oban. www.sams.ac.uk/research/coastal%20imapcts/ecol.htm.

Brooks, K. M. and Mahnken, C. V. W. 2003. Interactions of Atlantic salmon in the Pacific Northwest environment III. Accumulation of zinc and copper. *Fisheries Research* 62: 295–305.

Brooks, K. M., Stierns, A. R., Mahnken, C. V. W. and Blackburn, D. B. 2003. Chemical and biological remediation of the benthos near Atlantic salmon farms. *Aquaculture* 219: 355–377.

Brooks, K. M., Stierns, A. R. and Backman, C. 2004. Seven year remediation study at the Carrie Bay Atlantic salmon (Salmo salar) farm in the Broughton Archipelago, British Columbia, Canada. *Aquaculture* 239: 81–123.

Capone, D. G., Weston, D. P., Miller, V. and Shoemaker, C. 1996. Antibacterial residues in marine sediments and invertebrates following chemotherapy in aquaculture. *Aquaculture* 145: 55–75.

Chopin, T., Buschmann, A. H., Halling, C., Troall, M., Kautsky, N., Neori, A., Kraemer, G. P., Zertuche-Gonzalez, J. A., Yarish, C. and Neefus, C. 2001. Integrating seaweeds into marine aquaculture systems: a key towards sustainability. *Journal of Phycology* 37: 975–986.

Cook, E. J. and Kelly, M. S. 2007. Enhanced production of the sea urchin Paracentrotus lividus in integrated open-water cultivation with Atlantic salmon Salmo salar. *Aquaculture* 273: 573–585.

Cromey, C. J., Nickell, T. D. and Black, K. D. 2002a. DEPOMOD – modelling the deposition and biological effects of waste solids from marine cage farms. *Aquaculture* 214: 211–239.

Cromey, C. J., Nickell, T. D., Black, K. D., Provost, P. G. and Griffiths, C. R. 2002b. Validation of a fish farm waste resuspension model by use of a particulate tracer discharged from a point source in a coastal environment. *Estuaries* 25: 916–929.

Dean, R. J., Shimmield, T. M., Black, K. D. 2007. Copper, zinc and cadmium in marine cage fish farm sediments: an extensive survey. *Environmental Pollution* 145: 84–95.

Findlay, R. H. and Watling, L. 1997. Prediction of benthic impact for salmon net-pens based on the balance of benthic oxygen supply and demand. *Marine Ecology Progress Series* 155: 147–157.

Folke, C., Kautsky, N. and Troell, M. 1994. The costs of eutrophication from salmon farming – implications for policy. *Journal of Environmental Management* 40: 173–182.

Folke, C., Kautsky, N. and Troell, M. 1997. Salmon farming in context: response. *Journal of Environmental Management* 50: 95–103.

Furness, B. W. 1996. Interactions between seabirds and aquaculture in sea lochs. In *Aquaculture and Sea Lochs.*, (ed. K. D. Black), pp. 50–55. The Scottish Association for Marine Science, Oban.

Gowen, R. J. and Bradbury, N. B. 1987. The ecological impact of salmonid farming in coastal waters: a review. *Oceanography and Marine Biology Annual Reviews* 25: 563–575.

Hall, P. O. J., Holby, O., Kollberg, S. and Samuelsson, M. O. 1992. Chemical fluxes and mass balances in a marine fish cage farm. 4. Nitrogen. *Marine Ecology Progress Series* 89: 81–91.

Hall-Spencer, J., White, N., Gillespie, E., Gillham, K. and Foggo, A. 2006. Impact of fish farms on maerl beds in strongly tidal areas. *Marine Ecology-Progress Series* 326: 1–9.

Handy, R. D. and Poxton, M. G. 1993. Nitrogen pollution in mariculture – toxicity and excretion of nitrogenous compounds by marine fish. *Reviews in Fish Biology and Fisheries* 3: 205–241.

Heilskov, A. C. and Holmer, M. 2001. Effects of benthic fauna on organic matter mineralization in fish-farm sediments: importance of size and abundance. *Ices Journal of Marine Science* 58: 427–434.

Heuch, P. A. and Mo, T. A. 2001. A model of salmon louse production in Norway: effects of increasing salmon production and public management measures. *Diseases of Aquatic Organisms* 45: 145–152.

Heuch, P. A., Revie, C. W. and Gettinby, G. 2003. A comparison of epidemiological patterns of salmon lice, Lepeophtheirus salmonis, infections on farmed Atlantic salmon, Salmo salar L., in Norway and Scotland. *Journal of Fish Diseases* 26: 539–551.

Holby, O. and Hall, P. O. J. 1991. Chemical fluxes and mass balances in a marine fish cage farm. 2. Phosphorus. *Marine Ecology Progress Series* 70: 263–272.

Holmer, M. and Kristensen, E. 1992. Impact of marine fish cage farming on metabolism and sulphate reduction of underlying sediments. *Marine Ecology Progress Series* 80: 191–201.

Johnston, D. W. 2002. The effect of acoustic harassment devices on harbour porpoises (Phocoena phocoena) in the Bay of Fundy, Canada. *Biological Conservation*, 108: 113–118.

Kaiser, M. J., Laing, I., Utting, S. D. and Burnell, G. M. 1998. Environmental impacts of bivalve mariculture. *Journal of Shellfish Research* 17: 59–66.

Lepper, P. A., Turner, V. L. G. Goodson, A. D. and Black., K. D. 2004. Source levels and spectra emitted by three commercial aquaculture anti-predation devices, Seventh European Conference on Underwater Acoustics 5–8 July, Delft, The Netherlands.

McGinnity, P., Prodohl, P., Ferguson, K., Hynes, R., O'Maoileidigh, N., Baker, N., Cotter, D., O'Hea, B., Cooke, D., Rogan, G., Taggart, J. and Cross, T. 2003. Fitness reduction and potential extinction of wild populations of Atlantic salmon, Salmo salar, as a result of interactions with escaped farm salmon. *Proceedings of the Royal Society of London Series B-Biological Sciences* 270: 2443–2450.

McKibben, M. A. and Hay, D. W. 2004. Distributions of planktonic sea lice larvae Lepeophtheirus salmonis in the inter-tidal zone in Loch Torridon, Western Scotland in relation to salmon farm production cycles. *Aquaculture Research* 35: 742–750.

Montero, D., Kalinowski, T., Obach, A., Robaina, L., Tort, L., Caballero, M. J. and Izquierdo M. S. 2003. Vegetable lipid sources for gilthead seabream (Sparus aurata): effects on fish health. *Aquaculture* 225: 353–370.

Mourente, G. and Dick, J. R. 2002. Influence of partial substitution of dietary fish oil by vegetable oils on the metabolism of 1-C-14 18: 3n-3 in isolated hepatocytes of European sea bass (Dicentrarchus labrax L.). *Fish Physiology and Biochemistry* 26: 297–308.

Mourente, G., Good, J. E. and Bell, J. G. 2005. Partial substitution of fish oil with rapeseed, linseed and olive oils in diets for European sea bass (Dicentrarchus labrax L.): effects on flesh fatty acid composition, plasma prostaglandins E-2 and F-2 alpha, immune function and effectiveness of a fish oil finishing diet. *Aquaculture Nutrition* 11: 25–40.

Naylor, R. L., Goldburg, R. J., Primavera, J. H., Kautsky, N., Beveridge, M. C. M., Clay, J., Folke, C., Lubchenco, J., Mooney, H. and Troell, M. 2000. Effect of aquaculture on world fish supplies. *Nature* 405: 1017–1024.

Opstvedt, J., Aksnes, A., Hope, B. and Pike, I. H. 2003. Efficiency of feed utilization in Atlantic salmon (Salmo salar L.) fed diets with increasing substitution of fish meal with vegetable proteins. *Aquaculture* 221: 365–379.

Pauly, D., Christensen, V., Guenette, S., Pitcher, T. J., Sumaila, U. R., Walters, C. J., Watson, R. and Zeller, D. 2002. Towards sustainability in world fisheries. *Nature* 418: 689–695.

Pearson, T. H. and Rosenberg, R. 1978. Macrobenthic succession in relation to organic enrichment and pollution of the marine environment. *Oceanography and Marine Biology Annual Reviews* 16: 229–311.

Pearson, T. H. and Black, K. D. 2001. The environmental impact of marine fish cage culture. In *Environmental Impacts of Aquaculture*, (ed. K. D. Black), pp. 1–31. Sheffield Academic Press, Sheffield.

Penston, M. J., Mckibben, M. A., Hay, D. W. and Gillibrand, P. A. 2004. Observations on open-water densities of sea lice larvae in Loch Shieldaig, Western Scotland. *Aquaculture Research* 35: 793–805.

Pereira, P. M. F., Black, K. D., Mclusky, D. S. and Nickell, T. D. 2004. Recovery of sediments after cessation of marine fish farm production. *Aquaculture* 235: 315–330.

Revie, C. W., Gettinby, G., Treasurer, J. W. and Rae, G. H. 2002. The epidemiology of the sea lice, Caligus elongatus Nordmann, in marine aquaculture of Atlantic salmon, Salmo salar L., in Scotland. *Journal of Fish Diseases* 25: 391–399.

Ross, B. P., Lien, J. and Furness, R. W. 2001. Use of underwater playback to reduce the impact of elders on mussel farms. *Ices Journal of Marine Science* 58: 517–524.

Sanderson, J. C., Cromey, C. J., Dring, M. J. and Kelly, M. S. 2008. Distribution of nutrients for seaweed cultivation around salmon cages at farm sites in north-west Scotland.

Silvert, W. and Sowles, J. W. 1996. Modeling environmental impacts of marine finfish aquaculture. *Journal of Applied Ichthyology* 12: 75–81.

Silvert, W. and Cromey, C. J. 2001. Modelling impacts. In *Environmental Impacts of Aquaculture*, (ed. K. D. Black), p. 214. Sheffield Academic Press, Sheffield.

Strain, P. and Hargrave, B. 2005. Salmon aquaculture, nutrient fluxes and ecosystem processes in southwestern New Branswick. In *Environmental Effects of Marine Finfish Aquaculture. The Handbook of Environmental Chemistry (volume 5): Water Pollution*, (ed. B. Hargrave). Springer Verlag.

Troell, M., Kautsky, N. and Folke, C. 1999a. Applicability of integrated coastal aquaculture systems. *Ocean & Coastal Management* 42: 63–69.

Troell, M., Ronnback, P., Halling, C., Kautsky, N. and Buschmann, A. 1999b. Ecological engineering in aquaculture: use of seaweeds for removing nutrients from intensive mariculture. *Journal of Applied Psychology* 11: 89–97.

Whitmarsh, D. J., Cook, E. J. and Black, K. D. (2006). Searching for sustainability in aquaculture: an investigation into the economic prospects for an integrated salmon-mussel production system. *Marine Policy* 30: 293–298.

Willis, K. J., Gillibrand, P.A., Cromey, C. J. and Black, K. D. (2005). Sea lice treatments at a salmon farm have no adverse effects on zooplankton communities: a case study. *Marine Pollution Bulletin* 50: 806–816.

Chapter 8
Ethics, Governance and Regulation

Marc A. Saner

Introduction

Black's paper provides a broad, yet succinct survey of the environmental aspects of aquaculture. I will attempt to enrich this survey by first providing some analysis on the scope, completeness and coherence of the task that Black set out to do. In a second step, I will discuss implications of Black's descriptions and propositions from the perspectives of ethics, governance and regulation.

Scope

Black chose a humble title – the environmental *aspects* of aquaculture – but actually provides more. In the introduction, he alludes to his plan to take on more. He touches on the concept of the *sustainability* of aquaculture (a concept which encompasses issues beyond environmental aspects) and also points out that well-designed *regulations* (a governance instrument) are of central importance in addressing the results of his survey.

Black follows through on these three different topics. First and foremost, his survey encompasses the main impacts of aquaculture on the physical environment (qualitatively and occasionally quantitatively). Second, it addresses the broader impacts of aquaculture on society and the economy, as one would expect in a paper that touches on the sustainability of aquaculture. Third, Black puts forward a proposition for a regulatory approach – the "limiting factor approach" – which focuses on regulation for the control of the impacts of aquaculture on the physical environment.

Such a broad scope is beyond what can be fully addressed in a single paper, but this perspective provides us with a thread, that goes from a strong focus on the

M.A. Saner
Director of Research, Regulatory Governance, Carleton University, Ontario, Canada
e-mail: marc@saner.ca

K. Culver, D. Castle (eds.), *Aquaculture, Innovation and Social Transformation*,
© Springer Science+Business Media B.V. 2008

impacts of aquaculture on the physical environment, all the way to an interesting proposition to focus the regulation of aquaculture on the management of these impacts.

Completeness and Coherence

In general terms, there are two principal ways to approach a survey on the impacts of a human activity on the physical environment. Either one can emphasize the temporal scale by carrying out *life-cycle assessments* (LCA) of an activity or one can emphasize the spatial scale in *environmental assessments* (EA). A full assessment of environmental risks (quantity) and hazards (quality) would cover, ideally, both temporal and spatial scales.

Black chose to emphasize the spatial scale with a particular attention to the direct environmental impacts of *key effluents*. Effluents from aquaculture can reach first and foremost the seabed, the coast and the open waters – all of which Black discusses. It would be possible to also evaluate the effects effluents have on air quality, the climate, and the ozone layer (e.g., from the operation of machinery, the use of pesticides, or fermentation) or the deep sea and fresh water bodies (e.g., via current or migrating organisms), and so forth. However, in the context of aquaculture, they appear to be minor concerns – although Black does mention the production of methane (a greenhouse gas).

The primary effluents discussed in Black's paper could be categorized into three types – organic chemical effluents (Section 2), commercial chemical effluents (Sections 3 and 4), and living biological effluents (Sections 5 and 6). This approach provides a fairly complete and coherent list of potential environmental impacts.

However, only one type of *indirect* environmental impact is discussed, namely the environmental impact associated with the production of fish food as a key input into the aquaculture systems. It would be possible to discuss the environmental impacts associated with other inputs, pesticides for example, but these may well be less significant when compared to the effects associated with fish food production.

It would be interesting to add to this spatial approach a geographical survey of the main current and potential developments in aquaculture together with a discussion on the nature and vulnerability of habitats. This would improve the ability of the reader to think in terms of *comparative environmental impact analysis*. There is no doubt that aquaculture will alter the environment – but how should this change be judged when compared to other practices such as the building of harbors and the transportation of goods, the fishing in the coastal areas and the open seas, the production of protein through agriculture, and so forth? It is implied, however, that the habitats currently affected are sufficiently vulnerable and valuable to give rise to serious concerns.

It would also be interesting to further complete the discussion of the main impacts of aquaculture on the physical environment with a stronger attention to the temporal component, for example, by explicitly discussing the impacts from "cradle-to-grave." This is a difficult task, but experience in other industrial sectors have illustrated that this type of thinking is key to the successful protection of the environment. Temporal thinking opens the door for a thorough discussion of site development and rotation practices and highlights which portions of the impacts of aquaculture are persistent and which are transient. The smothering of reefs and clearing of mangroves are briefly mentioned towards the end of the paper, and Black addresses environmental persistence in some contexts, but no clear sense of the persistence of environmental impacts overall emerges. The long-term impacts on coral reefs, for example, is not trivial as stated in the Reuters news article from Israel entitled "Red Sea corals close to extinction" (as a consequence of aquaculture) (Goldin 2004).

I do not want to fault Black for limits in his argumentative thread moving from a thorough description of environmental impacts, to a brief discussion of the other pillars of sustainability and towards a conclusion that contains recommendations for regulatory concepts and action. It is notoriously difficult to convincingly move an argument from the description of matters-of-fact to prescriptions of appropriate values, standards or goals. Nevertheless, I have to point out that a *complete* argument for the limiting factor approach would require a new paper and I do hope that Black will write a "justification and feasibility of the limiting factor approach" at a later occasion. In the meantime, I am grateful that this innovative idea has been proposed for discussion as an opinion of an expert on aquaculture.

Ethics

Earlier I stated that the discussion of ethics, governance and regulation represents the second step of my commentary. I chose to put it that way for two reasons. First, it allows me to emphasize the move from the descriptive and predictive realm of science into the prescriptive realm of values, politics and public administration. I believe that it is worthwhile to emphasize this move because of the difficulty of combining the two realms that are at some distance, both philosophically and culturally (as I mentioned earlier). Second, it also delineates the move from straight commentary to an increasing input of own ideas and views.

Every applied scientific assessment operates within an ethical framework. It is worthwhile making this framework explicit because it fosters the understanding of the recommendations that flow from it. The implicit ethical framework in Black's paper is *enlightened anthropocentric*, meaning that the concern for the environment is driven by the utility of ecosystems to humans while fully recognizing the important current and future utility of these ecosystems. In this view, it is critically important to respect the carrying capacity of the ecosystems. It is not important from this point of view, however, to focus on issues of animal welfare, endangered species, or the protection of the environment for its own sake.

The position of enlightened anthropocentrism integrates well with the concept of sustainability. The Brundtland Report (WCED 1987) that publicized the concept of sustainable development, for example, explicitly rejects a concern for the environment that is independent of current or future human utility. The defense of enlightened anthropocentrism over ethical alternatives (for example, biocentrism, the Land Ethic, or ecocentrism in general) is quite difficult in philosophical terms and usually resorts to an appeal to persistent public popularity of anthropocentrism and the resulting political practicality. In other words, the best argument for enlightened anthropocentrism is that it represents the only position in environmental ethics that can have a practical policy impact. However, this does not mean that other ethical positions (such as the concern for animal welfare or endangered species) will not surface in the debate over aquaculture – these other viewpoints still could leave a mark on the design of regulatory approaches.

There is utility in understanding the ethical positions that stakeholders embrace. Such understanding is not only a requirement for the design of strategic communications by proponents of aquaculture; much more importantly, it is a requirement for true dialogue, which is the foundation of a stable governance system. I provide an example for each of these two points.

On the first point, proponents of aquaculture should respect good risk communication practices when designing strategic communications. One such practice is for communicators to inform early, clearly and honestly. Honesty is not just an ethical prescription here – it is simply a requirement for the building of trust – a very valuable commodity for industry and regulators alike. As part of good risk communication one would avoid, for example, to label broad-spectrum insecticides as "medicines" as Black proposes (or "therapeutics" – another label used by authors writing in the aquaculture context). "Medicine" is a highly unusual label for a household insecticide like Cypermethrin, which is used to kill cockroaches and is judged "highly toxic to fish" by the United States Environmental Protection Agency (NPTN 1998) (toxicity depends on concentration, of course). The use of such an unusual term has the flavour of manipulation and will likely have a negative impact on trust. Similarly, professional risk communicators would also see the similarity between the production of "reproductively incompetent fish" (to stop genetic pollution) and the recent public outcry over the so-called "Terminator Technology", which has been proposed for the use in transgenic crops. Reproductive incompetent fish are not only a benefit in terms of environmental risk, they also may represent a shift in the way fish stocks are owned and associated social issues.

On the second point, I would argue that dialogue with stakeholders is absolutely essential if an understanding of key social and economical impacts is needed – and such an understanding is indeed needed for a full evaluation of all pillars of the sustainable development concept (and for a meaningful interpretation of the enlightened anthropocentric position). For example, it is next to impossible to take the preferences and concerns of Aboriginals into account without dialogue. Openness to the culture and ethics (and indeed aesthetics) of Aboriginals, environmental activists, fishermen and aquaculturists, consumers and the public at large is essential to the development of defensible policies, regulatory systems and decision-making processes (See Felt, Shepert and Ryan, this volume).

Governance

The transition from a discussion on ethics to one on governance is almost imperceptible because the idea of good governance contains defensible ethics at its very core. This can be illustrated, for example, by the similarity of the points made above under ethics to the conception of "Modern Governance" in the Aquaculture Policy Framework of the Government of Canada (DFO 2002):

> In addition to supporting the department's regulatory function, this policy framework must set the stage for creative governance approaches in priority areas such as site access (e.g., proactive planning), scientific research and development (R&D), and government programs enabling industry development. These approaches must include mechanisms to engage aquaculturists, Aboriginal groups and stakeholders in decision-making and increase their accountability in relation to key sustainability objectives.

Achieving good governance, however, is not a simple matter. A good governance approach to the development and implementation of aquaculture regulations requires paying attention to principles such as those proposed by the United Nations Development Programme (UNDP) and further developed by the Institute On Governance (Table 8.1). These principles show that good governance requires more than a clear argument for a strategic direction. Instead, it is a matter of creating a system that provides long-term legitimacy, accountability, justice and performance. As a result, the development and implementation of new regulations (including the balancing between mandatory and voluntary approaches) is a complex and often slow process.

Regulation

Black concludes his evaluation of environmental impacts with a call for a "limiting factor approach" to regulation. His justification for this approach could be rendered more explicit and summarized as follows:

- All environmental impacts of aquaculture are scale dependent.
- Market forces alone will likely lead to scales of development of aquaculture (and associated environmental impacts) that would push ecosystems beyond their carrying capacity.
- If the regulatory system were capable of identifying and controlling the type of environmental impact that threatens the carrying capacity first, then the ecosystem would be safe.
- The ecosystem ought to be safe.
- Therefore, a regulatory system should be devised that is capable of identifying the environmental issue that poses the greatest risk (the "limiting factor") at a given location. The system should then cap the scale of development based on that limiting factor – even if the carrying capacity of the ecosystem has not yet been reached for other environmental stressors.

Table 8.1 Five principles of good governance (Graham et al. 2003)

Good governance principles	UNDP principles and related UNDP text on which they are based
1. Legitimacy and Voice	**Participation** – all men and women should have a voice in decision-making, either directly or through legitimate intermediate institutions that represent their intention. Such broad participation is built on freedom of association and speech, as well as capacities to participate constructively. **Consensus orientation** – good governance mediates differing interests to reach a broad consensus on what is in the best interest of the group and, where possible, on policies and procedures.
2. Direction	**Strategic vision** – leaders and the public have a broad and long-term perspective on good governance and human development, along with a sense of what is needed for such development. There is also an understanding of the historical, cultural and social complexities in which that perspective is grounded.
3. Performance	**Responsiveness** – institutions and processes try to serve all stakeholders. **Effectiveness and efficiency** – processes and institutions produce results that meet needs while making the best use of resources.
4. Accountability	**Accountability** – decision-makers in government, the private sector and civil society organizations are accountable to the public, as well as to institutional stakeholders. This accountability differs depending on the organizations and whether the decision is internal or external. **Transparency** – transparency is built on the free flow of information. Processes, institutions and information are directly accessible to those concerned with them, and enough information is provided to understand and monitor them.
5. Fairness	**Equity** – all men and women have opportunities to improve or maintain their well-being. **Rule of Law** – legal frameworks should be fair and enforced impartially, particularly the laws on human rights.

The strength and weakness of this approach is its simplicity. If the long-term preservation of the ecosystem is the ultimate goal then it makes sense to put all emphasis of regulation on the first threat that could cause the ecosystem to collapse. We do need the long-term function of these ecosystems to preserve the future of aquaculture and for many other reasons (including the desire to preserve these systems for their own sake). Thus, the proposition is meaningful and powerful in its clarity and simplicity.

Nevertheless, the proposition can be criticized. First, when considering all three pillars of sustainable development, the environment, the social fabric, and the economy (also known as planet, people and profit), the question of trade-off becomes very pressing. It is possible to gain on the economic and social sides in such a tremendous way that it "makes up" for the environmental loss. This trade-off can be expressed in economic terms by using the so-called "capital approach." For example, if a decline in natural capital (ecosystem stability) leads to an important increase in terms of social capital and the economy, then decision-makers may become convinced that the project should go forward. They can argue that the new social and economic capital will ultimately lead to a betterment of the physical environment because destruction is prevented elsewhere and because the social capital will translate into curative environmental inventions. In the same vein, the environmental destruction caused by aquaculture might be judged relatively minor, when compared to fishing operations or protein production in agriculture based on a comparative risk assessment. I am ready to believe that these arguments will ultimately fail – I only mention them here to illustrate that the limiting factor does not address several arguments that one could expect to come from advocates of aquaculture.

The limiting factor approach is also somewhat blind to arguments that one would expect to come from the critics of aquaculture. Habitat destruction associated with the creation of aquaculture sites may be overlooked by an emphasis on effluents, for example. Equally, aesthetic and cultural objections cannot be discussed within the limiting factor concept.

I conclude with the observation that the limiting factor approach represents an important contribution to the development of regulatory approaches. However, it will first have to pass the test of comparison with alternative options in a discussion that will, hopefully, be informed by good governance principles.

References

Fisheries and Oceans Canada (DFO) – Communications Branch. 2002. DFO's Aquaculture Policy Framework. DFO/6228. Cat. No. Fs23-411/2002.

Goldin, M. 2004. Red Sea Corals Close to Extinction. *PLANET ARK – World Environment News (Reuters New Service)*, May 7, 2004. http://www.planetark.org/avantgo/dailynewsstory.cfm?newsid=25839. Accessed: July 18, 2007.

Graham, J., Amos, B. and Plumptre, T. 2003. Principles for Good Governance in the 21st Century. Policy Brief #15, Institute On Governance, Ottawa, available at www.iog.ca.

National Pesticide Telecommunications Network (NPTN). 1998. Cypermethrin (factsheet). Oregon State University. http://npic.orst.edu/factsheets/cypermethrin.pdf. Accessed: July 18, 2007.

World Commission on Environment and Development (WCED). 1987. Our Common Future. Oxford University Press.

Chapter 9
A Synopsis of Environmental Issues Associated with Salmon Aquaculture in Canada

Fiona Cubitt, Kevin Butterworth and Robert Scott McKinley

Introduction

World-wide, aquaculture currently produces 43% of the world's food for human consumption. In 2004, 106 million tonnes of food were consumed, of which 45.5 million tonnes and a value of $63 billion (US) was attributed to aquaculture (FAO 2006). Canada is the world's 6th largest exporter of aquaculture products with a gross output of $785 million (CDN) in 2005 (Statistics Canada 2006). The contribution from British Columbia's salmon aquaculture industry was 53,800 tonnes of Atlantic salmon, and 16,800 tonnes of Pacific salmon (British Columbia Ministry of Environment (BCMOE) and British Columbia Ministry of Agriculture and Lands (BCMAL, 2005).

British Columbia's capacity for aquaculture production has been heralded by its proponents as a potential saviour of impoverished coastal communities struggling as a result of the decline of the capture fisheries and other traditional resource industries. Yet salmon aquaculture suffers from a particularly poor public image. In 1995, the provincial government of British Columbia, in consultation with Fisheries and Oceans Canada, implemented a moratorium on expansion of the salmon farming industry. This moratorium, which was in place for 7 years, was in response to public concern about the effects of the industry on wild salmon stocks and the marine environment (Report of the Auditor General 2006). Currently, perceived issues include disease outbreaks (viral, bacterial and parasitic), fish quality (contaminants, colour additives, antibiotic use), wild salmon returns (pink salmon in the Broughton Archipelago), fish waste (uneaten feed, faeces, mortalities, anti-fouling agents), predators (killing and harassment devices), farm escapes (exotic species introduction, ecology, disease, genetic dilution) and feeder fish population declines. Together, these issues contribute to an overall poor perception of open net-pen farming. The discussion on these issues is complex and ongoing, ranging from daily newspapers (e.g. Simpson 2007), weekly free press (e.g. Cox 2004), and websites

F. Cubitt
UBC-DFO Centre for Aquaculture and Environmental Research, University of British Columbia, 4160 Marine Drive, West Vancouver, BC V7V 1N6
e-mail: kfcubitt@interchange.ubc.ca

K. Culver, D. Castle (eds.), *Aquaculture, Innovation and Social Transformation,*
© Springer Science+Business Media B.V. 2008

associated with advocacy groups (e.g. Society for Positive Aquaculture Awareness (SPAA) 2007; Coastal Alliance for Aquaculture Reform (CAAR) 2007). Some of this discussion has even resulted in successful court cases (e.g. British Columbia Supreme Court 2007).

Three of the issues surveyed above are particularly prominent in public discussion of aquaculture in British Columbia: (1) the potential risk posed to wild salmon stocks by escaped aquaculture salmon, (2) the ecological 'footprint' of salmon farms, and (3) the risk posed to migrating juvenile wild salmon by sea lice hosted by farmed salmon.

The purpose of this chapter is to provide a Canadian contrast to the perspective offered by Black, and to present balanced scientific information on the regulation of aquaculture relative to coastal community and environmental sustainability. This chapter will focus on British Columbia as it is Canada's largest aquaculture producer (46% of annual revenue (Statistics Canada 2006)). Information presented here is the result of collaboration with the aquaculture industry, NGOs, First Nations, and provincial and federal governments.

The Regulatory Environment for Aquaculture in Canadian Provinces

The *Canadian Federal Fisheries Act* (1985) does not contain a definition of the term 'aquaculture.' The *British Columbia Fisheries Act* defines aquaculture as the "growing and cultivation of aquatic plants... or fish, for commercial purposes, in any water environment or in human made containers of water, and includes the growing and cultivation of shellfish on, in or under the foreshore or in water." In Canada's eastern-most province, Newfoundland and Labrador, the *Aquaculture Act* (1990), defines aquaculture as "the farming of fish, molluscs, crustaceans, aquatic plants and other aquatic organisms with an intervention in the rearing process to enhance production by regular stocking, feeding, and protection from predation, and includes fallowing and processes to mitigate environmental degradation and the placement of necessary gear and equipment." Finally, in New Brunswick, the *Aquaculture Act* (1988), defines aquaculture as "the cultivation of aquatic plants and animals, but does not include the cultivation of aquatic plants and animals in a laboratory for experimental purposes or in an aquarium."

In British Columbia, aquaculture is regulated through the Aquaculture Regulation (B.C.Reg 78/2002), under the *Fisheries Act* ([RSBC 1996] Chapter 149). BCMAL is the lead agency for aquaculture compliance, but some key authorities and functions reside with BCMOE. As a result, BCMAL inspection staff are responsible for assessing industry compliance. On the other hand, BCMOE are the lead for enforcement and are also responsible for monitoring compliance with environmental requirements.

Prior to the completion of the Salmon Aquaculture Review in 1997 (EAO 1997), the Aquaculture Regulation under the *Fisheries Act* (BC) was general in its approach, requiring only that a licence holder for example "take reasonable precautions to prevent the escape" of fish. In 2000, legislation came into play that for example, set and allowed enforcement of specific escape prevention and recapture requirements. Subsequently, the British Columbia Government developed regulatory standards, a result of two major revisions to the Aquaculture Regulation. In 2002/03 the Finfish Aquaculture Waste Control Regulation (FAWCR) was introduced. This regulation required the development of management practices which addressed environmental concerns including environmental monitoring underneath fish farms.

Escaped Farmed Salmon

Introduction

Accidental releases or 'escapes' of farmed salmon have been said to wreak havoc on wild fish (Langer 2003; Volpe 2001). Escape incidents have occurred as a result of adverse weather, predation or human error. The potential effects of escaped farmed fish on wild fish can be split into three categories: ecology, genetics and disease transmission (Hutchison 1997; Mathews et al. 2000; Butler 2002; Otteraa et al. 2004). Concerns regarding escapes have gained greater attention as farmed salmon production in British Columbia, Canada, has increased by 44% in the last 5 years, from 49,000 tonnes in 2000 to 70,600 tonnes in 2005, with a wholesale value of CAD$319 million (BCMOE and BCMAL 2005).

Legislation & Farm Practice

Under the Aquaculture Regulation, an operator must take precautions to prevent escapes. If they do occur, the operator must control, mitigate, remedy, confine and report escapes. Reporting of escape incidences (including those that are only suspected) is mandatory within 24 hours of discovery. In 2005, the government found no evidence of unreported escapes or suspected escapes (BCMAL and BCMOE 2006). A total of 64 escaped fish were reported in that year.

It is important to stress that the loss of farmed fish represents a substantial economic cost to the fish farmer, therefore measures are in place to minimise escapes. Predator nets surround the entire pen or site to prevent seals and sea lions coming into contact with the nets holding the fish. Shark guards are an additional layer used on the base of each pen to prevent dog fish coming into close contact with the fish. When fish are transferred from one pen to another, for example during transfer in or out of the net-pens, or during grading (which ensures fish in the same pen are of a similar size) nets are used to prevent accidental escapes. In addition, staff are trained to prevent escapes and during the transfer of fish, individuals are assigned the responsibility of locating potential areas of escape and actual escapes. Routine dive inspections are conducted (a minimum of every 90 days is required), increasing in frequency if there is evidence of predator activity or high fish mortality. Should an escape incident occur, all companies have nets on hand to recapture fish and boats on contract that can be brought into play to ensure a timely recapture of escaped fish. To this end, the Department of Fisheries and Oceans (DFO) issues a special 'ZZA' permit, in advance of an escape, to allow companies to recapture escaped Atlantic salmon without delay.

As the freshwater stage of farmed fish occurs in hatcheries, by the time the fish are transferred to sea, they represent an extensive investment in terms of man-hours and feed costs. As a result of the number of fish housed in each cage, escape incidents can result in the release of a few fish or of tens of thousands, depending on the cause.

The Extent of Escapes

Atlantic salmon dominate salmon production in British Columbia. In 2005, 53,800 tonnes of Atlantic salmon accounted for 76% of the total provincial production compared with 16,800 tonnes of chinook, coho, sockeye and marine trout combined (24% of total production; BCMOE and BCMAL 2005).

The first introductions of Atlantic salmon to the Canadian west coast occurred in 1904 with the aim of establishing a new population (Coleman and Rasch 1981). In British Columbia 8.6 million juvenile Atlantic salmon were released by the Department of Marine and Fisheries predominantly on the east coast of Vancouver Island and the Fraser River between 1905 and 1935 (MacCrimmon and Gots 1979; Ginetz 2002). The salmon were shipped as eyed eggs from the Miramichi River in New Brunswick, and the Thurso River, in Scotland and were introduced as eggs, fry, fingerlings and yearlings into more than 60 lakes and rivers in British Columbia (Ginetz 2002). These releases were an attempt to bring a highly prized and therefore economically valuable sport fishery to the Canadian west coast. All introduction attempts failed in Canada and the United States. Interestingly, of the non-native salmonids introduced at this time; brown trout (*Salmo trutta*), Atlantic salmon, and brook trout (*Salvelinus fontinalis*), Atlantic salmon were the only species *not* successfully introduced (Wydoski and Whitney 1979).

The numbers of escaped fish have ranged from 89,000 fish reported in 1998 to 34 fish reported in 2003 (DFO Atlantic Salmon Watch Program (ASWP) 2005). Although the number of Atlantic salmon produced in British Columbia has increased radically since their introduction to this area (from 33,100 tonnes in 1998 to 53,800 tonnes in 2005), the number of escapes has not. Figure 9.1 shows the number of escaped Atlantic salmon in British Columbia from 1991 until 2005 in relation to the number of Atlantic salmon produced. (Note: estimated by dividing the total production weight by an estimated market weight of 4.25 kg (Gross 1998)).

Many observers within the scientific community and the wider public have raised the question of the effects of these escapes. Juvenile offspring of escaped adults have been observed in rivers in British Columbia (Volpe et al. 2000; Harrower 2005 *pers*

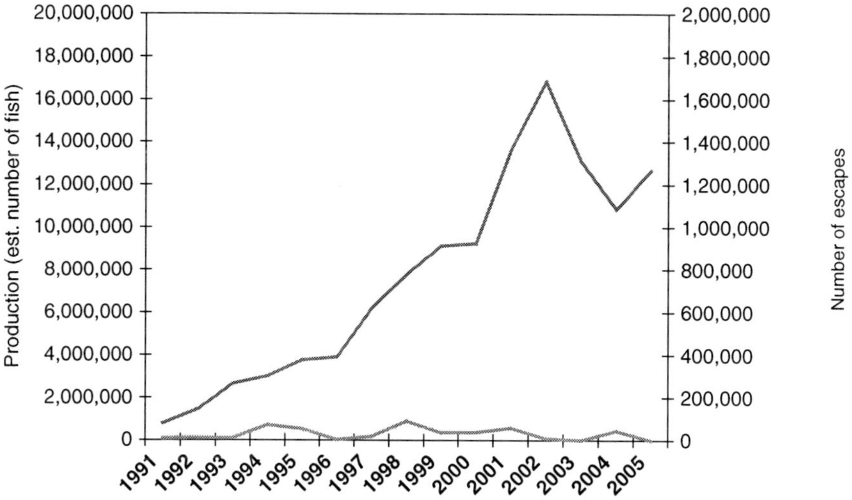

Fig. 9.1 Atlantic salmon production* (*dark grey*) and reported escapes (*light grey*) in British Columbia from 1998 to 2005 (DFO ASWP 2005; BCMAL and BCMOE 2006). Note ten-fold difference in scales
*Estimated by dividing the total production weight (BCMOE and BCMAL 2005) by an estimated market weight of 4.25 kg (Gross 1998)

com). However there is no evidence to date, of escaped Atlantic juveniles returning to rivers to spawn *after* migrating to sea.

As Atlantic salmon are a distinct species from Pacific salmon, the genetics of the Pacific species could only be affected through the interbreeding (hybridisation) of Pacific and Atlantic species. To date, there are no reported cases of hybridisation between Atlantic and Pacific species on the Pacific Northwest in the wild. Furthermore, laboratory studies carried out in controlled and optimal conditions have failed to cross Atlantic salmon with the Pacific pink, chum and coho salmon (Chevassus 1979; Loginova and Krasnoperova 1982; Gray et al. 1993). Therefore, it seems highly unlikely that Atlantic salmon have the potential to hybridise with, and thus influence the genetics of, Pacific species in the natural waters of British Columbia.

The Pacific species produced by aquaculture in British Columbia are chinook, coho, sockeye and marine trout (BCMOE and BCMAL 2005). The production of Pacific species is of a much lesser extent than that of Atlantic salmon. In 2005, 16,800 tonnes, of Pacific salmon was produced, compared with 53,800 tonnes of Atlantic, and farm-gate values of $66.5 million (CDN) and $252.2 million (CDN) respectively. As such, the production of Pacific salmon is much smaller than that of Atlantic salmon and, as shown in Fig. 9.2, the number of escapes is also lower for Pacific compared to Atlantic salmon.

Unlike the escaped Atlantic salmon, any environmental impacts from farmed Pacific salmonid escapes are likely to be dwarfed by the number of releases by public hatcheries breeding native, wild species. Public hatcheries are part of the

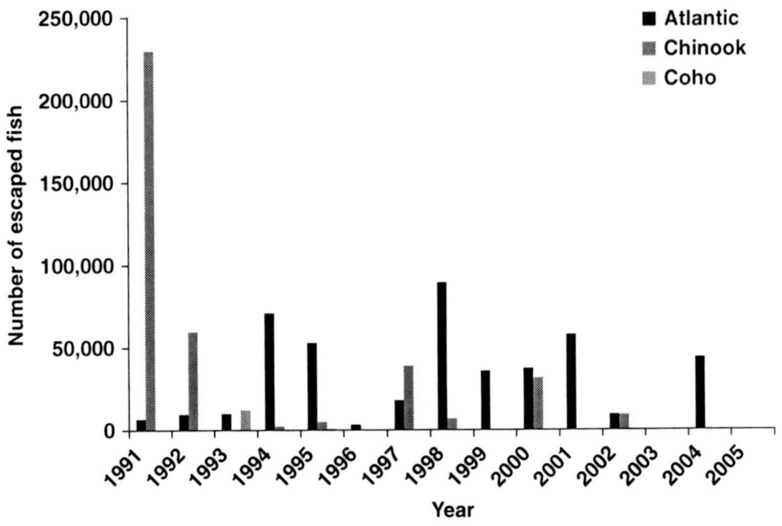

Fig. 9.2 Number of escapes of the three salmonid species farmed in British Columbia from 1991 to 2005 (Antaya 2006, *pers com*)

Fisheries and Oceans Canada's strategy to enhance salmon stocks. These facilities fertilise and then incubate eggs of local fish stocks, rearing them until the fish are released into the wild (DFO 2005b). In 2005, for each escaped chinook there were 9 million public hatchery fish released and for each escaped coho, there were 246,000 million fish released (DFO ASWP 2005; Lehmann and Irvine 2004).

Disease

While documentation of disease transmission between farmed and wild fish does exist (Munro et al. 1976), the direction of disease transmission between wild and farmed fish is particularly difficult to ascertain. It is likely that most fish diseases originate in wild populations (Olivier 2002). Therefore, wild fish have evolved mechanisms to deal with these diseases.

Disease introduction from hatcheries is rare but can occur. The monogean parasite, *Gyrodactylus salaries,* was introduced into northern Norwegian rivers during public hatchery fish stocking programs (Johnsen and Jensen 1991). However, the volume of smolts released from provincial, first nations and co-operative hatcheries is so large, the risk of disease introduction is much greater from these sources than from aquaculture hatcheries. As we have observed above, hatcheries annually release millions of fish while annual escapes are beneath one hundred.

In both commercial and public fish rearing facilities outbreaks of disease are treated quickly, and as a result, the probability of diseased fish escaping is very small. Furthermore, severely diseased fish are unlikely to survive for any length of time. Stephen and Ribble (1995) observed that 68% of clinically ill net pen fish died within 48 hours of first observation. Diseased fish do not feed or behave normally, and as a result, are less likely to school or come in close contact with other fish in the wild (Barber and Huntingford 1996), or even in enclosed aquaculture cages (Stephen and Ribble 1995). As a result of abnormal behaviour, diseased individuals are more prone to predation (Wolf 1985; Landeau and Terborgh 1986), which decreases the likelihood of disease transmission between fish.

Summary

The potential of escaped farmed Atlantic salmon to detrimentally affect Pacific salmonids does not appear to be realised in British Columbia. Although production of Atlantic salmon in British Columbia has increased rapidly, the number of escapes has not. Furthermore any effects of escaped farmed Pacific salmonids are likely to be dwarfed by extensive hatchery releases in this area. The likelihood of disease transmission between farmed and wild fish appears to be small. The occurrence of disease in aquaculture is very rare. Furthermore, diseased individuals behave abnormally and are more likely to get caught by a predator. Therefore, the chance of them associating with other salmonids is small. In conclusion, from the research currently available, the risk of escaped salmon detrimentally affecting wild stocks in British Columbia appears to be low.

Salmon Aquaculture's Environmental Footprint

Introduction

Ideally, a combination of good fish husbandry techniques, a balanced feeding regime, and favourable water conditions facilitate the production of stress-free, healthy salmon. Stress free fish grow faster, are less susceptible to disease, are of a higher quality, and hence represent a better return for the farmer. It is therefore in the best interests of the fish farmer to produce healthy, stress free salmon. Unfortunately, adverse environmental conditions such as elevated temperatures and algal blooms, husbandry practices (e.g. transportation), and occasionally lapses in husbandry standards can lead to stressed salmon which are more susceptible to disease (Pickering and Pottinger 1985; Ellsaesser and Clem 1986; Angelidis et al. 1987; Maule et al. 1989; Salonius and Iwama 1993; Espelid et al. 1996). Stressors are unavoidable in the intensive culture of animals, terrestrial and aquatic. However, best practises ensure this is minimized to ensure the health of the animals.

Disease Control

Animal Welfare

The majority of international standards, even the organic standards (e.g. IFOAM; EU regulation 1804/99 5.7), indicate that appropriate medicine (usually in the form of antibiotics) must be administered to sick animals as a matter of respect for animal welfare. 'Animal welfare' is a difficult term to define precisely. Different definitions focus on an animal's condition, on its subjective experience of that condition and/or on whether it can lead a natural life. These various definitions are not right or wrong, but simply highlight different aspects of welfare. Probably the most commonly referred to definition consists of 'five freedoms' recognizing five domains in which welfare may be compromised (Brambell Committee Report 1965):

1. Freedom from thirst, hunger and malnutrition by ready access to fresh, clean water and adequate, nutritious food to maintain full health and vigour.
2. Freedom from discomfort by providing an environment suitable to their species, including adequate shelter and a comfortable resting area.
3. Freedom from pain, injury and disease by prevention, rapid diagnosis and treatment.
4. Freedom to express normal behaviour by the providing sufficient space, proper facilities and company of the animal's own kind.
5. Freedom from fear and distress by ensuring living conditions which avoid mental suffering

These domains were developed for terrestrial farm animals, particularly mammals and to a lesser extent birds, but fish were considered later (FAWC 1996). However, fish are different from mammals in fundamental ways. For example, they do not maintain a warm body temperature, so effects of food deprivation are more subtle, and many species occur naturally in large shoals, so low densities may be more harmful than high densities.

In fish hatcheries, many fish are reared together, increasing the amount of close contact between the fish. This can potentially increase pathogen exposure. Antibiotic treatment within hatcheries that rear salmon for wild enhancement and for aquaculture may therefore create conditions for the production of antibiotic resistant disease strains. In the US states of Alaska, Washington, Oregon, and Idaho antibiotic-resistant strains of bacterial fish pathogens have been observed in Pacific Northwest salmon hatcheries for over 40 years. However, there have been no adverse impacts observed on wild salmon associated with these events (Waknitz et al. 2003). Furthermore, there is no information on antibiotic resistant disease strains having an impact on wild populations. This could be due to a lack of research or as a result of the selective pressures, such as predation, that are present in nature but absent in farm situations (McVicar 1997).

Antibiotics and Therapeutants in Salmon Farming

Disease identification, treatment and prevention are key aspects of any aquaculture venture, with prevention playing an increasingly important role as culture densities increase. Hence farmers maintain close contact with veterinarians to facilitate the shortest response time to any disease outbreak possible. Vaccinations, if available, are used when appropriate to prevent disease outbreaks. If an outbreak occurs, medication can be applied orally in the feed, in a bath treatment, or more unusually, via injection. The feed companies can have the medication added to the feed which is subsequently supplied to the farmer. For example, SLICE which is used in British Columbia for the treatment of sea lice is an in-feed treatment. Bath treatments, although commonly used to treat sea lice infestation in Norway and Chile, are relatively rare in British Columbia. The exceptions to this are salt and formalin treatments used during the freshwater period of the salmon life history. This is prior to the smolting process and transfer to net-pens in the marine environment for the grow-out phase.

Probably the most important point to remember about antibiotic use in salmon grown for human consumption in Canada is that antibiotics are only administered by veterinary prescription in response to disease. In terrestrial-based, meat producing farming, such as beef, poultry and pork, antibiotics are administered prophylactically to prevent disease before it occurs, promote pro-biotic function, and enhance growth e.g. monensin. In this respect salmon aquaculture compares very favourably with other agricultural meat producing industries from a consumer perspective. The reactive use of antibiotic therapy in response to a disease outbreak is not necessarily a good thing, however. Sick fish do not eat very much if at all, which presents logistical difficulties with the delivery of antibiotics mixed in with the normal salmon feed. If the ailing salmon are not ingesting sufficient amounts of the medicated feed the salmon will not respond to the treatment and may die. Additionally, it can create conditions favourable to the development of antibiotic resistance.

There are seven therapeutic products approved by Health Canada for use in Canadian salmon aquaculture, one anaesthetic, two anti-fungal treatments and four antibiotics (Table 9.1). All the antibiotic treatments are administered as in-feed supplements, and must be veterinary prescribed. The exception is Terramycin-Aqua (oxytetracycline), which can be obtained without a prescription under certain circumstances. Additionally, Health Canada can issue temporary authorizations to use drugs not approved in Canada through Emergency Drug Releases and Experimental Studies Certificates. Finally, veterinarians can take the initiative and issue prescriptions for therapeutants authorized for animals other than fish. These are called 'off-label' prescriptions.

The uses of these veterinary drugs are monitored by the Ministry of Agriculture and Lands under the *Pharmacists, Pharmacy Operation and Drug Scheduling Act* and Medicated Feeds regulations. Once the salmon are harvested, monitoring of drug residues falls under the remit of the Canadian Food Inspection Agency (CFIA). The CFIA is responsible for the monitoring of all agricultural food and

Table 9.1 Therapeutic products approved for use by Health Canada in salmon farming. Adapted from Health Canada Bureau of Veterinary Drugs

Brand names	Active ingredients	Compliance	Species	Technical use	(AMRL)
Romet 30 (Hoffman-LaRoche)	Sulpha-dimethoxine Ormeto-prim	1990	Salmonids	Antimicrobial	0.1 ppm
Tribrissen 40 (Coopers Agropharm)	Sulphadiazine Trimetho-prim	1992	Salmonids	Antimicrobial	0.1 ppm 0.1 ppm
Terramycin-Aqua (Pfizer Inc.)	Oxytetracycline	1989	Salmonids Lobster	Antimicrobial	0.1 ppm
Aquaflor (Schering-Plough)	Florfenicol	1996	Salmonids	Antimicrobial	0.8 ppm
Aqua Life-TMS (Syndel Labs)	Tricaine-methane-sulfonate	1995	Salmonids	Anaesthetic	Not Established
Parasite-S (Western Chemicals)	Formaldehyde	1994	Salmonids	Antifungal	Not Needed
Perox-Aid (EKA Chemicals)	Hydrogen Peroxide	1998	Salmonid Eggs	Antifungal	Not Needed

meat products and conducts spot audits on farmed salmon for the presence of drug residues.

It is obvious that the presence of therapeutant drug residues represents a key potential hazard in farmed salmon products. With this in mind the CFIA has instituted a Quality Management Program (QMP) as a control mechanism to monitor the quality and safety of seafood, including farmed salmon. This program is modelled on Hazard Analysis Critical Control Point (HACCP) principles. As part of the program, the CFIA monitors the therapeutant drug residue levels in aquaculture products in Canada. The monitoring procedure consists of certification programs and documentation by the growers and the industry, system and procedure audits, and third-party verification. Additionally, laboratory testing provides a useful method of verifying the checks and balances described above, which is fundamental to ascertaining the success of the HACCP principles and hence the Quality Management Program. These on-going tests are conducted by the CFIA.

There are various reports of the improvements made by salmon farmers over the last decade in curbing antibiotic use. BCMAL reports that antibiotic use dropped from 516 grams per metric tonne in 1997 to 165 grams per metric tonne in 2001. This translates to about 2.5% of all feed being medicated each year (BC Salmon Farmers Association 2007), mostly on fish not yet at harvestable size. Hence salmon farming is the world's largest protein-producing sector with the smallest use of antibiotics (Willoughby 1999; Waknitz et al. 2003). While this attests to the increasingly effective good husbandry techniques used by farmers, the potential risk of antibiotic residues in the final product remains of concern.

For each antibiotic, the Veterinary Drugs Program (VDP) (formerly the Bureau of Veterinary Drugs) of Health Canada responsible for approving drugs used by the aquaculture industry, stipulates a withdrawal time. The withdrawal time is the time that it takes the fish to metabolise or breakdown the drug to concentrations that no longer represent a potential health hazard to humans, and is somewhat temperature dependant. During the withdrawal time fish cannot be harvested for human consumption. Therefore adherence to the antibiotic withdrawal times will facilitate the production of farmed salmon that are deemed safe for human consumption by Health Canada. Organic standards decree a somewhat more cautious withdrawal time of twice (e.g. EU Regulation 1804/99 5.7; IFOAM draft standards 2005) or three times (NASAA standards, Australia) the conventional time.

Environmental Impact

In addition to concerns on the potential risks of antibiotic use in agriculture to human health, there is the potential impact on the environment. This is particularly relevant in the marine habitat where there is the potential that the theraputant can enter the food chain via other organisms inhabiting the vicinity of salmon farms, such as oysters, Dungeness crab, and red rock crab.

Overall, the use of antibiotics approved by Health Canada for salmon aquaculture appear to pose little risk to the environment, as these antibiotics are degraded within reasonable time limits. Relatively, there is a higher potential risk from Terramycin-Aqua due to the penchant of this product for marine sediments below the net-pens. Oxytetracycline HCl, the active ingredient in Terramycin-Aqua, is more persistent and can still be detected in the deeper layers of sediment 180 days after the cessation of use (Hektoen et al. 1995). This is 100 days longer than the withdrawal time at temperatures below 10°C. The levels in the upper layers of sediment rapidly reduced, a phenomenon the authors attributed to leaching and redistribution. The area of sediments that contain measurable oxytetracycline residues appears to be localized, with residues detectable to a distance of 30 m from the cage, but absent at 100 m from the cage (Capone et al. 1996). In wild fish associated with the farms, Oxytetracycline was detected up to only 13 days after cessation of treatment (Bjorklund et al. 1990), 67 days less than the withdrawal time. Only trace amounts of oxytetracycline were found in oysters (*Crassostrea gigas*) or Dungeness crab (*Cancer magister*) associated with heavily treated farms. Half the red rock crab (*Cancer productus*) collected under the cages during and within 12 days of oxytetracycline treatment contained high levels of oxytetracycline (Capone et al. 1996). Unfortunately the authors did not continue to monitor the red rock crabs to record just when these levels dissipated.

Reductions in antibiotic use in salmon aquaculture in BC over time as discussed above is likely to further reduce this potential risk. Equally, Health Canada appear to have adequate controls in place to negate the risk posed by antibiotic residues to human health in salmon aquaculture products.

Sea Lice Therapeutants

In Canada there are two available sea lice treatments, Calicide® and SLICE®. These are both in-feed treatments which have some advantages over bath treatments such as Excis® (Cypermethrin) and Salmosan® (Azamethipos), which are used in Europe. In-feed treatments are non-weather dependant and easy to administer, unlike bath treatments where re-infestation can occur within a short period, and are only effective against the mobile stages of lice. Additionally, and not surprisingly, immersing salmon in a bath treatment of therapeutant stresses the fish.

Of the two therapeutants available in Canada, SLICE® is 100% effective against all parasitical stages of the sea louse, whereas Calicide® exerts its effects at the moulting stages, and so must be used before adult lice appear. As it is the number of adult female lice that is used to trigger sea lice treatments, Calicide has a very limited practical application. Hence, it is almost exclusively SLICE that is used to treat sea lice infestations in British Columbia. Both of these therapeutants are classified as drugs, not pesticides. The distinction is due to the mode of delivery which is through the feed. Pesticides are delivered in a bath treatment like Excis® and Salmosan® discussed above. *Calicide*®: The active ingredient is teflubenzuron, a benzoylphenylurea (BPU) insecticide. Calicide® is marketed by Trouw Aquaculture, a subsidiary of Nutreco. (In Canada it is marketed by Skretting.) Other (BPU) insecticides are diflubenzuron, triflumuron, hexaflumuron, lufenuron, chlorfluazuron, flufenoxuron and flucycloxuron (Hiemstra et al. 1999). Teflubenzuron has been used as an insect growth inhibitor and has been principally studied in its role as an insecticide in the terrestrial environment.

There is very little information available on the environmental fate and ecological effects of teflubenzuron in aquatic environments. The available ecotoxicological information (Anon 1997; Trouw 1999; Jenkins 1995; Baird et al. 1997; Baird 1998) on the possible risks of Calicide concludes that there is a negligible risk to zooplankton in the water column. The main environmental risk from teflubenzuron will come through faeces and waste feed. Hence there is a moderate risk of build up in the sediments through repeated applications. This can be reduced by careful management strategies. Unfortunately, it is difficult to predict the ecological risk of teflubenzuron to the marine environment because of the current lack of research data. However, although it has low toxicity to fish, mammals and birds, it would still be prudent to monitor the effects on crustacean near the cage sites.

SLICE®: The active ingredient in SLICE® is emamectin benzoate, a member of the avermectin family, which are insecticidal, acaricidal, and antihelminthic agents that have been isolated from the fermentation products of *Streptomyces avermitilis*, a member of the actinomycete family of soil bacteria.

Emamectin benzoate (SLICE®, Proclaim®, Denim®) is produced by the same fermentation system as abamectin. The mode of action of emamectin benzoate is to block the neurotransmitter γ,γ–aminobutyric acid (GABA) at the neuromuscular junction. This results in the cessation of activity, such as feeding, due to paralysis, and death occurs with a number of days. When used to treat sea lice infestation SLICE® is highly effective at removing all stages of sea lice, an improvement on Calicide®, discussed above.

Here in Canada, SLICE®, marketed by Shering-Plough, is currently under review by the Veterinary Drugs Directorate (VDD), a part of Health Canada. To date, the VDD have approved SLICE® for use by veterinarians through the Emergency Drug Release (EDR) process. As part of the assessment process by the VDD, Shering-Plough has made the environmental data on the safety of the compound available for scrutiny (Technical Report 2001). There have been studies looking at the effects in fish, crustacea, mollusks, mammals and birds (Technical Report 2003). There is some evidence that emamectin benzoate causes premature moulting in female American lobsters, *Homerus americanus,* and a concurrent loss of eggs if the females were gravid at the time of exposure (Waddy et al. 2002). Additionally, there was a toxic effect observed in copepods.

Specific to British Columbia, there was also research carried out on the effects of emamectin benzoate on the dungeness crab and the spot prawn, two commercially important species in the Pacific Northwest (Van Aggelen et al. 2003). There was no toxicity shown in either of these species.

Overall, from the available data, SLICE appears to be an effective treatment for sea lice that has minimal impacts on the environment and poses very little risk to the consumer.

Summary

It is in the interests of any farmer to ensure a high standard of animal health. This includes managing and, when necessary, treating disease outbreaks. When animals are kept in close proximity, an increase in pathogen exposure can potentially occur. In Canadian aquaculture, vaccinations are used to prevent disease outbreaks, and fish are treated upon the outbreak of a disease. Antibiotics are only administered by a veterinary in response to a disease. Unlike other industries, antibiotics are not used prophylactically. As a result salmon farming is the world's largest protein-producing sector with the smallest use of antibiotics. Drug use is controlled by BCMAL, whereas the issue of residue falls under the remit of the federal CFIA. Each antibiotic used has a stipulated withdrawal time.

The use of approved antibiotics for salmon aquaculture appears to pose little risk to the environment, as these antibiotics are degraded before they cause harm. Health Canada appears to have adequate controls in place to negate the risk posed by antibiotic residues to human health in salmon aquaculture products.

In Canada, the predominant sea lice treatment is SLICE®. This treatment appears to be an effective treatment for sea lice that has minimal impacts on the environment and poses very little risk to the consumer.

Salmon Lice (*Lepeophtheirus salmonis*)

Introduction

In the North Atlantic, *L. salmonis* has been a severe problem for both farmed and wild salmonids, with an estimated cost of between €23.9 million and €59.7 million in Norway alone (Boxaspen and Næss 2000; Fast et al. 2002; Stone et al. 2002; Glover et al. 2003, 2004; Heuch et al. 2003; Wagner et al. 2003, Wagner and McKinley 2004). However, the situation in the North Pacific along British Columbia's coast is rather different. For example, instead of one species of migrating salmon there are five species, each of which has a different level of susceptibility to *L. salmonis*.

The sea lice that naturally affect salmon and trout in the marine environment belong to the family *Caligidae* and to the genera, *Caligus* and *Lepeophtheirus*. In British Columbia, 14 species (2 species of *Caligus* and 12 species of *Lepeophtheirus*) of sea lice parasitize many different species of marine fish (Kabata 1973). These 14 species of sea lice have a similar body shape (Fig. 9.3). Additionally,

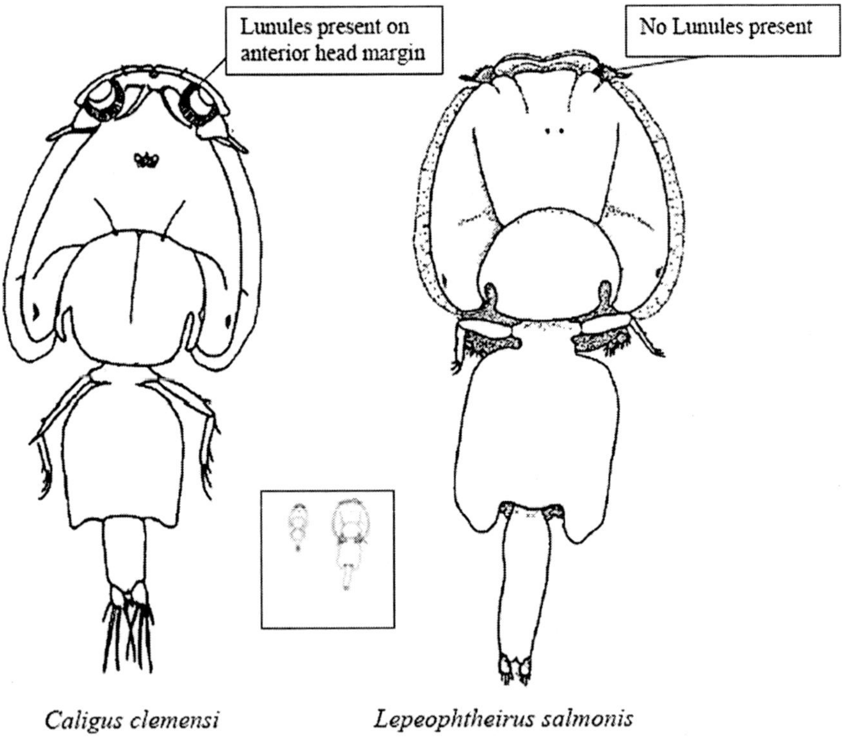

Fig. 9.3 External anatomy of the distinguishing features between the genus Caligus and the genus Lepeophtheirus. Dorsal view shown. Small inset shows relative size difference (Butterworth et al. 2005)

differences within species of each of these two genera are small, and difficult to identify without the assistance of a magnifying glass and some taxonomic training. Only two species, *Lepeoptheirus salmonis* and *Caligus clemensi*, pose a potential threat to both farmed and wild salmon.

In the context of salmon aquaculture and wild salmon in British Columbia, *Lepeophtheirus salmonis* is the important species. Also commonly known as the salmon louse, *L. salmonis* is a parasitic caligid copepod (Johnson and Albright 1991; Butterworth et al. 2004). Ubiquitous in the North Pacific and Atlantic Oceans, *L. salmonis* is a parasite on both farmed and wild salmon and on sea-run trout (Kabata 1973; Pike 1989; Johnson et al. 1996; Bjørn and Finstad 1998). On salmon, high infection intensities lead to stress, impaired performance, reduced physiological ability, and in extreme cases, death through either primary lesions or secondary infections (Bjørn and Finstad 1997; Pike and Wadsworth 1999; Bowers et al. 2000; Finstad et al. 2000; Wagner et al. 2003).

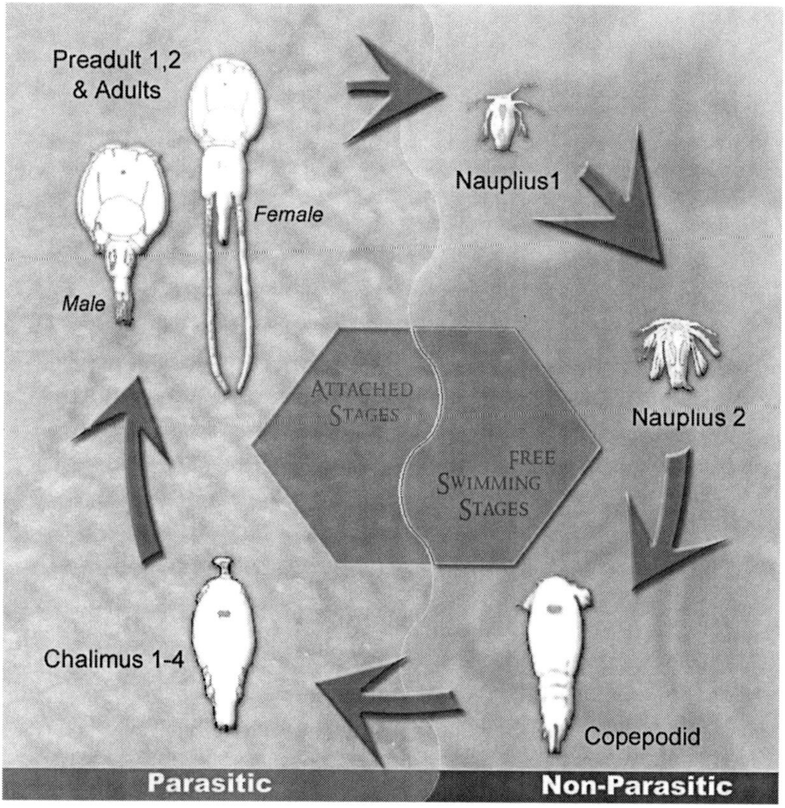

Fig. 9.4 Life cycle of L. salmonis showing nauplii, copepodid, chalimus, pre-adult and adult life stages
Source: Catherina Murphy, AquaNet Canada, re-printed with permission

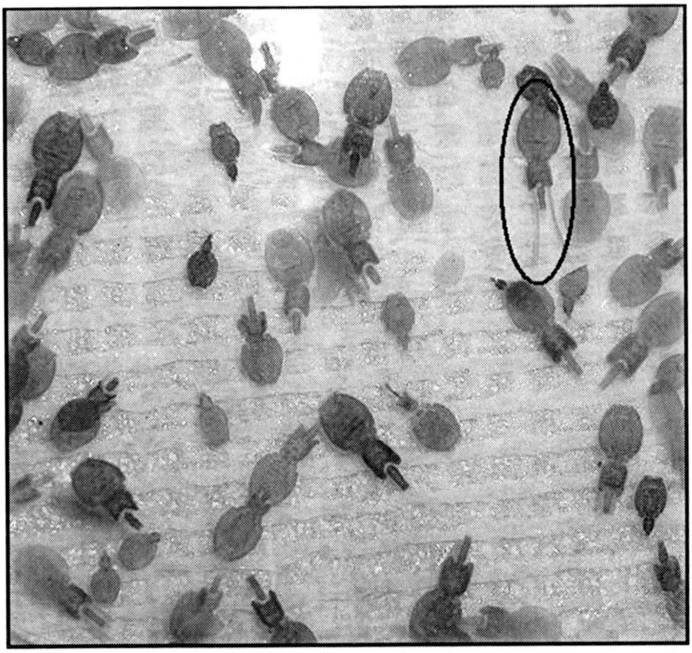

Fig. 9.5 Adult male and female L. salmonis. The females are slightly larger with elongated genital segments. Inside the black oval on the top right of the photograph developing egg strings are visible on a female
Source: Photograph by K.G. Butterworth

Life Cycle

L. salmonis has 11 life stages, starting from the egg through to the adult stage (Fig. 9.4). The first three stages are free swimming and non-parasitic. The larval sea lice then finds a host and progresses through 7 parasitic life stages before the next generation of eggs are produced (Fig. 9.5).

Factors that Influence Salmon Lice Development

Environmental factors that have the greatest impact on the development and success of the infectious larval stage (the copepodid stage), are temperature and salinity. Depending on the combination, these two factors can either promote swift growth and survival of salmon lice, or retard their development and severely reduce their survival.

At temperatures below 7°C, salmon lice in the free swimming copepodid stage are less able to take the next step and settle onto a host (Fig. 9.6), than when the water temperature is warmer (Tucker et al. 2000). The impact of temperature on the overall generation time for salmon lice is equally pronounced. At 7.5°C, the generation time is 106 days, but at 14°C, this decreases to 36 days (Tully 1992). Such

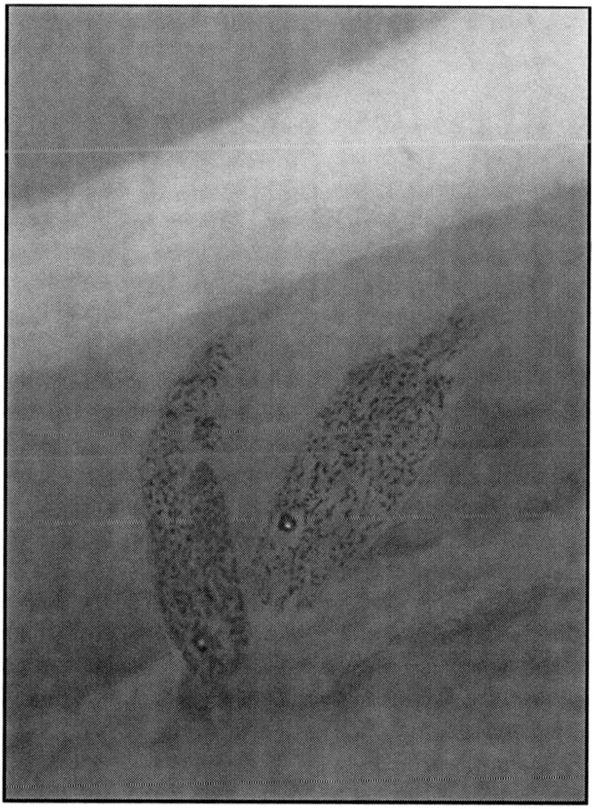

Fig. 9.6 Successfully settled and moulted and feeding; close-up photograph of chalimus stages of *L. salmonis* attached to a juvenile pink salmon smolt
Source: Photograph by K.G. Butterworth

temperature-dependant growth rates can significantly impact population densities of the copepodid life stage (settlement stage) of *L. salmonis* in shallow coastal waters, which are prone to warming. As a result, warmer summers are likely to result in an increased number of sea lice copepods in these waters.

Salinity plays a very important part in the life cycle of *L. salmonis*. Successful development of the copepodid stage has been reported to occur only at salinities above 30 (Pike and Wadsworth 1999). In British Columbia, successful copepodid development and subsequent host settlement have been achieved at salinities as low as 28% (Butterworth 2005, *pers com*). Once at this stage, copepodids actively avoid sea water with salinities below 20% (Heuch 1995) and their optimal survival is at 30% (Johnson and Albright 1991). Hence, lower salinities such as those commonly recorded in British Columbia inshore waters, could have a significant damping effect on *L. salmonis* distribution and population sizes.

Differences Between Populations in the North Pacific and Atlantic Oceans

In the North Atlantic, *L. salmonis* has been a severe problem for both farmed and wild salmonids (Fast et al. 2002; Stone et al. 2002; Glover et al. 2003, 2004; Heuch et al. 2003; Wagner et al. 2003, Wagner and McKinley 2004). Atlantic salmon make up the bulk of finfish aquaculture in the North Atlantic and *L. salmonis* have had a large impact on wild Atlantic salmon and sea-run brown trout (*Salmo trutta*). The population of wild salmon stocks in the North Atlantic have and continue to decrease (Hiscock et al. 2005). This decrease, alongside the poor ability of Atlantic salmon to re-establish populations in traditional spawning streams (Cubitt et al. 2006) have led to farmed Atlantic salmon outnumbering wild Atlantic salmon in some locations.

The situation in the North Pacific along British Columbia's coast is rather different. First, instead of one species of migrating salmon, there are five that undertake long migrations. To complicate the picture further, *L. salmonis* shows differential levels of prevalence on different species of Pacific salmon. In the Pacific Ocean, the highest levels of infection have been reported on Pink salmon (*Oncorhynchus gorbuscha*) and rainbow trout (*O. mykiss*) (Nagasawa 2001). Lower infection levels were found on Coho salmon (*O. kisutch*), Chum salmon (*O. keta*) and Chinook salmon (*O. tshawytscha*). The Sockeye salmon (*O. nerka*) was found to have the lowest infection levels, but this relationship is poorly understood as subsequent surveys found high levels of salmon lice on sockeye salmon in excess of those found on pink salmon (Beamish et al. 2005; Nagasawa 2001).

In British Columbia, wild salmon vastly outnumber farmed salmon. There are 128 salmon farm tenures in British Columbia, compared to more than 9,600 distinct stocks of wild Pacific salmon identified on the British Columbia coast. All of these distinct stocks are considered to be separate populations, subject to unique environmental and anthropogenic pressures. Additionally, over-wintering wild coho and chinook salmon (Healey 1991; Robinson et al. 1992) and schools of wild sticklebacks (Jones et al. 2006) in coastal waters provide an ideal stock of potential hosts upon which the salmon lice can over-winter, ready to infect out-migrating wild smolts in the spring. Hence, there is a large potential reservoir of salmon lice associated with wild fish. This is the opposite of the North Atlantic, where due to the severe depletion of wild fish stocks, salmon farms contain the largest pool of potential hosts upon which salmon lice can over-winter.

The Impact of Salmon Lice on Salmon Health

To date, the bulk of research has focussed on the impact of salmon lice infestations on Atlantic species of salmonid (Stone et al. 2002; Glover et al. 2003, 2004; Heuch et al. 2003; and Wagner et al. 2003, Wagner and McKinley 2004) and not on Pacific salmon. Research on Atlantic salmon is very useful to scientists studying Pacific salmon, as it provides core methodology and insights on some of the physiological mechanisms that are affected by salmon lice infestation. This is called the August Krogh principle, and is the underlying ethos behind the field of comparative physiology and biochemistry. Although the mechanisms may be the same, the levels of susceptibility and response vary between species. Hence, the number of salmon lice that cause mortality in Atlantic salmon cannot be assumed to cause mortality in Pacific salmon. Each species needs to be examined individually.

Research has shown that while Atlantic salmon have little resistance to salmon lice infestation, this resistance can be strengthened by selective breeding (Kolstad et al. 2005). Atlantic salmon and sea trout develop lesions when infected with salmon lice, and appear to have very little defence against the infestation aside from turning away from the sea and heading back into fresh water streams. This causes the lice, which are intolerant to low salinities, to drop off the afflicted salmon.

In July and August of 2003 and 2004, Fisheries and Oceans Canada conducted a study of skin damage caused by salmon lice to returning wild Pacific salmon (Beamish et al. 2004, 2005). Of 1,046 wild Pacific salmon found infected with salmon lice, the authors reported that there were a small number of sockeye and pink salmon with lesions present (skin removed and muscle exposed or skin partially removed exposing necrotic tissue) and haemorrhaging at margins of lesions (Beamish et al. 2004, 2005). However, the authors stressed that this was a rare occurrence.

The majority of the pink salmon and some of the sockeye salmon had subcutaneous haemorrhaging ranging from mild red discolouration to moderate over the area only half the size of the anal fin. Although these data would suggest that the Pacific salmon appear to be more resistant to skin damage from salmon lice than their Atlantic counterparts, there are reported cases where Pacific salmon have had severe lesions from salmon lice infestation (Kabata 1970; Johnson et al. 1996).

Perceived differences in the severity of skin damage between Pacific and Atlantic species may be in part due to the lack of epithelial hyperplasias and the inflammatory response of Altantic salmon to an infestation of salmon lice (Johnson and Albright 1992; Johnson 1993). It is important to note that even though the Pacific salmon species appear to be more resistant to salmon lice infestation, there is as yet no scientific assessment of the impact of infestation intensity on the general health of these salmon. Thus, there is no scientific data available on the number of sea lice needed to have a detrimental effect on Pacific salmon.

The Interaction of Wild and Farmed Salmon

The alleged role of commercial salmon farms as a possible source of salmon lice infections in passing wild salmon has received much attention from both the scientific community and the popular press. The debate in British Columbia has focussed on pink salmon stocks in the Broughton Archipelago (for a comprehensive review of this issue, see Brooks 2005). The controversy is based on evidence that correlates higher salmon lice infestation intensities in areas of British Columbia with salmon farms, as opposed to areas without salmon farms (Morton et al. 2004, 2005; Morton and Routledge 2006). An increase in sea lice with proximity to farms has also been reported in the Atlantic Ocean (Costelloe et al. 1996, 1998; Bjørn et al. 2001; Penston et al. 2002; McKibben and Hay 2002). The correlation compares two variables (proximity to salmon farms and the number of salmon lice on wild salmon) that are changing in a similar manner and therefore appear to be linked. There is no evidence of a link (cause and effect) between the two variables. Hence, the variables, in this case proximity to salmon farms and the number of salmon lice on wild salmon, may or may not be related.

Although it has not been possible to date to establish a direct causal link between the decline of wild salmon stocks and the expansion of the salmon aquaculture industry in British Columbia, it is of obvious importance to establish whether salmon farms are actually significantly contributing to salmon lice prevalence among wild Pacific salmon (Butterworth et al. 2004). If the salmon farms are not contributing significantly to the problem, then salmon lice on farmed salmon is a farm management problem, not a potential interaction issue between wild and farmed salmon. Were this separation to be proven, it would facilitate the development of more specific management policies for the effective control of salmon lice on salmon farms. More recent research suggests that it is possible for farmed and wild salmon to co-exist in a sustainable manner in the same habitat (Beamish et al. 2006). In the Pacific, salmon lice monitoring programs have reported the occurrence of salmon lice on 91–100% of salmon sampled in areas with and without salmon farms (Nagasawa 2001; Beamish et al. 2004, 2005).

Complicating the picture is the occurrence of *L. salmonis* on the three-spine Stickleback *Gasterosteus aculeatus*, a species common to inshore waters in British Columbia (Jones et al. 2006). Additionally, given the rates of dispersal of the lice in their larval stages (O'Donoghue et al. 1998) by dynamic flow fields caused by changing tides, currents, and local shifts in wind direction, there is a huge potential for larval dispersal (Asplin et al. 1999, 2004). Therefore, more conclusive research is needed before a cause and effect relationship can be demonstrated between salmon lice present on salmon farms, and infection levels among wild Pacific salmon in British Columbia.

Successful management at a culture facility depends on the development and implementation of a fish health management plan that mitigates risk factors of disease. To this end, the BC government requires all private companies and public fish culture facilities to develop and maintain an up-to-date Fish Health Management Plan (FHMP) specific to their facility. For private companies, the FHMP will be enforced as a Term & Condition of an aquaculture licence. The FHMP must contain specific components, for example, identification and management of fish health risks, use of drugs and chemicals reduction of exposure to, or spread of, disease causing agents. The health status of the fish needs to be routinely assessed, fish health records must be maintained, and an action plan needs to be in place to control and prevent the spread of disease in the event of an outbreak. There must also be provisions for the detection, investigation and management of any potential disease outbreaks, including sea lice.

Summary

Sea lice biology and therefore prevalence, is greatly dependent on environmental variables. For example temperature and salinity have a large impact on *L. salmonis* growth rates and survival. In particular, free swimming sea lice larval stages are intolerant of low salinities such as those found in British Columbia's inshore waters. Sticklebacks and wild salmon could provide a potential host for sea lice. Much of the information currently available on the impact of sea lice on health is from Atlantic salmon, not Pacific salmon. For example, it is not known at this time what intensity (concentration) of sea lice has on the health of Pacific salmon, nor what level would cause mortality. Because there is as yet no answer to either question, it is currently not possible to conclude that sea lice on British Columbia salmon farms are having a significant impact on the parasite loads of the wild Pacific salmon.

Biological Waste

Introduction

In the Pacific Northwest, Europe and Chile, salmon are transferred to net-pens situated in coastal waters following freshwater stages. As these net-pens are not fully enclosed, this raises concerns over the potential for biological aquaculture waste to pollute the surrounding environment and also to detrimentally affect the aquaculture industry via through reduced growth and survival (Tveterås 2002).

In rough terms, biological waste consists of nutrients. If an ecosystem is lacking in the particular nutrients that are introduced as biological waste, this introduction can result in extensive changes to the surrounding ecosystem. For example, additional nitrogen can result in the rapid production of plant life such as algal blooms. The increase in plant life rapidly depletes available dissolved oxygen which is detrimental to other inhabitants of that area.

The majority of waste from fish farms consists of waste feed and metabolic by-products from fish. The main constituents of this waste are carbon, nitrogen and phosphorus (Beveridge et al. 1991). Lesser wastes from salmon farming operations include substances related to auxiliary salmon farm operations, including sewage from human workers, net cleaning operations, mortality disposal, bleeding and offal from product preparation, and garbage. These additional contributors will not be discussed within the present paper.

Legislation

In British Columbia, legislation regarding waste from fish farms falls under several acts. Aquaculture licenses are granted by the British Columbia Ministry of

Agriculture and Lands, under the *Fisheries Act*, ([RSBC 1996] Chapter 149, Sections 13(5) and 14(2)). Species and production limits are conditions of these licenses (Section 16(d)). *The Environmental Management Act*, ensures that any waste discharges are associated with a permit, approval or an applicable regulation. The Finfish Aquaculture Waste Control Regulation (FAWCR; B.C. Reg. 256/2002, includes amendments up to B.C. Reg. 321/2004), ensures that information regarding the surrounding environment, species, stocking and feeding levels are declared. For example, as part of this regulation, reports must be compiled at peak biomass and prior to stocking, and as a result of changes of 20% or more by the licensee.

According to the Annual Inspections on Marine Finfish Aquaculture Sites for 2004 and 2005 Inspection Cycles, conducted by BCMAL and BCMOE (BCMAL and BCMOE 2006), compliance was high overall (Fig. 9.7). In particular, the industry had a compliance level of over 97% or over for all issues that directly impact aquaculture waste. Currently, BCMAL inspectors visit all operating farms at least once yearly. In some cases repeated inspections are necessary to ensure compliance or to investigate additional reports of potential non-compliance. In addition, other provincial and federal authorities such as BCMAL Fish Health technicians, BCMOE Waste Biologist staff, Fisheries and Oceans Canada and Worker's Compensation Board visit sites, resulting in an average of at least three to four visits per year (BCMAL and BCMOE 2006).

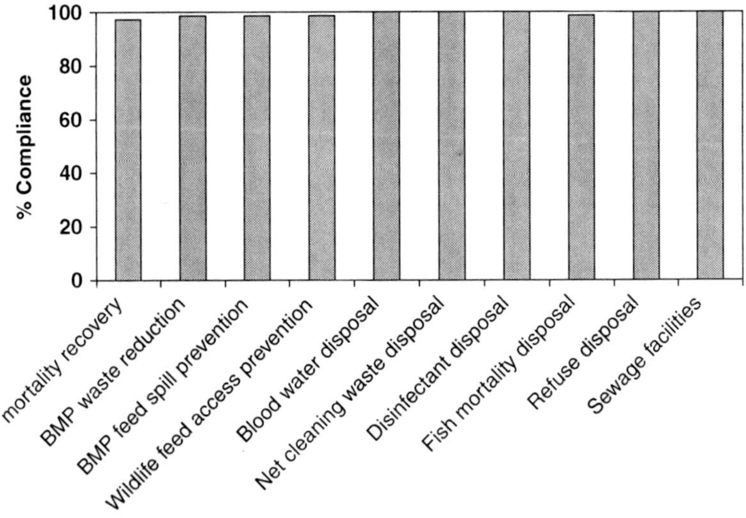

Fig. 9.7 Percentage compliance with issues that directly impact waste from aquaculture sites in 2005. Percentages are the number of sites in compliance in relation to the 75 sites in operation (BCMAL and BCMOE 2006)

Uneaten Feed

The fact that feed costs account for 35–60% (Anon 2002; DFO 2005a) of production costs, coupled with the wide range of potential environmental effects, means that it is in the interest of the aquaculture industry to minimise the loss of waste feed into the surrounding environment. Calculations of feed loss in Europe and the North American East Coast range from 5–30% (Beveridge et al. 1991; Gowen and Bradbury 1987; Gowen et al. 1991; Findlay and Watling 1994; Pearson and Black 2001; Strain and Hargrave 2005; Weston 1986). The majority of the earlier estimates were taken when moist feed, which creates higher levels of waste food, was used. In British Columbia dry feed has been used for more than 10 years, therefore these figures are undoubtedly an over-estimation.

Available estimates from the Pacific Northwest include 1%, 3%, 5% or less, 9%, 15% and 17% (Sams 2005, *pers com*; DFO 1997; Brooks and Mahnken 2003; Alive and Inseparable 2006). The amount of feed lost depends primarily on feed composition, efficiency of feeding regimes and environmental conditions. Using these estimates of waste along with those obtained from the Ministry of Environment (data from each of the marine aquaculture companies on an annual basis as part of their reporting requirements under the Finfish Aquaculture Waste Control Regulation (FAWCR)) one can calculate a best and worst case scenario for the total weight of uneaten feed pellets generated by the British Columbia salmon farming industry in a particular year. In 2005, the British Columbia salmon farming industry used 109,663 tonnes of feed (White 2006, *pers com*). Therefore, one can estimate the total amount of uneaten feed to be between 1,097 tonnes (using 1% feed waste) and 18,643 tonnes (using 17% waste feed).

For Atlantic salmon farmed species in British Columbia, the rate of feed intake has been shown to tail off rapidly (Talbot et al. 1999, see Fig. 9.8), so it is particularly important to decrease feeding at this point to prevent high levels of feed waste. To this end, the majority of fish farms in British Columbia monitor fish feeding and pellet wastage using underwater camera systems, demonstrating the importance of careful feed management. Data from Talbot et al. (1999) was collected without the use of underwater camera systems in net-pens of similar densities to

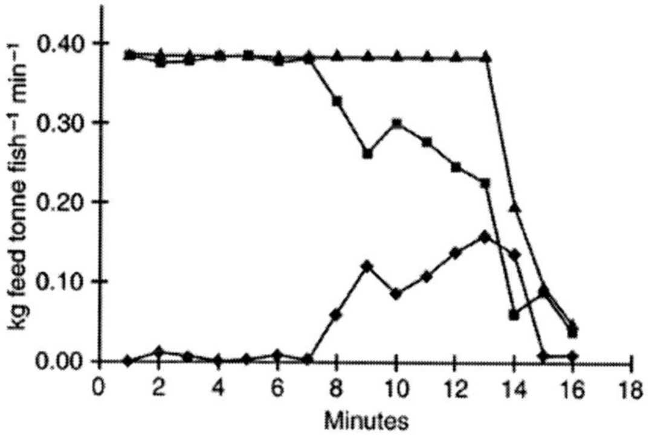

Fig. 9.8 From Talbot et al. 1999. Example of the pattern of feed intake in a cage containing 17.4 tonnes of Atlantic salmon with an average weight of 0.7 kg. The water temperature was 12°C (▲, feed delivery rate; ■, feed ingestion rate; ♦, feed wastage rate; all units in kg feed tonne fish^{-1} min^{-1})

those used in British Columbia, but proportionally smaller ($12 \times 12 \times 10$ m compared with $30 \times 30 \times 20$ m predominantly used in British Columbia). Many studies have observed little or no feed underneath fish farms in British Columbia (Brooks 2000a – f, cited in Brooks and Mahnken 2003). Scientific research on ways to minimise feed waste is ongoing (Cho and Bureau 1997, 2001; Cubitt et al. 2006; McFarlane et al. 2004).

Waste Effects

In order to grasp the extent of the waste from aquaculture, it would seem logical to ascertain the total output of waste from aquaculture either in British Columbia collectively, or by individual farm. As farms do not have measures of actual waste output, this is not a straightforward measure to attain. It is particularly difficult to calculate the actual biological output from the aquaculture industry in British Columbia. Waste varies because the output varies from farm to farm. Factors affecting organic accumulation rates include fish biomass and feeding rates; fish food and faecal material particle sizes and densities; net-pen configuration; water depth; current speeds; and the degradation rate of sedimented carbon. In addition, there is an increase in organic waste when fish are stressed or there is a high mortality at the farm (Brooks 2001). As a result legislation, in part, focuses on effects of organic waste such as levels of sulphur, an element indicative an environmental imbalance of nutrients.

In British Columbia, sulphide has been shown to increase with an increase in salmon biomass (Brooks 2001). Brooks also observed three regimes regarding the effect that sulphide had on species within the sediment in the vicinity of fish farms. Firstly, these species were influenced by the presence of between 200 and 300μM of sulphur. Secondly, six opportunists were observed to proliferate at levels up to 5,000 μM of sulphur, but at sulphur levels greater than 5,000 μM, the presence of infauna decreased radically.

Nitrogen is excreted by fish in particulate and dissolved matter. The discussion below includes dissolved excreted nitrogen. Salmon feeds are purported to contain between 6.1 and 7.7% nitrogen (Alive and Inseparable 2006). Recent studies report levels of 7.2% (Gillibrand et al. 2002), between 6.3 and 8.7% (DFO 2005a), 6.6% (Higgs et al. 2006). Just less than half the amount of nitrogen found in fish feed (between 6% and 8.5% of feed) is found in fish faeces. Thus, nitrogen makes up between 3% and 4% of faeces (D. Higgs 2006, *pers com*; Buryniuk et al. 2006; McGhie et al. 2000).

By quantifying waste, estimates of released nitrogen have ranged from 20 kg to 43 kg of nitrogen for each tonne of Atlantic salmon produced (Johnsen and Wandsvik 1991; Johnsen et al. 1993). Given that 46,100 tonnes of Atlantic salmon were produced in British Columbia in 2004 (BCMOE and BCMAL 2005), this would amount to between 922 tonnes and 1,982 tonnes of nitrogen released per year. By using back calculations from amount of fish produced, feed given, feed conversion ratio and feed composition (Note: if there are 70,600 tonnes of salmon produced, using a 1.2 feed conversion ratio (FCR), 84,720 tonnes of feed was used. At a rate of 17% waste feed, this gives 14,402 tonnes waste feed. If we assume 40% of feed is protein this gives 5,761 tonnes of protein. Lastly, if we assume 16% of the protein is nitrogen this leaves 2,304 tonnes of nitrogen released from waste feed alone.), Higgs (UBC-DFO Centre for Aquaculture and Environmental Research, *pers com* 2006), calculated a value of 1,987 tonnes of nitrogen.

The report, *Alive and Inseparable, British Columbia's Coastal Environment* (2006), provides calculations and results on the nitrogen output from salmonid aquaculture. The result for nitrogen output was between 3,342 and 3,605 tonnes in 2003. Using similar methodology and feed usage numbers provided from the BC Ministry of Environment (White 2006, *pers com*) yields values of between 3,783 and 4,080 tonnes in 2004 and between 4,125 and 4,449 tonnes in 2005.

In 1997, the amount of dissolved nitrogen released from salmon farms was found to be a small fraction of the amount released from riverine sources and sewage sources (Burd 1997). More recent figures document 16,601 tonnes of nitrogen produced by land agriculture (poultry, diary, hog and beef farming; based on the 2001 Census of Agriculture) in the lower mainland (Timmenga and Associates 2003). Furthermore, although the amount depends on the source of information, Victoria, British Columbia's capital city, releases between 34,000,000 m^3 (Sierra Legal Defence Fund 2004; MacLeans 2005) and 94,000,000 m^3 (Green Party 2005) of raw sewage into the Juan de Fuca Straight each year. Taking the lower of the two values together with average values observed in domestic wastewater (Metcalf & Eddy Inc. 1991), between 8,500 and 17,000 tonnes of nitrogen are released from Victoria each year. Therefore, in comparison to both city domestic sewage output and other food production industries, both of which end up in the sea, it would appear that the amount of nitrogen released by aquaculture is small.

Although nitrogen release may be small, it still has the potential to impact the environment that it is released into. Studies carried out by in Washington's Puget Sound, (Brooks 1994a, 1994b, 1995a, 1995b) found a statistically significant increase in dissolved organic nitrogen 6m down current at one of eight farms measured. This increase was not found at 30 m down current. The highest observed level of toxic unionized ammonia (NH^3) was 0.0004 mgl^{-1} which is lower (by a factor of 87.5) than the United States Environmental Protection Agency (EPA) chronic exposure (4 days) concentration limit of 0.035 mgl^{-1}, at a temperature of 15°C, when sensitive salmonid species are present. Typical permissible ammonia levels in the UK 0.4–0.5 mgl^{-1}) in the UK (Weston et al. 1996). Also in Washington, at a poorly flushed site, a level of 0.0023 mgl^{-1} was found in the centre of the sea cage facility, 7.6 times higher than the corresponding reference level of 0.0003 mgl^{-1}. Therefore, it would appear that although nitrogen levels increased within the farm vicinity, the effect was localised.

Fish excrete carbon as faecal particulate waste, as dissolved carbon dioxide and as bicarbonate via the gills. The particulate fraction is addressed here singularly as the amounts of dissolved waste added are insignificant compared to the extensive carbonate-buffering system of the ocean (EAO 1997). Scientific studies concluded that the carbon content of salmonid feed was between 44% and 57% (Buryniuk et al. 2006; DFO 2005a; Findlay and Watling 1994; Gowen and Bradbury 1987; Gillibrand et al. 2002; McGhie et al. 2000; Silvert 1994). Furthermore, 14% of the consumed carbon was used for growth (after Pérez et al. 2002) and 60% was respired or excreted as urine (Gowen et al. 1991). The remaining 26% was assumed to be incorporated into the faeces. In comparison, Higgs et al. (unpub data, UBC-DFO Centre for Aquaculture and Environmental Research, for Alive and Inseparable 2006) found that 9.25% of carbon was indigestible and therefore was excreted as fish faeces. Previous studies (Buryniuk et al. 2006; McGhie et al. 2000) demonstrated that between 55% and 76% of the carbon from feed was found in faeces.

The report, *Alive and Inseparable, British Columbia's Coastal Environment (2006)*, provides calculations and results on the carbon output from salmonid aquaculture as a result of faeces and from waste feed. This report integrated information on the proportion of carbon in feed, the amount assimilated by fish, the amount of feed used and also the amount of waste feed. The resultant figure for particulate carbon output was between 6,354 and 10,540 tonnes in 2003. Using similar methodology and feed usage numbers provided from the British Columbia Ministry of Environment (White 2006, *pers com*) yields values of between 7,192 and 11,929 tonnes in 2004 and between 7,842 and 13,007 tonnes in 2005. In comparison, the Fraser River, is estimated to contribute 171,000 tonnes of particulate organic carbon per year (Johannessen et al. 2003), 13 times that released by salmon aquaculture.

The addition of the carbon results in the most noticeable impact of the nutrients mentioned. As detailed in the section on eutrophication, the biological breakdown of organic carbon in bottom sediments can eliminate oxygen and produce toxic substances, eliminating all organisms aside from certain, predominantly opportunistic, anaerobic bacteria.

Reduction and Recovery

To reduce the potential effect of waste from salmon aquaculture, sites are selected with particular characteristics, for example high water flow. There are many other influential variables. The total volume introduced into the system is important. Brooks (2001) derived a model of dispersion, feed (fed to fish on the farm) and distance which explained 70% of aggregated data on TVS[1] deposition. This model suggested that dispersion rates had a small effect on deposition. In contrast, feed had a larger effect on deposition, but farms became more efficient with higher numbers of fish.

Although many scientific papers address the effect of fish farms, there is relatively little information on the recovery from these effects. In Scottish sea lochs, no significant benthic macrofaunal recolonization was apparent 1 year after fish

[1] Total Volatile Solids, weight loss upon ignition to 550°C of a sample of solids, primarily organic matter.

production ceased (Gowen et al. 1988; Lumb 1989). In Tasmania, a 7-week recovery period from moderately disturbed to undisturbed sediment conditions was observed (Ritz et al. 1989). Also in Tasmania, after 12 months fallowing, fish-farm-derived organic matter in surface sediment at the centre of the cage remained greater than at 30 m distance, but redox potentials indicated that normal 'oxic' conditions had returned (McGhie et al. 2000) the total leased site was 14 ha and annual production was based on 260,000 smolts. In Norway, recovery of the macrobenthic community was slow following the removal of a salmon farm. The community was still highly dominated by *Capitella capitata* 1 year later (Johannessen et al. 1994). On the Canadian east coast, the benthic macrofaunal community had not recovered 1 year after fish farming ceased (Pohle et al. 2001).

As commercial aquaculture sites are typically located in areas with good access and reasonable flushing, one may assume that site recovery, or remediation, upon fallowing would be similar from place to place. In keeping with this, Brooks et al. (2003) predicted chemical remediation of less than 6 months for four sites in the Broughton Archipelago, a similar length of time to the 85 day period reported by Woodward et al. (1992), based on benthic respiration rates, and that reported by Andersen and Kristensen 1992 who studied the effects on the benthos. Brooks and Mahnken (2003) noted that changes imposed as a result of aquaculture lasted from several months to several years. A condensed period of rapid recovery has been observed, proceeded by a more extensive period of slower recovery (Mahnken 1993). It should be noted that Brooks et al. (2004) observed a long remediation time of between 4 and 7 years at a heavily impacted site in British Columbia. Overall, remediation times are proportional to at least the depth of accumulated waste, sediment and water physicochemical characteristics and water depths and current speeds.

Fallowing, the removal of all fish and nets from sea cage sites can be used to reduce the accumulation of sediment beneath the cages (Beveridge 1987; Lumb 1989). Studies have shown that sediments can respond rapidly to decreases in feed levels, even when large biomasses remain. Brooks (2001) found that sediment sulphide concentrations decreased by 2.5 as a result of 3 months of lower feed levels. In British Columbia, the 7 year moratorium (1995–2002) of BC fish farming had an effect on the ability of farmers to fallow sites as no new sites were granted within this time, arguably affecting fish farms and surrounding environment. Currently, fallowing is achieved either by repositioning the cages or by alternately farming at two separate sites, usually for 12–24 months (one growing cycle). The frequency of rotation and the length of the fallowing period are at the operator's discretion and vary considerably depending on company policy and production pressure.

Research to Minimise Waste

In Canada it is illegal to combine different kinds of farming, such as salmon and mussel or salmon and kelp. It is a condition of the licence granted by BCMAL that, amongst other minimum separations, finfish aquaculture sites must be 125 m away from shellfish operations. Recent research has focussed on this area as a method of reducing waste and improving the efficiency of the aquaculture system.

In the Bay of Fundy researchers demonstrated that blue mussels (*Mytilus edulis*) at commercial salmon farms in the Bay of Fundy grew 20% faster than those at reference sites further away (Barrington 2002). On the West Coast of Scotland, growth of mussels at a site in close proximity (10 m) to a salmon farm was significantly higher than at a control site some 500 m distant. The increased growth in these systems has been attributed to utilisation of organic waste. Aside from the demonstrated environmental benefit of integrated aquaculture, the organisms that have been tested all have a viable market currently in existence, so there is a powerful economic incentive. In a study investigating the economics of integrative aquaculture, Whitmarsh et al. (2006) hypothesised that combining mussel with salmonid aquaculture would yield an 'economy of integration', a 9% increase in profit over farming each species separately. As a result of the delicate balance between the cost of capital and the value of the products, it was predicted that operation costs and future price trends were very important, such that a decrease in salmon prices of 2% would wipe out the economy of integration.

Canada's west coast may be equally as suitable as its east coast for integrated aquaculture. The Pacific SEA-Lab Research Society is currently developing and testing a small commercial-scale SEA-System (integrated multi-trophic aquaculture configuration) with the support of government, industry and local First Nations. The farm is designed to take advantage of the organic waste streams from traditional fish culture systems by integrating a kelp component to assimilate the dissolved nutrients, a downstream shellfish component to address the fine particulate fractions and sea cucumber beneath the fish to utilize the solid waste material. Initial research has shown that these groups of organisms have the ability to use these wastes without developing a negative impact to seafood safety and product quality. The current research, initiated in 2007, will continue over the next 5 years and provide scientific data at a commercial scale (Cross 2007, *pers com*).

Closed containment aquaculture is hailed as an ideal mitigative strategy by NGO's (David Suzuki Foundation 2007; Friends of Clayoquot Sound 2007; Georgia Straight Alliance and CAAR 2005; National Environmental Trust and Pure Salmon Campaign 2005; Raincoast Conservation Society 2007; T. Buck Suzuki Foundation 2007). Closed containment provides a barrier between the cultured fish and the environment. Therefore it is theoretically possible to totally remove any interaction of sea lice, disease or waste between the farmed and wild fish. In addition, solid waste could be collected and transported to a land facility for further use.

There are substantial costs associated with closed containment systems. In British Columbia, there have been a number of pilot programs which failed (British Columbia Salmon Farmers Association 2007). The impacts of this technology on the health and welfare of the farmed fish and any potential environmental impacts are unknown. Further research is needed to prove that it is a viable option.

Research is being carried out to develop more environmentally friendly feeds. For example Sarker et al. (2005) demonstrated reduced phosphorus excretion and improved nitrogen retention with the addition of specific diet supplements and Tantikitti et al. (2005) reported reduced levels of nitrogenous waste when fish protein was partially substituted by soybean protein. Substitution of fish oil and meal in commercial feed becomes increasingly likely as prices rise worldwide Hardy et al. (2001) and studies show that are no deleterious effects on the quality of the fish (Higgs et al. 2006).

Summary

Waste released from fish farms is controlled in British Columbia by the BC Ministry of the Environment, under the *Fisheries Act*, the *Environmental Management Act* and the Finfish Aquaculture Waste Control Regulation and includes multifaceted monitoring, management plans and cut-off levels of key pollutants. Fisheries and Oceans Canada is involved in farm siting criteria and conducts research in this area.

The proportion of feed that is uneaten by farmed fish has decreased over the past 15 years alongside increases in feed efficiency, and therefore waste. As fish feed is one of the industry's key production costs, these trends are likely to continue. The main components of waste from aquaculture are carbon, nitrogen, phosphorus and sulphur. Carbon is a major component of waste from salmon farms in British Columbia, but the volume of carbon released is dwarfed by other biological systems such as the Fraser River. Similarly, it would appear that a great deal more nitrogen is released by other industries than from the finfish aquaculture sector in British Columbia. Further, as much of the nitrogen released from fish farms is likely to travel on ocean currents its effect is likely to diminish over a widespread area. Although sulphur, which is used as a regulatory indicator, can result in extensive changes to local fauna (a decrease non-opportunist species and an increase in opportunistic species) the levels associated with fish farms appear to be relatively low.

Field studies have revealed extensive information on the fate and effects of waste from fish farms. Although recovery from such effects can be rapid, it is highly dependant on the site characteristics and history. Very little research has been carried out on methods of impact mitigation and further research on this issue is needed. Although, the recovery of sediment from organic enrichment is relatively well investigated, site specific environmental complexities result in a lack of general extrapolations or trends. Further investigation is needed, in the form of carefully designed and executed laboratory studies.

Presently, cautious investigation of multi-trophic integrated aquaculture is underway. There are many environmental benefits associated with this methodology, but further research is needed to assess the environmental benefits and commercial profitability of full repeated production cycles. Closed containment aquaculture appears at first to offer a solution to waste output from aquaculture, but in reality it would result in concentrated waste which would subsequently need to be processed. Although this may provide an effective way of recycling ecosystem nutrients, this technology does not appear to be economically viable at this time. Further research is needed to ascertain its costs and benefits.

In conclusion, aquaculture waste in British Columbia is tightly monitored and there is no scientific evidence of the extensive all-pervasive environmental impacts often asserted in the media. Near field effects of aquaculture waste are reasonably well known and documented and recently, they have been successfully predicted. The potential far-field effects of organic enrichment associated with net-pen aquaculture are more difficult to document and have not been adequately studied. Research is ongoing in an attempt to increase our understanding of the characteristics and behaviour of waste in the far-field.

Overall Conclusion

Overall, aquaculture in British Columbia is tightly regulated, perhaps more so than any other jurisdiction. The Salmon Aquaculture Review, presented in 1997, prompted a change from recommended strategy and guidance legislation and enforcement (EAO 1997).

The much-discussed potential of escaped farmed Atlantic salmon to detrimentally affect Pacific salmonids has not happened in British Columbia. Although production of Atlantic salmon in British Columbia has increased rapidly in the past decade, the number of escapes has not. Furthermore, the likelihood of disease transmission between farmed and wild fish appears to be small.

In Canadian aquaculture, vaccinations are used to prevent disease outbreaks, and fish are treated upon the outbreak of a disease. Antibiotics are only administered by a veterinary in response to a disease, not prophylactically, which is an important point in contrast with other industries. As a result salmon farming is the world's largest protein-producing sector with the smallest use of antibiotics. Drug use is controlled by BCMAL, whereas the issue of residue falls under the remit of the federal CFIA. Each antibiotic used has a stipulated withdrawal time.

The use of approved antibiotics for salmon aquaculture appears to pose little risk to the environment, as these antibiotics are degraded within reasonable time limits. Health Canada appears to have adequate controls in place to negate the risk posed by antibiotic residues to human health in salmon aquaculture products. The predominant treatment for sea lice in Canada is SLICE®. This treatment appears to be an effective treatment for sea lice that has minimal impacts on the environment and poses very little risk to the consumer.

Sea lice biology and therefore prevalence, is greatly dependent on environmental variables. For example temperature and salinity have a large impact on *L. salmonis* growth rates and survival. In particular, free swimming sea lice larval stages are intolerant of low salinities such as those found in British Columbia's inshore waters. Sticklebacks and wild salmon could provide a potential host for sea lice.

Much of the information currently available on the impact of sea lice on health is from Atlantic salmon, not Pacific salmon. For example, it is not known at this time what intensity (concentration) of sea lice has on the health of Pacific salmon, nor what level would cause mortality. Because there is as yet no answer to either question, it is currently not possible to conclude that sea lice on British Columbia salmon farms are having a significant impact on the parasite loads of the wild Pacific salmon.

Biological waste is perhaps the most tightly controlled aspect of aquaculture in British Columbia, including multifaceted monitoring, management plans and cut-off levels of key pollutants. Nutrients such as carbon, nitrogen and phosphorus are released and result in changes to the ocean floor in the farm vicinity. Relative to other nutrient releases, these inputs are small, but the analytical detection of changes is highly site dependent and can occur for periods of weeks to years. Although research has been conducted on impact mitigation, such as environmentally friendly feeds in integrated multi-trophic aquaculture, further research is needed on this topic.

In conclusion, aquaculture waste in British Columbia is tightly monitored and there is no scientific evidence of the extensive all-pervasive environmental impacts often asserted in the media.

Since its inception in British Columbia, discussion concerning aquaculture has often generated conflicting information. Unfortunately, it is not feasible to postulate the effect that this may have had on British Columbia legislation of aquaculture. The information presented here provides evidence of the ability of the British Columbia government to evaluate, review and revise legislation accordingly.

References

Alive and Inseparable. 2006. British Columbia's Coastal Environment. BC Government Publication, Victoria, B.C., p. 388.

Andersen, F.O. and Kristensen, E. 1992. The importance of Benthic Macrofauna in decomposition of microalgae in a coastal marine sediment. *Limnol Oceanogr* 37(7): 1392–1403.

Angelidis, P., Baudin-Laurencin, F., and Youinou, P. 1987. Stress in rainbow trout, *Salmo gairdneri*: effects upon phagocyte chemiluminescence, circulating leucocytes and susceptibility to *Aeromonas salmonicida. J Fish Biol* 31(Suppl. A): 113–122.

Anon, 1997. Application for the marketing authorisation in the UK of Calicide (teflubenzuron) for the treatment of sea lice infestations of Atlantic Salmon. Part III A5 *Ecotoxicity*, 8 Volumes.

Anon, 2002. Lønnsomhetsundersøkelse for matfiskproduksjon, laks og ørret (in Norwegian). Økonomiske Analyser Fiskeoppdrett nr. 1/2002, Fiskeridirektoratet.

Antaya, Y. 2006. British Columbia Ministry of Agriculture and Lands, personal communication.

Aquaculture Act, R.S.N.L. 1990, c. A-13.

Aquaculture Act, S.N.B. 1988, c. A-9.2.

Asplin, L., Salvanes, A.G.V., and Kristoffersen, J.B. 1999. Non-local wind-driven fjord-coast advection and its potential effect on plankton and fish recruitment. *Fish Oceanogr* 8: 255–263.

Asplin, L., Boxaspen, K., and Sandvik, A. D. 2004. Modelled distribution of sea lice in a Norwegian fjord. *ICES C.M. 2004/P*, 11: 12.

Baird, D., Telfer, T.C., and Jenkins, W.R. 1997. Ecotoxicity Expert Report on Calicide. Final Report to Nutreco ARC.

Baird, D. 1998. Expert Report Addendum. Report to Nutreco ARC June 1998.

Barber, I. and Huntingford, F.A. 1996. Parasite infection alters schooling behaviour: Deviant positioning of helminth-infected minnows in conspecific groups. *Proc R Soc Lond B Biol Sci* 263: 1095–1102.

Barrington, K.A., MacDonald, B.A., and Robinson, S.M.C. 2002. Feeding behavior of blue mussels living within an Atlantic salmon aquaculture site. *Bulletin - Aquaculture Association of Canada*. 102(3): 109–110.

Beamish, R.J., Jones, S., Dawe, S., Gordon, E., Sweeting, R.M., Neville, C.M., Johnson, S., Trudel, M., MacDonald, T., and Ambers, N. 2004. Prevalence, intensity and life history strategy of sea lice on adult Pacific salmon returning to the spawning areas in the Central Coast of British Columbia. Fisheries and Oceans Canada.

Beamish, R., Neville, C.M., Sweeting, R.M., and Ambers, N. 2005. Sea lice on adult Pacific salmon in the coastal waters of British Columbia, Canada. *Fish Res* 76: 198–208.

Beamish, R., Jones, S., Neville, C.M., Sweeting, R.M., Karreman, G., Saksida, S., and Gordon, E. 2006. Exceptional marine survival of pink Salmon that entered the marine environment in 2003 suggests that farmed Atlantic salmon and Pacific salmon can coexist successfully in a marine ecosystem on the pacific coast of Canada. *ICES J Mar Sci* 63(7): 1326–1337.

Beveridge, M.C.M. 1987. Cage Aquaculture. Fishing News Books, Farnham, Surrey.

Beveridge, M.C.M., Phillips, M.J., and Clarke, R.M. 1991. A quantitative and qualitative assessment of wastes from aquatic animal production. In: Brune, D., and Tomasso, J.R. (Eds.), *Aquaculture and Water Quality: Advances in World Aquaculture*. World Aquaculture Society, Baton Rouge, LA, 506–533.

Bjorklund, H., Bondestam, J., and Bylund, G. 1990. Residues of oxytetracycline in wild fish and sediments from fish farms. *Aquaculture* 86: 359–365.

Bjørn, P.A. and Finstad, B. 1997. The physiological effects of salmon lice infection on sea trout post smolts. *Nord J Freshw Res* 73: 60–72.

Bjørn, P.A. and Finstad, B. 1998. The development of salmon lice (*Lepeophtheirus salmonis*) on artificially infected post smolts of sea trout (*Salmo trutta*). *Can J Zool*, 76, 970–977.

Bjørn, P.A., Finstad, B,. and Kristoffersen, R. 2001. Salmon lice infection of wild sea trout and Arctic char in marine and freshwaters: the effects of salmon farms. *Aquacult Res* 32: 947–962.

Bowers, J.M., Mustafa, A., Speere, D.J., Conby, G.A., Brimacombe, M., Sims, D.E., and Burka, J.F. 2000. The physiological response of Atlantic salmon, *Salmo salar* L., to a single experimental challenge with sea lice *Lepeophtheirus salmonis*. *J Fish Dis*, 23: 165–172.

Boxaspen, K. and Næss, T. 2000. Development of eggs and the planktonic stages of salmon lice (*Lepeophtheirus salmonis*) at low temperatures. *Contrib Zool*, 69(1/2).

Braaten, B.N.D. and Joyce, J. 1991. Status of pollution from aquaculture in six Nordic countries, release of pollutants, effects and wastewater treatment. In: *Aquaculture and the Environment*. Special Publication of the European Aquaculture Society.

British Columbia Ministry of Agriculture and Lands (BCMOE) and British Columbia Ministry of Environment (BCMAL), 2005. The 2005 British Columbia Seafood Industry Year in Review. Victoria, Government of British Columbia.

British Columbia Ministry of Agriculture and Lands (BCMAL) and British Columbia Ministry of Environment (BCMOE), 2006. Annual Inspections on Marine Finfish Aquaculture Sites for 2004 and 2005 Inspection Cycles, 85. Available at: http://www.agf.gov.bc.ca/fisheries/aqua_report/2004-5/2004-5_Report.pdf; accessed: August 26, 2007.

British Columbia Salmon Farmers Association, 2007. Report Card: Special Committee on Aquaculture. Digital publication: http://www.salmonfarmers.org/files/hottopic_05_16_07_e.pdf; accessed August 26, 2007.

British Columbia Supreme Court, 2007. Creative Salmon Company Ltd. v. Staniford, 2007 BCSC 62. Date: 20070115. Docket: S053729. Registry, Vancouver.

Brooks, K.M. 1994a. Environmental Sampling at Sea Farm Washington, Inc., Net-pen Facility II in Port Angeles Harbor, WA, During 1994. Available from Washington Department of Natural Resources, Olympia, WA.

Brooks, K.M. 1994b. Environmental Sampling at Global Aqua USA, Inc., Saltwater II Salmon Farm Located in Rich Passage, WA, 1994. Global Aqua USA, Inc., Bainbridge Island, WA.

Brooks, K.M. 1995a. Environmental Sampling at Sea Farm Washington, Inc., Net-pen Facility II in Port Angeles Harbor, WA, During 1995. Available from Washington Department of Natural Resources, Olympia, WA.

Brooks, K.M. 1995b. Environmental Sampling at Global Aqua USA, Inc., Saltwater II Salmon Farm Located in Rich Passage, WA, 1994. Global Aqua USA, Inc., Bainbridge Island, WA.

Brooks, K.M. 2000a. Sediment Concentrations of Zinc Near Salmon Farms in British Columbia, Canada During the Period June Through August 2000. British Columbia Salmon Farmers Association, Campbell River, BC, Canada.

Brooks, K.M. 2000b. Database Report to the Ministry of Environment Describing Sediment Physicochemical Response to Salmon Farming in British Columbia, 1996 Through April 2000. British Columbia Salmon Farmers Association, Campbell River, BC, Canada.

Brooks, K.M. 2000c. Report to the Ministry of Environment on the Determination of Copper Loss Rates from Flexgard XITM Treated Nets in Marine Environments and Evaluation of the Resulting Environmental Risks. British Columbia Salmon Farmers Association, Campbell River, BC, Canada.

Brooks, K.M. 2000d. Report to the Ministry of Environment on Sediment Concentrations of Sulfides and Total Volatile Solids Near Salmon Farms in British Columbia, Canada, During the Period June Through August 2000, and Recommendations for Additional Sampling. British Columbia Salmon Farmers Association, Campbell River, BC, Canada.

Brooks, K.M. 2000e. Report to the Ministry of Environment on the Results of the June 2000 Interim Salmon Farm Monitoring at Stolt Sea Farm, Inc., Salmon Aquaculture Tenures Located in British Columbia. Stolt Sea Farm, Inc., Campbell River, BC, Canada.

Brooks, K.M. 2000f. Literature Review and Model Evaluation Describing the Environmental Effects and Carrying Capacity Associated with the Intensive Culture of Mussels (*Mytilus edulis galloprovincialis*). Taylor Resources, Shelton, WA.

Brooks, K.M. 2001. An Evaluation of the Relationship Between Salmon Farm Biomass, Organic Inputs to Sediments, Physicochemical Changes Associated with Those Inputs and the Infaunal Response—with Emphasis on Total Sediment Sulfides, Total Volatile Solids, and Oxidation–Reduction Potential as Surrogate Endpoints for Biological Monitoring. Available from Technical Advisory Group, Ministry of Environment, Nanaimo, BC, Canada.

Brooks, K.M. and Mahnken, C.V.W. 2003 Interactions of Atlantic salmon in the Pacific Northwest environment II. Organic wastes. *Fish Res*, 62, 255–293.

Brooks, K.M., Stierns, A.R., Mahnken, C.V.W., and Blackburn, D.B. 2003. Chemical and biological remediation of the benthos near Atlantic salmon farms. *Aquaculture* 219; 255– 377.

Brooks, K.M., Stierns, A.R., and Backman, C. 2004. Seven year remediation study at the Carrie Bay Atlantic salmon (*Salmo salar*) farm in the Broughton Archipelago, British Columbia, Canada. Aquaculture. *Aquaculture* 239, 81–123.

Brooks, K.M. 2005. The effects of water temperature, salinity, and currents on the survival and distribution of the infective copepodid stage of sea Lice (*Lepeophtheirus Salmonis*) originating on Atlantic salmon farms in the Broughton Archipelago of British Columbia, Canada. *Rev Fish Sci* 13: 177–204.

Burd, B. 1997. Waste discharges. In Salmon Aquaculture Review Report – Technical Advisory Team Discussion Papers Volume 3, Part D.

Buryniuk, M., Petrell, R.J., Baldwin. S., and Lo, K.V. 2006. Accumulation and natural disintegration of solid wastes caught on a screen suspended below a fish farm cage. *Aquac Eng* 35: 78–90.

Butler, J.R.A. 2002. Wild salmonids and sea louse infestations on the west coast of Scotland: sources of infection and implications for the management of marine salmon farms. *Pest Manag Sci* 58: 595–608.

Butterworth, K.G., Li, W., and McKinley, R.S. 2004. Carbon and nitrogen stable isotopes: a tool to differentiate between *Lepeophtheirus salmonis* and different salmonid host species? *Aquaculture* 241(1–4): 529–538.

Butterworth, K.G. 2005. Personal communication with K. Butterworth (January). Vancouver: Centre for Aquaculture and Environmental Research, University of British Columbia.

Butterworth, K.G., Ronquillo, J.D., and McKinley, R.S. 2005. Simplified illustrated sea lice identification guide for *Lepeophtheirus salmonis* and *Caligus clemensi* in British Columbia, Canada. *ACC Spec Publ* 9: 101–103.

Capone, D.G., Weston, D.P., Miller, V., and Shoemaker, C. 1996. Antibacterial residues in marine sediments and invertebrates following chemotherapy in aquaculture. *Aquaculture* 145(1–4): 55–75.

Chevassus, B. 1979. Hybridization in salmonids: results and perspectives. *Aquaculture* 17: 113–128.

Cho, C.Y. and Bureau, D.P. 1997. Reduction of waste output from salmonid aquaculture through feeds and feeding. *Prog Fish Culturist* 59: 155–160.

Cho, C.Y. and Bureau, D.P. 2001. A review of diet formulation strategies and feeding systems to reduce excretory and feed wastes in aquaculture. *Aquac Res* 32: 349–360.

Coastal Alliance for Aquaculture Reform (CAAR), 2007. Website: www.farmedanddangerous.org; accessed: August 26, 2007.

Coleman, P. and Rasch, T. 1981. A detailed listing of the liberations of salmon into the open waters of the State of Washington during 1980, Progress Report No. 132. Washington Department of Fisheries, Olympia, WA.

Costelloe, M.J., Costelloe, G., and Roche, N. 1996. Planktonic dispersal of larval salmon-lice, *Lepeophtheirus salmonis*, associated with cultured salmon, *Salmo salar*, in Western Ireland. *J Mar Biol Assoc U.K.* 76: 141–149.

Costelloe, M., Costelloe, J., O'Donohoe, G., Coghlan, N.J., Oonk, M., and Van der Heijden, Y. 1998. Planktonic distribution of sea lice larvae, *Lepeophtheirus salmonis*, in Killary Harbour, West Coast of Ireland, *J Mar Biol Assoc U.K.* 78: 853–874.

Cox, S. 2004. Opponents of proposed sablefish farms charge that disease and parasites will devastate wild stocks. Sound familiar? Georgia Straight [Vancouver] 16 December, 2004.

Cross, S.F. 2007. University of Victoria. Personal Communication.

Cubitt, K.F., Butterworth, K.G., Finstad, B., Huntingford, F., and McKinley, R.S. 2006. Escaped farmed salmon; a threat to BC's wild salmon? *Fraser Institute Alert Series*. Vancouver, Canada.

Cubitt, K.F., Williams, H.T., Rowsell, D., McFarlane, W.J., Gosine, R.G., and McKinley, R.S. 2006. Development of an intelligent reasoning system to distinguish hunger states in rainbow trout (*Oncorhynchus mykiss*). *Comp Electron Agr*.

David Suzuki Foundation. 2007. Salmon Aquaculture Solutions. http://www.davidsuzuki.org/Oceans/Aquaculture/Salmon/Solutions.asp; accessed August 26, 2007.

DFO. 1997. Waste discharge. In: Salmon Aquaculture Review. *Environmental Assessment Office*, Stn. Prov. Govt., Vic., BC, Canada, WD1–47.

DFO Atlantic Salmon Watch (ASWP). 2005. Data provided by Atlantic Salmon Watch Program, Department of Fisheries and Oceans, Victoria, British Columbia, Canada.

DFO. 2005a. Biotechnology to help develop better salmon feeds. Aquaculture, current topics http://www. dfo-mpo.gc.ca/science/aquaculture/biotech/fact3_e.htm; accessed: August 26, 2007.

DFO – Fisheries and Oceans Canada. (2005b). Hatcheries, Fishways and Channels. Available online at: http://www-heb.pac.dfo-mpo.gc.ca/facilities/hat-fw-sc_e.htm; accessed August 26, 2007.

Ellsaesser, C.F., Clem, L.W. 1986. Haematological and immunological changes in channel catfish stressed by handling and transport. *J Fish Biol* 28: 511–521.

Environmental Assessment Office (EAO). 1997. Salmon aquaculture review. www.intrafish.com/laws-and-regulations/report_bc/; accessed: August 26, 2007.

Espelid, S., Lokken, G.B., Steiro, K., and Bogwald, J. 1996. Effects of cortisol and stress on the immune system in Atlantic salmon (*Salmo salar* L.). *Fish Shell Immunol* 6: 95–110.

Fast, M.D., Ross, N.W., Mustafa, A., Sims, D.E., Johnson, S.C., and Conboy, G.A. 2002. Susceptibility of rainbow trout oncorhynchus mykiss, atlantic salmon salmo salar and coho salmon oncorhynchus kisutch to experimental infection with sea lice lepeophtheirus salmonis. *Dis Aquat Organ* 52(1): 57–68.

FAWC. 1996. Report on the welfare of farmed fish. Farm Animal Welfare Council (FAWA), UK.

Findlay, R.H. and Watling, L. 1994. Toward a process level model to predict the effect of salmon net-pen aquaculture on benthos. In: Hargrave, B.T., (Ed.), Modelling Benthic Impacts of

Organic Enrichment from Marine Aquaculture *Canadian Technical Report of Fisheries and Aquatic Sciences*, Canada, 47–48.

Finstad, B., Grimnes, A., Bjørn, P.A., and Hvidsten, N.A. 2000. Laboratory and field investigations of salmon lice [*Lepeophtheirus salmonis* (Krøyer)] infestation on Atlantic salmon (*Salmo salar* L.) postsmolts. *Aquacult Res* 31: 795–803.

Fisheries Act, R.S. 1985, c. F-14, s. 1.

Fisheries Act, R.S.B.C. 1996, c. 149.

Food and Agriculture Organisation (FAO). 2006. State of world aquaculture. Fisheries Department, FAO Fisheries Technical Paper 500, 134p. FAO Rome.

Friends of Clayoquot Sound. 2007. Close the Cages or Close Down! http://www.focs.ca/ fishfarming/solutions.asp; accessed August 26, 2007.

Georgia Straight Alliance and CAAR. 2005. New Aquaculture Technology: Closing in on Solutions. http://www.georgiastrait.org/Articles2005/salmon_newtech.php; accessed August 26, 2007.

Gillibrand, P.A., Gubbins, M.J., Greathead, C., and Davies, I.M. 2002. Scottish executive locational guidelines for fish farming: Predicted levels of nutrient enhancement and benthic impact. *Scottish Fisheries Research Report* 63. Fisheries Research Services, Marine Laboratory, Aberdeen, Scotland.

Ginetz, R.M.J. 2002. On the risk of colonization by Atlantic salmon in BC waters. On special assignment to the B.C. Salmon Farmers Association from Fisheries and Oceans Canada. Available online at: http://www.salmonfarmers.org/attachments/colonization.pdf; accessed August 26, 2007.

Glover, K.A., Skaala, O., Nilsen, F., Olsen, R., Teale, A.J., and Taggart, J.B. 2003. Differing susceptibility of anadromous brown trout (salmo trutta L.) populations to salmon louse (lepeophtheirus salmonis (kroyer, 1837)) infection. *ICES J Mar Sci,* 60(5): 1139–1148.

Glover, K.A., Hamre, L.A., Skaala, O., and Nilsen, F. 2004. A comparison of sea louse (lepeophtheirus salmonis) infection levels in farmed and wild atlantic salmon (salmo salar L.) stocks. *Aquaculture,* 232(1–4): 41–52.

Gowen, R.J. and Bradbury, N.B. 1987. The ecological impact of salmonid farming in coastal waters: a review. *Oceanogr Mar Biol Annu Rev* 25: 563–575.

Gowen, R.J., Brown, J.R., Bradbury, N.B., McLusky, D.S. 1988. Investigations into benthic enrichment, hypernutrification and eutrophication associated with mariculture in Scottish coastal waters (1984–1988). Department of Biological Sciences, University of Stirling, 289.

Gowen, R.J., Weston, D.P., and Ervik, A. 1991. Aquaculture and the benthic environment: A review. In: Cowey, C.B., and Cho, C.Y. (Eds.), *Nutritional Strategies and Aquaculture Waste.* Proc. 1st Int. symp. Nutritional Strategies in Management of Aquaculture, U. Guelph, Ontario.

Gray, A.K., Evans, M.A., and Thorgaard, G.H. 1993. Viability and development of diploid and triploid salmonid hybrids. *Aquaculture* 122: 125–142.

Green Party. 2005. Harris calls for tougher waste-water treatment standards. Press release, 2005-12-22.

Gross, M.R. 1998. One species with two biologies: Atlantic salmon (Salmo salar) in the wild and in aquaculture. *Can J Fish Aquat Sci* 55: 131–144.

Hardy, R.W., Higgs, D.A., Lall, S.P., and Tacon, A.G.J. 2001. Alternative dietary protein and lipid sources for sustainable production of salmonids. *Fisken og Havet* 8: 44.

Harrower, B. 2005. Personal communication with B. Harrower, June, 2005. British Columbia Ministry of Agriculture and Lands, Victoria, British Columbia, Canada.

Healey, M.C. 1991. The life history of chinook salmon. In: Groot C., and Margolis, L. (Eds.), *Pacific Salmon Life Histories*. U.B.C. Press, 311–393.

Heuch, P.A. 1995. Experimental evidence for aggregation of salmon louse copepodids (*Lepeophtheirus salmonis*) in step salinity gradients. *J Mar Biol Assoc U.K.* 75(4): 927–939.

Heuch, P.A., Revie, C.W., and Gettinby, G. 2003. A comparison of epidemiological patterns of salmon lice, lepeophtheirus salmonis, infections on farmed atlantic salmon, salmo salar L., in Norway and Scotland. *J Fish Dis* 26(9): 539–551.

Hektoen, H., Berge, J.A., Hormazabal., V., and Yndestad, M. 1995. Persistence of antibacterial agents in marine sediments. *Aquaculture* 133(3–4): 175–184.
Higgs, D.A. 2006. UBC-DFO Centre for Aquaculture and Environmental Research. Personal communication.
Hiemstra, M., Toonen, A., and DeKok, A. 1999. Determination of benzoylphenylurea insecticides in pome fruit and fruiting vegetables by liquid chromatography with diode array detection and residue data obtained in the Dutch national monitoring program. *J AOAC Int* 82(5): 1198–1205.
Higgs, D.A., Belfry, S.K., Oakes, J.D., Rowshandelil, M., Skura, B.J., and Deacon G. 2006. Efficacy of an equal blend of canola oil and poultry fat as an alternate dietary lipid source for Atlantic salmon (*Salmo salar* L.) in sea water. I: effects on growth performance, and whole body and fillet proximate and lipid composition. *Aquac Res* 37: 180–191.
Higgs, D.A. 2007. UBC-DFO Centre for Aquaculture and Environmental Research. Personal communication.
Hiscock, K., Sewell, J., and Oakley, J. 2005. *Marine Health Check 2005*. Marine Life Information Network, Marine Biological Association of the United Kingdom. World Wildlife Fund-United Kingdom.
Hutchison, P. (ed.). 1997. Interactions between salmon culture and wild stocks of Atlantic salmon: the scientific and management issues. Proceedings of ICES/NASCO Symposium, Bath, England. *ICES J Mar Sci* 54.
IFOAM. 2005. The IFOAM norms for organic production and processing. IFOAM Germany. ISBN: 3-934055-58-3, 33–34.
Jenkins, W.R. 1995. Ecotoxicity – Expert Report on Ektobann. Pharmaco LSR Ltd.
Johannessen, P. J., Botnen, H. B., and Tvedten, Ø. F. 1994. Macrobenthos: before, during and after a fish farm. Aquaculture and Fisheries Management, 25: 55–66.
Johannessen, S.C., Macdonald, R.W., and Paton, D.W. 2003. A sediment and organic carbon budget for the greater Strait of Georgia. *Estuary Coast Shelf Science* 56: 845–860.
Johnsen, B.O. and Jensen, A.J. 1991. The *Gyrodactylus* story in Norway. *Aquaculture* 98: 289–302.
Johnsen, R.I., Grahl-Nelson, O., and Lunestad, B.T. 1993. Environmental distribution of organic waste from a marine fish farm. *Aquaculture* 118: 229–244.
Johnsen, F. and Wandsvik, A. 1991. The impact of high energy diets on pollution control in the fish farming industry. In: Cowey, C.B., and Cho, C.Y. (Eds.), *Nutritional Strategies and Aquaculture Waste*. Proceedings of the First International Symposium on Nutritional Strategies in Management of Aquaculture Waste, University of Guelph, Ontario, 51–62.
Johnson, S.C. and Albright, L.J. 1991. The development stages of *Lepeophtheirus salmonis* (Krøyer, 1837) (Copepoda: Caligidae). *Can J Zool* 69: 929–950.
Johnson, S.C. and Albright, L.J. 1992. Comparative susceptibility and histopathology of the response of naive Atlantic, Chinook and Coho salmon to experimental infection with *Lepeophtheirus salmonis* (Copepoda: Caligidae). *Dis Aquat Organ* 14(3): 179–193.
Johnson, S.C. 1993. A comparison of development and growth rates of *Lepeophtheirus salmonis* (Copepoda: Caligidae) on naïve Atlantic (*Salmo salar*) and chinook (*Oncorhynchus tshawytscha*) salmon. In: Boxshall, G. A., and Defaye, D. (Eds). *Pathogens of Wild and Farmed Fish: Sea Lice*, Chichester, Ellis Horwood 68–82.
Johnson, S.C., Blaylock, R.B., Elphick, J., and Hyatt, K.D. 1996. Disease induced by the sea louse (*Lepeophtheirus salmonis*) (Copepoda: Caligidae) in wild sockeye salmon (*Oncorhynchus nerka*) stocks of Alberni Inlet, British Columbia. *Can J Fish Aquat Sci* 53: 2888–2897.
Jones, S.R.M., Prosperi-Porta, G., Kim, E., Callow, P., and Hargreaves, B. 2006. The Occurrence of *Lepeophtheirus salmonis* and *Caligus clemensi* (Copepoda: Caligidae) on three-spine Stickleback *Gasterosteus aculeatus* in Coastal British Columbia. *J Parasit* 92(3): 473–480.
Kabata, Z. 1970. Diseases of Fishes. In Snieszko, S.F. and H.R. Axelrod, eds. *Book 1: Crustacea as Enemies of Fishes*. New Jersey: T.F.H. Publications.
Kabata, Z. 1973. The species of *Lepeophtheirus* (Copepoda: Caligidae) from fishes of British Columbia. *J Fish Res Board Can* 30(6): 729–759.
Kolstad, K., Grisdale-Heiland, B., Meuwissen, T.H.E., and Gjerde, B. 2005. Family differences in feed efficiency of Atlantic salmon (Salmo salar): a pilot study. *Aquaculture* 247(1–4): 21.

Landeau, L. and Terborgh, J. 1986. Oddity and the 'confusion effect' in predation. *Anim Behav* 34: 1372–1380.
Langer, O.E. 2003. Is there a bottom line in the wild salmon – farmed salmon debate? A technical opinion. Pub. The David Suzuki Foundation. Vancouver, BC. pg.7.
Lehmann, S. and Irvine, J.R. 2004. Canadian Enhanced Salmonid Production During 1978–2003 (1977–2002 brood years). NPAFC Doc. 801, Fisheries and Oceans Canada.
Loginova, G.A. and Krasnoperova, S.V. 1982. An attempt at crossbreeding Atlantic salmon and pink salmon (preliminary report). *Aquaculture* 27: 329–337.
Lumb, C.M. 1989. Self-pollution by Scottish Salmon Farms? *Mar Pollut Bull* 20: 375–379.
MacCrimmon, H.R. and Gots, B.L. 1979. World distribution of Atlantic salmon, Salmo salar. *J Fish Res Board Can* 36: 423–457.
MacLeans. 2005. From sea to stinking sea by Ken MacQueen. October 17, 2005.
Mahnken, C.V.W. 1993. Benthic faunal recovery and succession after removal of a marine fish farm. PhD thesis, University of Washington, Seattle, p. 290.
Mathews, M.A., Poole, W.R., Thompson, C.E., McKillen, J., Ferguson, A., Hindar, K. and Wheelan, K.F. 2000. Incidence of hybridization between Atlantic salmon, Salmo salar L., and brown trout, Salmo trutta L. *Ireland, Fish Manag Ecol* 7: 337–347.
Maule, A.G., Tripp, R.A., Kaattari, S.L., and Schreck, C.B. 1989. Stress alters immune function and disease resistance in chinook salmon (*Oncorhynchus tshawytscha*). *J Endocrinol* 120: 135–142.
McFarlane, W.J., Cubitt, K.F., Williams, H., Rowsell, D., Moccia, R., and McKinley, R.S. 2004. Can feeding status and stress level be assessed by analyzing patterns of muscle activity in free swimming rainbow trout (*Oncorhynchus mykiss* Walbaum)? *Aquaculture* 239: 467–484.
McGhie, T., Crawford, C., Mitchell, I., and O'Brien, D. 2000. The degradation of fish-cage waste in sediments during fallowing. *Aquaculture* 187: 351–366.
McKibben, M.A. and Hay, D.W. 2002. Planktonic distribution of sea lice *Lepeophtheirus salmonis* larvae in intertidal plankton samples in Loch Shieldaig, Western Scotland in relation to local salmon farm production cycles. *ICES CM 2002/T:06.*
McVicar, A.H. 1997. Disease and parasite implications of the coexistence of wild and cultured Atlantic salmon populations. *ICES J Mar Sci* 54(6): 1093–1103.
Metcalf & Eddy Inc. 1991. *Wastewater Engineering, Treatment, Disposal, and Reuse*, third ed. McGraw-Hill Inc., New York, p. 1333 (revised by Tchobanoglous, G. and Burton, F.L.).
Morton, A., Routledge, R., Peet, C., and Ladwig, A. 2004. Sea lice (*Lepeophtheirus salmonis*) infection rates on juvenile pink (*Oncorhynchus gorbusha*) and chum (*Oncorhynchus keta*) salmon in the nearshore environment of the British Columbia Coast, Canada. *Can J Fish Aquat Sci* 61: 147–157.
Morton, A., Routledge, R.D., and Williams, R. 2005. Temporal patterns of sea louse infestation on wild Pacific salmon in relation to the fallowing of Atlantic salmon farms. *N Am J Fish Manage* 25(3): 811–821.
Morton, A. and Routledge, R. 2006. Fulton's condition factor: is it a valid measure of Sea Lice impact on Juvenile Salmon? *N Am J Fish Manage* 26: 56–62.
Munro, A.L.S., Liveredge, J., and Elson, K.G.R. 1976. The distribution and prevalence of infectious pancreatic necrosis in wild fish in Loch Awe. *Proc R Soc Edinb-B* 75: 223–232.
Nagasawa, K. 2001. Annual changes in the population size of the salmon louse *Lepeophtheirus salmonis* (Copepoda:Caligidae) in a high-seas Pacific salmon (*Oncorhynchus spp.*), and relationship to host abundance. *Hydrobiologia* 453/454: 411–416.
National Environmental Trust and Pure Salmon Campaign. 2005. Closed Containment: The Best Solution. http://www.puresalmon.org/solutions.html; accessed August 26, 2007.
O'Donoghue, G., Costello, M., and Costello, J. 1998. Development of a management strategy for the reduction/elimination of sea lice larvae *Lepeophtheirus salmonis* parasites of salmon and trout. *Mar Res Ser* 6: 1–51. The Marine Institute, Dublin, Ireland.
Olivier, G. 2002. Disease interactions between wild and cultured fish – Perspectives from the American Northeast (Atlantic Provinces). *Bull Eur Assoc Fish Pat* 22: 103–109.
Otteraa, H., Skilbrei, O., Skaala, O., Boxaspen, K., Aure, J., Taranger, G.L., Ervik, A., and Borgstroem, R. 2004. The Hardanger Fjord—Salmonid Aquaculture and effects on wild

salmonid populations. Fisken og Havet [Fisken Havet]., 43 pg. 2004. Published by the Institute of Marine Research. Bergen, Norway.
Pearson, T.H. and Black, K.D. 2001. The environmental impacts of marine fish cage culture. In: Black KD (ed) *Environmental impacts of aquaculture*. CRD, Boca Raton, Florida, p 213.
Penston, M.J., McKibben, M., Hay, D.W., and Gillibrand, P.A. 2002. Observations of sea lice larvae distributions in Loch Shieldaig, Western Scotland. *ICES CM2002/T:09*.
Pérez, O.M., Telfer, T.C., Beveridge M.C.M., and Ross L.G. 2002. Geographical Information Systems (GIS) as a simple tool to aid modelling of particulate waste distribution at marine fish cage sites. *Estuar Coast Shelf Sci* 54: 761–768.
Pickering, A.D. and Pottinger, T.G. 1985. Cortisol can increase the susceptibility of brown trout, *Salmo trutta* L., to disease without reducing the white blood cell count. *J Fish Biol* 27: 611–619.
Pike, A.W. 1989. Sea lice – major pathogens of farmed Atlantic salmon. *Parasitol Today*, 5: 291–297.
Pike, A.W. and Wadsworth, S.L. 1999. Sea lice in salmonids: their biology and control. *Adv Parasit*, 44: 233–337.
Pohle, G., Frost, B., and Findlay, R. 2001. Assessment of regional benthic impact of salmon mariculture within the Letang Inlet, Bay of Fundy. *ICES J Mar Sci* 58 (2): 417–426.
Raincoast Conservation Society. 2007. Closed containment systems. http://www.raincoast.org/proj-aquaculture/proj-aqua-5.shtml; accessed August 26, 2007.
Report of the Auditor General of Canada. 2006. Fisheries and oceans — the effects of Salmon Farming in British Columbia on the management of wild salmon stocks. http://www.oag-bvg.gc.ca/domino/reports.nsf/html/0030ce.html; accessed August 26, 2007.
Ritz, D.A., Lewis, M.E., and Shen, M. 1989. Response to organic enrichment of infaunal macrobenthic communities under salmonid seacages. *Mar Biol* 103: 211–214.
Robinson, S., Lander, T., MacDonald, B., Barrington, K., Chopin, T., Martin, J.D., Bastarache, S., Belya, E., Haya, K., Sephton, D., Page, F., Martin, J.L., Eddy, S., and Sandercock, F.K. 1992. The life history of coho salmon. In: Groot, C., and Margolis, L., (Eds.), *Pacific Salmon Life Histories*. U.B.C. Press, 395–445.
Salonius, K. and Iwama, G.K. 1993. Effects of early rearing environment on stress response, immune function, and disease resistance in juvenile coho (*Oncorhynchus kisutch*) and chinook salmon (*O. tshawytscha*). *Can J Fish Aquat Sci* 50: 759–766.
Sams, L. 2005. Marine Harvest Canada. Personal communication.
Sarker, S.A., Satoh, S., and Kiron, V. 2005. Supplementation of citric acid and amino acid-chelated trace element to develop environment-friendly feed for red sea bream. *Pagrus major. Aquaculture* 248: 3–11.
Sierra Legal Defence Fund. 2004. The National Sewage Report Card: Grading the Sewage Treatment of 22 Canadian Cities, p. 76.
Silvert, W. 1994. Modelling benthic deposition and impacts of organic matter loading. In: Hargrave (Ed.), Modelling Benthic Impacts of Organic Enrichment from Marine Aquaculture. *Can Tech Rep Fish Aquat Sci* 1949: 1–18.
Simpson, S. 2007. Salmon farms get passing grade. Vancouver Sun, 27th June, 2007.
Society for Positive Aquaculture Awareness (SPAA). 2007. www.farmfreshsalmon.org; accessed August 29, 2007.
Statistics Canada. 2006. Aquaculture industry, by selected provinces. http://www40.statcan.ca/l01/cst01/prim56a.htm?sdi=aquaculture; accessed August 29, 2007.
Stephen, R.C. and Ribble, C.S. 1995. An evaluation of surface moribund salmon as indicators of seapen disease status. *Aquaculture* 133: 1–8.
Stone, J., Roy, W.J., Sutherland, I.H., Ferguson, H.W., Sommerville, C., and Endris, R. 2002. Safety and efficacy of emamectin benzoate administered in-feed to atlantic salmon, salmo salar L., smolts in freshwater, as a preventative treatment against infestations of sea lice, lepeophtheirus salmonis (Krøyer). *Aquaculture* 210(1–4): 21–34.
Strain, P.M. and Hargrave, B.T. 2005. Salmon aquaculture, nutrient fluxes and ecosystem processes in southwestern New Brunswick. Chapter 2 In: *The Handbook of Environmental Chemistry*.

Environmental Effects of Marine Finfish Aquaculture. Volume 5: Water Pollution. Springer, Berlin Heidelberg New York.

Sutherland, T.F., Martin, A.J., and Levings, C.D. 2000. The characterization of suspended particulate matter surrounding a salmonid net-pen in the Broughton Archipelago, British Columbia. Available from Department of Fisheries and Oceans, West Vancouver Laboratory, West Vancouver, BC, Canada.

T Buck Suzuki Foundation 2007. Salmon farming solutions. http://www.bucksuzuki.org/farming_solutions.htm; accessed 25th June, 2007.

Talbot, C., Corneillie, S., and Korsoen, O. 1999. Pattern of feed intake in four species of fish under commercial farming conditions: implications for feeding management. *Aquac Res* 30(7): 509–518.

Tantikitti, C., Sangpong, W., and Chiavareesajja, S. 2005. Effects of defatted soybean protein levels on growth performance and nitrogen and phosphorus excretion in Asian seabass (*Lates calcarifer*). *Aquaculture* 248: 41–50.

Technical Report. 2001. Potential environmental impact of emamectin benzoate formulated as SLICE for salmonids.

Technical Report. 2003. SLICE – Guidelines for the use of SLICE in aquaculture. Vol. 1.

Timmenga & Associates Inc. 2003. Report: evaluation of options for fraser valley poultry manure utilization. Timmenga & Associates Inc. Vancouver, Canada.

Trouw. 1999. Additional reports on environmental safety of Calicide. Report to SEPA by Trouw Aquaculture ARC, 2 Volumes.

Tucker, C.S., Sommerville, C., and Wootten, R. 2000. The effect of temperature and salinity on the settlement and survival of copepodids of *Lepeophtheirus salmonis* (Krøyer, 1837) on Atlantic salmon, *Salmo salar*, L. *J Fish Dis* 23: 309–320.

Tully, O. 1992. Predicting infestation parameters and impacts of caligid copepods in wild and cultured fish populations. *Invertebr Reprod Develop* 22: 91–102.

Tveterås, S., 2002. Norwegian salmon aquaculture and sustainability: the relationship between environmental quality and industry growth. *Mar Res Econ* 17: 121–123.

Van Aggelen, G.C., Linssen, M., and Endris, R. 2003. Toxicity of emamectin benzoate in commercial fish feed to adults of the spot prawn and dungeness crab. *Oceans* 3: 1205–1208.

Volpe, J.P., Taylor, E.B., Rimmer, D.W., and Glickman, B.W. 2000. Evidence of natural reproduction of aquaculture-escaped Atlantic salmon in a coastal British Columbia River. *Conserv Biol* 14: 899–903.

Volpe, J.P. 2001. Super Un-natural. Atlantic Salmon in BC Waters. Pub. The David Suzuki Foundation, Vancouver BC, p. 36.

Waddy, S.L., Burridge, L.E., Hamilton, M.N., Mercer, S.M., Aiken, D.E., and Haya, K. 2002. Emamectin benzoate induces molting in American lobster, *Homarus americanus. Can J Fish Aquat Sci* 59(7): 1096–1099.

Wagner, G.N., McKinley, R.S., Bjorn, P.A., and Finstad, B. 2003. Physiological impact of sea lice on swimming performance of Atlantic salmon. *J Fish Biol* 62(5): 1000–1009.

Wagner, G.N., and McKinley, R.S. 2004. Anaemia and salmonid swimming performance: The potential effects of sub-lethal sea lice infection. *J Fish Biol* 64(4): 1027–1038.

Waknitz, F.W., Iwamoto, R.N., and Strom, M.S. 2003. Interactions of Atlantic salmon in the Pacific Northwest, IV: Impacts on the local ecosystems. *Fish Res* 62: 307–328.

Weston, D.P. 1986. The environmental effects of floating mariculture in Puget Sound. Available from Washington Department of Fisheries and the Washington Department of Ecology, Olympia, WA.

Weston, D.P., Phillips, M.J., and Kelly, L.A. 1996. Environmental impacts of salmonid culture. *Dev Aquacult Fish Sci* 29: 919–967.

White, K. 2006. BC MAL, Personal communication. Data Submitted as Part of the Finfish Aquaculture Waste Control Regulation.

Whitmarsh, D.J., Cook, E.J., and Black, K.D. 2006. Short communication searching for sustainability in aquaculture: An investigation into the economic prospects for an integrated salmon–mussel production system. *Mar Policy* 30: 293–298.

Willoughby, S. 1999. *Manual of Salmonid Farming*. Fishing News Books, London.

Wolf, N.G. 1985. Odd fish abandon mixed-species groups when threatened. *Behav Ecol Sociobiol* 17: 47–52.

Woodward, I.O., Gallagher, J.B., Rushton, M.J., Machin, P.J., and Mihalenko, S. 1992. Salmon farming and the environment of the Huon Estuary, Tasmania. *Tech Rep-Sea-Fish Res Div Sea Fish Tas Mar Lab* 45: 59.

Wydoski, R.S. and Whitney, R.R. 1979. Inland Fishes of Washington. University of Washington Press, Seattle.

Part IV
The Interaction Between Traditional and Local Knowledge, and Modern Aquaculture

Introduction

Were it not for the cod, Europeans might not have settled North America, or at very least, settlement might have followed a much different pattern. Cod brought Europeans first to the Grand Banks off Newfoundland, then down the East Coast of North America past Cape Cod and on to Virginia and the Carolinas. European settlement began in seasonal outposts established to dry fish for shipment to Europe, and grew as settlers saw little reason to return to Europe from the rich fishing grounds. Throughout the four hundred years between discovery and exhaustion of the cod fisheries, knowing where to find the fish was key to fishers' prosperity – knowledge first kept secret by Portuguese fishers, gradually slipping out until the Grand Banks saw fleets from as far away as Russia. The collapse of the cod fisheries brought an end to that prosperity, as stocks once thought to be inexhaustible were rapidly overwhelmed by huge numbers of fishers using sophisticated fish-finding and harvesting technologies (Kurlansky 1998). Other species fared little better: the once abundant Atlantic salmon is now an endangered species in Maine (NOAA 2000), and while the West Coast salmon fishery has so far remained robust, signs of trouble are evident: early sockeye returns in 2007 are 10% of the expected numbers (CBC 2007).

The causes of the collapse of key capture fisheries are subject to heated debate. Fishers in Canada often point to government mismanagement allowing an unsustainable increase in numbers of fishers involved in the fishery, and to deficiencies in the scientific knowledge used in ecosystem-based management approaches. Fishers contend that among the errors found in government management approaches, ignorance and devaluing of traditional and local ecological knowledge represent a squandered opportunity to draw on coastal communities' memory of centuries of participation in the fishery. Local and traditional ecological knowledge may contribute to understanding of potential sites for aquaculture, interaction between aquaculture sites and the capture fisheries, and inter-species interaction in integrated multi-trophic aquaculture. Sources of traditional and local ecological knowledge vary across Canada, from the Newfoundlanders who carry forward the knowledge of their forebearers who first chose to settle in the bays of Newfoundland, to First Nations in British Columbia for whom the fisheries were historically their primary source of food protein.

While reports from these groups vary with their interests in particular fisheries, there is a common thread in their response to 'government science': government-employed scientists are not the only observers of our aquatic resources, and even well-trained government scientists can make mistakes in interpretation of the data they have gathered. A final observation on the relevance of local and ecological knowledge might also address the question of whether coastal communities have in some sense squandered their entitlement to a say in management of aquatic resources by having participated actively, even eagerly in overuse of aquatic resources. One response to that question might be to point again to the value of experience. Even if it is true that traditional fishers have been somehow complicit in mismanagement of our aquatic resources, even experience of how not to fish is relevant to understanding matters of great concern to aquaculture, from siting to species and diets and more. Successful aquaculture, like successful capture fisheries, is knowledge-based and must be careful to learn lessons from the past to enable a better future.

In this chapter's first contribution, Larry Felt discusses ecological knowledge in Newfoundland and Labrador. The story of the relationship is one of difference and controversy, an unfortunate outcome of what Felt characterizes as failed communication. Even where perspectives have been shared, difficulties in appreciation of others' ways of speaking and portraying knowledge have eroded rather than built points of commonality. The results have been predictably disappointing. In the absence of the means to integrate local fishers' knowledge with scientific approaches, fisheries management has not reflected local knowledge adequately and communities have felt excluded and devalued. This situation may yet change as Fisheries and Oceans Canada is increasingly aware of the importance of consulting coastal communities and other stakeholders regarding fisheries management. In the interim, however, there is a gap between holders of traditional and scientific knowledge, and that gap in our approach to the capture fisheries must be understood and given attention in our development of new approaches to management of aquaculture.

Teresa Ryan argues that in the case of the Pacific salmon fishery and any aquaculture, knowledge of fish ought to be considered a natural resource along with the fish themselves. Unfortunately, according to Ryan's assessment of the situation, the institutions charged with responsible management of capture fisheries and aquaculture are not using this knowledge. Rather, a 'top-down' management system leaves valuable knowledge resources underutilized, and so squanders an opportunity to understand better the nature of the resources being managed, and further squanders an opportunity to build First Nations' knowledge into practical decision-making regarding management decisions taken. Until these opportunities are taken up, we risk repeating the story of cod mismanagement in our management of Pacific salmon, and integration of the capture fisheries with aquaculture will remain a back-burner issue even as controversy over aquaculture continues.

Closing these gaps will be difficult because the relevant knowledge systems are incomplete and fragmented, not just within the 'top down' management frameworks operated by departments and agencies within governments, but within knowledge contributors including First Nations. Just as fishers in Newfoundland hold competing views, there is no single 'First Nations perspective,' a problem which Marcel Shepert portrays as a product of a lack of engagement in management decisions. In the absence of First Nations' collective support of a cohesive statement of perspective, the available First Nations views are too fragmented to have much chance of being taken up for integration with the already established, comparatively consolidated view of fisheries managers. Shepert argues for the implementation of systemic measures, such as formalized collaborative and information sharing activities, in order to bring about a system of governance which uses and respects traditional knowledge in governing both capture and culture fisheries. Shepert recognizes that this task will require that First Nations put their own house in order, a task which Shepert takes up in a carefully nuanced fashion, connecting, for example, the role of local knowledge in management of the capture fishery to repeated use of that knowledge in governance of aquaculture. Shepert emphasizes throughout his contribution the importance of knowledge, no matter what its source, to sound governance. As echoed by Felt and Ryan, traditional and local knowledge are different in origin from scientific knowledge, not different in kind or value.

References

Canadian Broadcasting Corporation (CBC) News. 2007. B.C. Salmon Numbers Cause for Concern. http://www.cbc.ca/story/canada/national/2004/09/20/salmon040920.html. Accessed September 13, 2007.

Kurlansky M. 1998. *Cod: A Biography of the Fish that Changed the World*. New York: Penguin.

National Oceanic and Atmospheric Administration (NOAA) National Fish and Wildlife Service. 2000. Wild Atlantic Salmon in Maine Protected as Endangered. http://www.publicaffairs.noaa.gov/releases2000/nov00/noaanfws1113.html. Accessed September 13, 2007.

Chapter 10
"It All Depends on the Lens, B'y"[1]: Local Ecological Knowledge and Institutional Science in an Expanding Finfish Aquaculture Sector

Larry Felt

Introduction

A palpable lesson of resource management in the last twenty-five years is the necessity of meaningfully involving resource users in all elements of the management process. Whether the process is termed 'co-management' (Felt et al. 1997), collaborative governance (Paquet 2000) or any of several related terms, managerial devolvement is necessitated by the inadequacy of states to impose management from afar. If legitimacy and hence increased rule compliance is to occur, local resource users must understand management initiatives, consider them beneficial as well as consistent with their knowledge of the resource and feel they have fully participated in their creation and adoption. An increasingly recognized benefit of such management 'partnerships' is the ability to utilize the extensive knowledge resource users have gained as a result of resource harvesting. Indeed, several jurisdictions have formally incorporated the obligation of governments to utilize local understanding, or as the literature references it, Traditional Ecological Knowledge (TEK) or Local Ecological Knowledge (LEK) in developing and assessing resource plans. While the distinction between LEK and TEK oft times overlaps, most commonly TEK refers to people drawing upon longer time scales frequently of several hundred years (such as with much aboriginal knowledge) and LEK for that acquired and codified over three to four generations as, for example, Eastern Canadian fishermen and women whose knowledge of fish and the sea might date from the four or five generations since European settlement. Following this distinction, the term TEK will be used in this chapter.

Academic and government recognition of local knowledge has led to a significant expansion in our understanding of its structure, content and forms of presentation (Neis and Felt 2000; Berkes 1999). As well, a sizeable literature has arisen comparing and contrasting it to institutionalized natural science typically undertaken

L. Felt
Department of Sociology, Memorial University of Newfoundland, Newfoundland, Canada
e-mail: lfelt@mun.ca

[1]Comment offered by a fisher in explaining the differences in seeing bottom rocks, shadows and fish in a shallow salt water inlet using different colored sunglasses.

© Springer Science+Business Media B.V. 2008

within state management bureaucracies (Felt 2005a; Neis and Felt 1995). Particularly significant in this regard are LEK's differences in temporal and spatial scales, anecdotal or story-like presentation and holistic or ecosystem-integrated form (Hutchings, Neis and Ripley 1996). The most formidable challenge lies in finding ways to utilize local knowledge alongside natural science for management development. While most scientists acknowledge and are sympathetic regarding utilization of local understanding, they are at considerable doubt regarding its utility and legitimacy. As a result, the most common forms of accommodation have involved utilizing the knowledge in ways that assist scientific inquiry rather than providing some degree of alternative to it. Examples include using resource users knowledge to help frame scientific research questions, or considering local knowledge as a useful though distinctly secondary substitute where natural science investigation is lacking. The result is that scientists, despite some desire to do more, treat LEK as a proxy for knowing until their knowledge is collected. While outright disdain is seldom communicated, it is clear that the two ways of knowing are not seen as being equal. A final form of utilization is to ask resource users for their assessments of scientific findings to potentially fill in missing gaps or clarify processes underlying findings. Typically overwhelmed by the maze of charts, graphs and equations and awed by the university pedigrees and technical language of the scientist deferment and deference typically rule (Fischer 2000).

These utilizations place local knowledge in a subsidiary relationship to bureaucratic science. Moreover, all inadequately deal with issues of interpretive dispute and disagreement. A particularly helpful contribution would be to suggest ways of comparing and contrasting differences in observation and interpretation drawing on the respective frameworks of each type of knowledge. Lack of such capability is arguably the most critical issue for achieving meaningful cooperative and collaborative resource management utilizing best practices from each form of knowledge.

To be successful, resolution will most likely need to be based on the mutually recognized legitimacy of different knowledge forms, a reasonably detailed understanding of how each is constructed and some way of preventing the more dominant and powerful claims of science from devaluing if not outright dismissing any competing claims. This chapter addresses these issues and the search for accommodation through focusing upon an emergent aquaculture industry in Newfoundland and Labrador, Canada's most Easternly province. Aquaculture is chosen as a case study due to its nascent status. Through examining an industry relatively early in its development that intrudes upon traditional users, it may be easier to see how conditions supportive of reconciliation and accommodation may be promoted.

Ongoing research (Felt and Oliver 2006) suggests this to be a superior strategy than examining more mature relationships in other resource areas where, virtually without exception, the rhetoric of mutual respect and legitimacy courted to alternative knowledge forms considerably exaggerates actual practice and where processes of knowledge erosion (Felt 1994) appear well-advanced. In this latter situation, possessors of local knowledge, normally through continuous exchange with natural scientists and managers in state-sponsored committees begin, over time, to internalize and verbalize the paradigm of natural science quite often to the devaluation of their own acquired knowledge. As fishers become 'co-opted', they, in effect, become amateur scientists. Over time, this may lead them to question their own understanding, assume tasks akin to research assistants for scientists and even, in more extreme cases, personally reject their TEK and champion fisheries science. New industries thus offer at least two advantages for understanding the relationship between local knowledge and scientific understanding. First, formalized, state-sponsored consultation mechanisms are not yet fully mature and second, there most likely does not yet exist a corpus of 'accepted scientific data' specific to the site to which both scientists and managers are committed to.

The chapter begins with a brief overview of the finfish aquaculture industry in the province juxtaposed with an overview of traditional fisheries in the area. Following this a brief comparison of LEK with what might be termed 'institutionalized fisheries science' relevant to aquaculture is presented. Next, using information from semi-structured interviews with fishers, several areas of disagreement with government and industry are presented along with the local knowledge upon which they are based. Since government and industry positions on these issues are based upon some degree of scientific support, such a focus allows us to more carefully analyze similarities and differences between LEK and fisheries science as a springboard for proposing strategies of accommodation and mutual respect.

Focusing upon four areas of contested knowledge, the chapter examines the claims and counter claims for each. The investigation focuses upon how each knowledge claim was assembled, what limitations might be applicable to each and how any disagreements might be best understood. Through this 'de-construction' of each claim, a clearer understanding of each is sought as well as strategies for reconciliation and accommodation. The chapter concludes with recommendations for better understanding the contribution each form of knowledge makes to aquaculture management and ways to more effectively integrate fishers and their knowledge into resource management.

A Note on Data and Methods

It has been recognized for some time that the extensive knowledge fishers possess is potentially highly relevant for various aspects of aquaculture development as well as ongoing monitoring (Calder 1997). A number of uses for the knowledge have been suggested including cage site selection, anticipating potential pollution, disease, and parasite infestation as well as aquaculture/wildfish interaction. As part of an earlier study dating to 1989, the author has studied the region's fisheries. Initial work focused on describing various fisheries, particularly those for Atlantic salmon, herring and lobster. Nineteen fishers were sampled initially to represent a reasonably representative cross section from the adjacent communities. Cataloguing and understanding LEK was a critical component of that earlier work (Felt 1994). Since that time, intermittent conversations have occurred with all original interviewees, usually at workshops and local fisher committee meetings.

As aquaculture developed, issues associated with fish farming gradually became part of ongoing conversations with fishers. As a result of publicly stated plans by government to expand aquaculture development in the area, during 2001 and 2002 more focused work on the role of LEK and cage rearing was initiated. The majority of the original fishers were contacted and asked to update general information concerning any changes to the local ecosystem as well as more specific questions considered relevant to aquaculture development and expansion. In the interviewing, it was clear that many had been consulted in various ways by government, aquaculture companies or consultants working on behalf of these two groups. As well, local fishers' committees made several presentations on their behalf.

During 2004 more focused telephone interviews were carried out with twelve of these fishers. These semi-structured interviews focused upon their view of the overall evolution of aquaculture in the region with particular emphasis upon the extent to which their views were accepted and implemented by either government or the industry following a 1997 government study on potential areas of conflict between traditional fishing and aquaculture in the province (Government of Newfoundland and Labrador, 1997).

Fishers were asked to explain in detail any areas of frustration and disagreement as well as explain why they felt as they did. Fishers drew extensively on their knowledge of the marine environment in responding to these questions. In addition to the interviews supplementary materials such as local fisher committee reports, media reports, scientific studies and government policy statements and reports were also examined. From these materials, four areas of disagreement between fishers and government were selected. In each case, some level of scientific evidence was available that challenged fishers' claims. These four include: the impact of aquaculture on parasite infestation on wild species, effects of cage rearing sites on local marine pollution in short and longer term, effects of cage site selection on local access to wild species such as scallops and the behavior of escaped aquaculture fish on local wild species such as salmon, capelin and lobster. Before examining these areas of dispute, a brief overview of the fish farming industry and region is useful.

Finfish Aquaculture in Newfoundland

Finfish aquaculture is relatively new to Eastern Canada. It is seen as a more sustainable alternative to the vagaries of ever-expanding, industrially-based wild fisheries (Wright 2001). Following the collapse of Northern cod (Gadus morhua) a seemingly inexhaustible species of groundfish that gave birth to European settlement in Eastern North America, fish husbandry took on more and more promise in Northeastern North America. To that end, the state of Maine and every Eastern Canadian province has undertaken one or more strategic development plans for various 'aquaculture friendly' species. The most extensively developed aquaculture sector in all these provinces is finfish, especially salmonids (salmon and trout). Dominant salmonid species include anadromous rainbow or 'steelhead' trout (*Oncorhynchus mykiss*) and Atlantic salmon (*Salmo salar*). In Newfoundland fish are grown in marine anchored cage sites almost exclusively on the province's South coast in the Central zone where non-lethal winter sea temperatures, and narrow, protective fjord-like bays offer protection from storms, and lack of ice allows year-round (see Map 1).

Steelhead trout were initially selected as the most lucrative species though since 2000 Atlantic salmon production has increased and as of 2005, the last year for which complete data is available, surpassed steelhead trout in both production volume (3000 to 2006 metric tones) and value ($16.4 million to $10.2 million) as can be seen in Table 10.1. The overwhelming majority of product is exported, primarily to the Northeastern United States. Approximately 200 workers are employed raising, processing and marketing these two species, the majority for most of the year. In response to stiff competition in recent years from offshore areas such as Chile, Peru and Argentina, considerable consolidation has occurred with four companies currently in operation in the area and one, North Atlantic Sea Farms, dominating

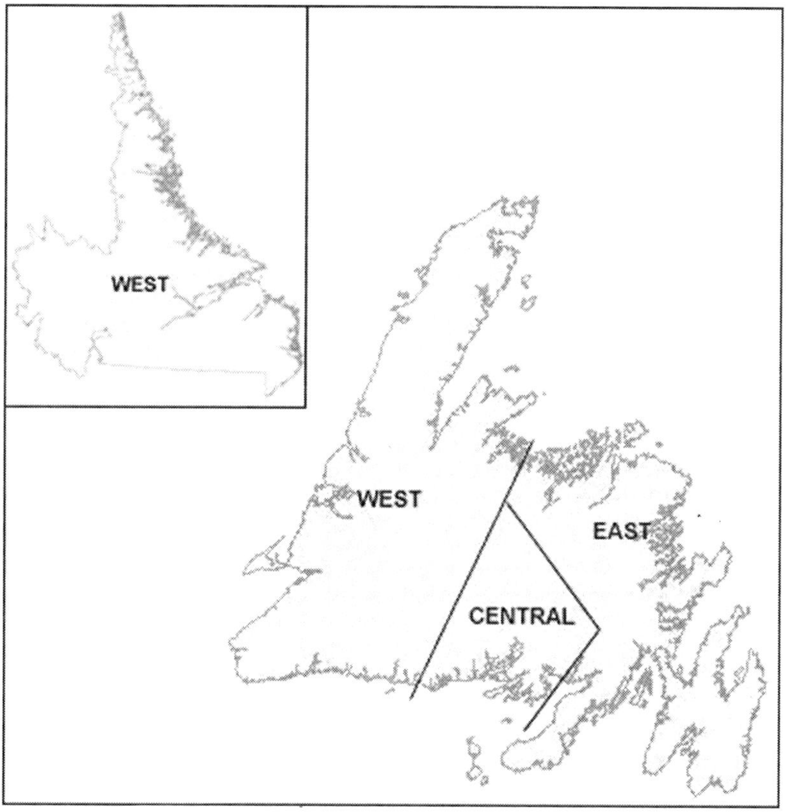

Aquaculture Management Zones

Table 10.1* Aquaculture production and value, 1998–2005

	1998	1999	2000	2001	2002	2003	2004	2005
Atlantic Salmon Production (tonnes)	401	399	670	1092	1270	1450	1990	3000
Atlantic Salmon Value ($'000)	2,924	2,462	4,962	5,200	6,132	6,670	9,500	16400
Steelhead Trout Production (tonnes)	1315	2078	842	1719	1600	1150	1339	2006
Steelhead Trout Value ($'000)	6,919	11,401	5,493	9,752	8,800	6,324	7,480	10200
Blue Mussels Production (tonnes)	946	1700	1051	1452	1700	1300	2298	3157
Blue Mussels Value ($'000)	815	3,799	2,699	3,929	5,500	2,280	5,055	7,000
Atlantic Cod Production (tonnes)	11	105	155	207	227	30	27	19
Atlantic Cod Value ($'000)	83	335	495	525	900	30	108	81

* Source: Government of Newfoundland and Labrador, Department of Fisheries and Aquaculture, Production and Value. http://www.fishaq.gov.nl.ca/Statistics/statistics.stm and 2005 year in review (http://www.fishaq.gov.nl.ca/publications/yir2005.pdf.

production. In November 2004, the provincial government was forced to provide additional supplies of working capital through an Aquaculture Working Capital Loan Guarantee (AWCLG) program as a means of responding to threats of bankruptcy and shortages of working capital for the industry (Government of Newfoundland and Labrador 2007). Despite often precarious financial conditions, the industry has continued to expand with a 50% expansion in production and a 56% increase in market value between 2004 and 2005 alone (Government of Newfoundland and Labrador 2005). Total production is still small compared to Canadian aquaculture production of salmonids at approximately 5–8%. In 2006, the provincial government announced a major expansion in salmonid aquaculture for the region through a joint partnership with Cooke Aquaculture, one of the largest integrated aquaculture industries in New Brunswick. Details of how this expansion will develop remain unclear up to the summer of 2007, however.

Traditional Fisheries in the Region

Until the groundfish collapse of 1992, the region within which finfish aquaculture is centered was an active multispecies fishery with inshore, small boat as well as offshore, deepwater sectors. At its peak, as many as two thousand men and women earned a livelihood from cod, Atlantic salmon, lobster, capelin, herring, redfish, turbot and several other species (DFO 1992). As with many other regions of the province, Atlantic cod (Gadus morhua) was the dominant species in both landed volume and value.

While crab and shrimp have provided some alternatives in recent years, the region has suffered arguably more than many others. For example, out migration from the roughly seven towns within the prime aquaculture area has been among the highest in the province. The two largest suffered a 21.9% and 12% loss respectively up to 2005 (Community Accounts 2007). As is typical, emigration was skewed towards younger, largely better educated, men and women. Fishing, including aquaculture, continues to be the economic foundation for the region though per capita and household incomes have failed to grow dramatically and several near-by communities remain in a state of crisis. Community Accounts data show, average family income, in real terms, has remained constant since 1994 at around $27,000 or roughly 70% of the provincial level and little more than 50% of the Canadian average. While it is difficult to estimate numbers of active fish harvesters and processors, present harvesting and processing employment is certainly no more than 80% and 50% of pre 1992 levels (DFO 2004). Crab and lobster dominate landed value.

Before discussing disagreements and alternative knowledge claims surrounding salmonid aquaculture in the province, it is useful to understand the processes of 'knowing' whereby fisheries science and TEK constructs their respective understanding.

Ways of Knowing: A Brief Overview of LEK and Fisheries Science

While social scientists do not normally reject notions of 'external reality' and 'proof' through the criterion of falsifiability, most support the perspective that all understanding is, in important respects, socially constructed and that this construction underlies what one knows and how one knows it. Groups create cultures and subcultures. Through socialization and experience, these cultures develop paradigms (Kuhn 1970) or integrated systems of values, norms (rules), terms and relationships that prioritize what is important and structure understanding. Such 'knowledge paradigms' mediate between groups and their physical worlds and provide both a context for new situations as well as a repository for integrating new with old. Note that a claim of social construction is not intended to devalue or 'bias' a particular view but rather to assist in how understanding is achieved (Hacking 1999). In this wider view, fisheries science in all its forms as well as LEK are social constructions based on different paradigms.

In the twenty-five years since the path breaking work of Robert Johannes (Johannes 1981, 1989) social scientists have documented the extensive traditional knowledge that resource users acquire in their harvesting of marine resources through their daily, long-term interactions in the ocean. Previous work in the region (Felt 2004) has documented a rich knowledge of local bays and species that inhabit it. This LEK shares many qualities with other works in its vernacular construction (use of locally specific, colorful terminology), integration or holism at the local marine ecosystem level, content, form of presentation, sharedness and localized spatial scale. Since it is drawn from the lived, shared experiences of its practitioners, it is cumulative, inductive, holistic and expressed in language and categories of the culture from which it arises. Since the primary purpose of such knowledge is highly instrumental, that is to catch more fish or lobster, it is continually being 'ground truthed' or tested and is highly sensitive to changes in species sought or the wider marine environment. Finally, knowledge is communicated to insiders and outsiders alike in non-numerical anecdotal (story-like) forms more often than not embedded in colourful humour, spicy irony and acquiescing fatalism.

Consider the seemingly disconnected comment I heard a fisherman make to his son and fishing partner in an enraged voice one evening in another area of the province: 'Take the bloods of bitches to the gallows! We got to get the blueberries off the nets.' The explanation was simple enough. The nets were to be taken to a three-pole tripod (gallows) to be dried in order to remove a parasite thought to infect codfish with bluish-black spots in their flesh (blueberries) and therefore fetch lower prices. To assemble such knowledge in a comprehensive and integrated fashion is no easy task given such subtlety and variety of expression. Fishermen easily utilize their knowledge to draw on paper, even paper bags, bathymetry (for example, known deep holes within the underwater trenches of the region) as well as other important physical and biological features including seasonal variations in fish, tide, water temperature, prey species weather patterns and numerous other ongoing features of everyday life.

The end result is a reasonably integrated understanding of living and physical factors that define the bay as a 'living system.' While general knowledge is widely shared, it would be an exaggeration to say that everyone possesses knowledge to the same extent and that there is no disagreement. Fishermen harvesting specific fish species generally know more about it than those for whom it is an incidental bycatch. Additionally, the type of gear and boat one uses mediates this knowledge to some extent with, for example, 'hook and line' (baited lines of hooks up to a mile long) and gillnet fishermen possessing more detailed knowledge of ocean bottom configurations than trap and seine fishermen. Nevertheless, a surprising consensus does exist on a wide variety of parameters of the local ecosystem.

As extensive and integrated as local knowledge is, it is also pervaded by uncertainty. Many things occur that do not fit past experiences. Until explanations can be found, they are thought to 'just' happen for no apparent reason. This uncertainty underscores much of the fatalism and irony that is forcefully communicated in stories and other means of knowledge transmission.

It is important to understand that LEK is working knowledge ultimately and palpably measured by the number of fish in the boat at the end of the day. As such it is continuously being retested and modified against the catch. If fishing is poor, alternative explanations from shared past experiences are considered. "Am I using all the knowledge relevant to the situation? Is something changing in the ecosystem to push or pull the fish somewhere else?" Scientists will easily recognize these as alternative hypotheses. Given the highly fluctuating nature of the marine environment where species and physical events typically occur at varying temporal and spatial scales, explanations are oft times highly elusive. Sometimes, a considerable amount of time must elapse before the connection between change and explanation is reached. Interestingly, like fisheries scientists, fishers usually require long periods of time before reaching conclusions. A fisher from another area expressed it simply:

> We've seen lots of things round here about cod. Sometimes they just about disappear for a few years. Other times they don't show up where you expect them too. Sometimes you find them somewhere else and other times there are just lean years. You got to keep looking and trying.

The general scientific paradigm underlying fisheries science is well enough known (Fischer 2000; Finlayson 1991; Hilborn and Walters 1982) that extensive discussion need not be given it here. Briefly, fisheries science, in its various forms, is driven by:

- Some form of the scientific paradigm requiring clear, hopefully universal, definition and measurement of relevant concepts
- Focus upon key, paradigm-derived research questions
- Rigorous research designs and sampling procedures
- Quantification
- Strongly comparative at large temporal and spatial scales
- The search for general laws and propositions
- Expression of knowledge in probabilistic, statistical statements
- Corroboration/falsification optimally occurs through replication by independent scientists.

These are, of course, ideal objectives with actual research often falling considerably short of their attainment as numerous studies of scientific effort have demonstrated (Latour 1999; Kuhn 1970; Strum and Fedigan 2000).

Similarities and differences between how fishers and scientists come to make knowledge claims have also been widely chronicled. While the majority of this work focuses upon obvious differences such as anecdotal verses quantified, local verses general spatial scales, and species verses ecosystem focused, none have directly examined what is arguably a fundamental problem if both forms are to be part of a 'knowledge dialogue.' This difference is partially, but only partially, captured in the frequently voiced invidious comparison between anecdotal and scientific information. A fisheries scientist summarized the issue in these words:

> ... I don't think anyone disputes the fact that fishers know a lot. The problem I have is consistency and generality. Science speaks with one voice and it seems as if every time I talk to a different fisher I get a slightly different message.

For the scientist, the very hallmarks of science seemed to be missing in LEK. But does science 'speak with one voice' and does LEK's anecdotal style using stories, moralisms and colorful localized language preclude consistency and generalization? The answer to both questions is clearly NO and understanding why is critical if local knowledge is to be treated as anything other than a cursory, supplemental addition to science incapable of standing very confidently on its own.

In theory, science should NOT speak with a single voice until numerous replications through independent research confirms the adequacy of the research and the interpretations drawn from it. This is ideally achieved through rigorous replication and peer review. Replication is preferably undertaken with both the original data collected as well as new data using the same measurement and sampling procedures. Commitment to a shared set of procedures and interpretive rules allows confirmation and acceptance to eventually prevail.

Probably a majority of actual research does not follow such a path for a host of largely practical reasons. Replicating completed research is financially and intellectually costly. Moreover, unless data is publicly collected it is often difficult to gain access to data for review and re-assessment. Intellectually, the rewards of replication are considerably less as anyone who has tried to publish an article concluding that previous work is largely correct will quickly attest. In those research areas where replication is the norm, for example in emergent areas of scientific knowledge, diversity of opinion and research conclusion is arguably no less varied than talking with a dozen fishers. The result, to return to our discussion of TEK and fisheries science, is that the comparison between the two is all too often between a single piece of scientific research on a subject compared to numerous fishers' conclusions on the same or a similar subject.

In addition to what we might term a highly skewed sampling issue, fisheries scientists largely share a similar language and understate in a variety of ways, the uncertainty and incompleteness of their understanding. A good example can be gleaned from stock assessment science, a fisheries science specialization in which the primary objective is estimating the current population of a stock of fish for allocation and management purposes. It is not uncommon for population estimates to provide a number, or more accurately a range of numbers expressed as a range at a given confidence level, in which the ratio from bottom to top estimate is 300–400 percent. In extreme cases, this might mean the difference between allowing a commercial fishery and complete closure. It would be interesting to see what the results might be if the same stock assessment were undertaken independently and findings discussed and debated among the researchers and other qualified researchers. It is arguable that variability might be no less than among fishers. Since this is not likely to happen and the language of confidence levels and unexplained variation tend to promote a more controlled sense of the problem, the actual variability between the groups is likely exaggerated (Finalyson 1991; Felt 2005).

Where fisheries science is undertaken as part of a state resource management bureaucracy, homogeneity of research results is arguably even more likely. While research inside resource management bureaucracies is exceedingly difficult to conduct, and becoming more so all the time, sufficient work exists, often from retired or previous bureaucratic scientists themselves, to suggest a tremendous set of largely informal and unwritten pressures to homogenize research results and minimize differences in interpretation and conclusion (Hutchings et al. 1997; Finalyson 1991; Felt 2005). Since the invidious comparisons made between fishers' knowledge and fisheries science are largely between fishers and institutionalized, bureaucratic science within a state agency, the difficulty in comparison becomes more acute.

Thus, the contrast between a consistent, single scientific voice and disparate, oft-times contradictory voices from fishers is more contrived than real. There are real bases for creating different knowledge or at least different conclusions such as gear type fished, degree of political involvement in which common posturing is felt required, and location fished among fishers that provide different emphases and potentially different conclusions. But whether this is any greater than occurs in a 'free wheeling' scientific investigation utilizing different research designs and asking slightly different questions is not easy to say. Unless these processes are understood for both sides, LEK is always likely to be devalued compared to bureaucratic or state science for its seemingly disparate, unconnected conclusions as well as the structure of its knowledge...

Having spent some time on apparent differences and suggested reasons for believing the gulf is not as distant as might be supposed, it is important before turning to some concrete areas of disagreement to restate the many similarities between the two. It is, after all, the similarities that provide the pillars for any bridge between the two. And the similarities are considerable, particularly if we consider how scientific knowledge is ideally to be achieved Although the specific processes and manner in which each is achieved are different, LEK and science share being empirical, falsifiable, replicable, dynamic and tolerant of uncertainty. Moreover, both, in theory, acknowledge and promote a holistic, ecosystem focus for understanding, though the extent to which this is practiced differs within as well as between the types.

Fishers and scientists share a clear commitment to evidence whether it results from highly rigorous probability sampling or fish in the boat. While fishers are more likely to create data inductively on a day-to-day, year-to-year or generation-to-generation basis, scientists are initially more deductive in establishing relevant sampling frames and data collection instruments then shifting to more inductive strategies to build time series data bases. At least in theory, both offer the possibility of being 'wrong', whether this be no fish in the boat or subsequently uncovered sampling errors or measurement biases that lead to inaccurate data and/or estimations and conclusions. Both are also committed to replication. For fishers, any understanding is only as good as it works time after time while for the scientist replication occurs as time series are created at annual or some other temporal scale. Both modes of understanding are also open and dynamic. By this, it is meant that they are constructed in ways that allow for change at any point in the framework. Thus, ideally, each knowledge framework is continuously open to testing, retesting and alteration and even transformation if necessary. Finally, both have agnosticism or the lack of knowledge built into them though they are expressed in dramatically different ways—fatalism verses wide confidence intervals.

Awareness of these similarities and differences are critical in understanding differences in conclusions and, more importantly, in preventing differences from precluding greater cooperation and understanding in future management.

Where Minds Diverge: Disagreements over Aquaculture Impacts

Interviews indicate little outright opposition to aquaculture development and some expansion in the region despite a number of specific disagreements on its effects. The major reason for this appears to be the comprehensive manner in which the provincial government has attempted to consult with fishers and develop comprehensive regulations in anticipation of problems that have plagued salmonid cage farming elsewhere in Canada, Norway, Ireland and Scotland (Porter 2005; Government of Newfoundland and Labrador 1997, 2005). In fact, provincial regulations appear to be as progressive and environmentally sound as any in the world and the clear leader in Canada. A detailed comparison of the NASCO (NASCO 2003) position paper on sustainable aquaculture practices and a recent evaluation undertaken by the World Wildlife Fund and the Atlantic Salmon Federation (Porter 2005) suggest that Newfoundland's salmonid aquaculture industry meets most criteria for sustainability and minimal impact on wild fish. In contrast, larger operations in other provinces have largely failed to match this level with the result that Canada as a whole is not given a very strong evaluation in Porter's report despite Newfoundland and Labrador's generally progressive regulations.

The four disagreements chosen for discussion illustrate most of the points thus far made. Although only single scientific studies have been undertaken for each disagreement, general differences in power and legitimacy of science combined with the tendency to emphasize variability rather than consensus amongst fishers has led to fishers' concerns being, if not trivialized, at least accorded little more than cursory attention. And yet, a more careful examination of the disagreements suggest there is little substantive reason to reject their view in preference to government claims. The first of these disagreements concerns possible parasite infestation resulting from the high fish density and insufficient dispersal of waste from cages into the open sea. While several parasites were mentioned by fishers, the most common were sea lice or small, snail-like copepods (crustaceans).

Case 1: Disagreement over Parasitic Infestation

Sea lice is a common term used for a grouping of small crustaceans occurring naturally throughout the world that may use several species of fish as hosts. In small numbers, sea lice typically pose little threat to fish health though in large numbers their blood extraction and feeding can cause damage and even mortality. Greatest danger occurs with young fish, particularly salmonids, in the area of the head where scale development is not present. While there are over 200 different species world-wide, only a few are found in the North Atlantic (Schering-Plough Animal Health 2005). The most common species affecting salmonid aquaculture is *Lepeophteirus salmonis* distributed throughout the North Atlantic and specific to salmonid species. Other species along the Atlantic coast include *Caligus elongates* and *Caligus curtus*. The latter two are not host specific and may infect a wide range of fishes including salmonids. Sea lice often undergo population explosions in the high density cage sited areas. For this reason, sea lice monitoring on wild fish as well as farmed ones is a routine activity. Good husbandry practices along with medication can usually keep lice epidemics from occurring.

Every fisher interviewed claimed to be able to recognize sea lice (though only one knew that there were several species of which only one was host specific to salmonids). Sea lice were perceived to have increased in the area during the last five to eight years by 91% (11 of 12) fishers recently interviewed. Herring, mackerel, escaped steelhead trout and wild Atlantic salmon were all identified as showing more infestation than in earlier years. Six fishers claimed that lice-like crustaceans were also more noticeable on cod and other ground fish close to cage sites while the other four claimed not to possess enough recent experience with groundfish to be certain of any changes. Infestation was said to be particularly high early in the year, typically when nets were set for lobster bait and later in the year when mackerel appeared.

Parasites are routinely monitored at the individual cage level as well as more generally by the provincial government. As well, Fisheries and Oceans Canada, in cooperation with the Conne River first nation, operate an Atlantic salmon-counting fence on an important nearby Atlantic salmon river. None of the three sources report increases in lice infestation. It is not unusual for adult wild Atlantic salmon to carry a few lice as they enter freshwater but daily inspections at the fence do not suggest any significant infestation (T. R. Porter, Personal communication). These observations are based on adult returns, however, leaving open some possibility that infestation might be a significant source of mortality for young, juvenile salmon or 'smolts' leaving rivers in Spring. Given declines in returns of adult wild Atlantic salmon to the area coincident to aquaculture development, juvenile infestation and associated mortality remains at least a possibility.

The Provincial government has stated that protocols in place (Government of Newfoundland and Labrador 2005) minimize the likelihood of infestation and, should it occur, provide countermeasures to address it. Some level of sea lice is seen as practically inevitable given the densities of fish in cage settings. Recent changes in allowable density, that is number of farmed fish per cage by the Provincial government are further thought to adequately address the issue. As well, farm operators report no unusually high levels of lice among caged fish.

Fishers were asked to respond to this information. Unfazed, all said it was largely a question of where and when one looked. One summarized nicely their colleagues views in making a comparison between sea lice and current disagreements over the abundance and distribution of cod:

> They (DFO) tell us there aren't many cod out there since they can't find them in their surveys. Maybe they aren't looking in the right places at the right time. I'm not saying sea lice are at some crisis stage or anything like that (though there are more of them) and I don't know how many are attached to fish in the cages but I see them on fish we catch and I certainly see them on the steelhead that have escaped from the cages that we get in our nets once in a while.

A logical resolution to this disagreement might be to extend sampling to other fish (or at least other life stages in the case of wild Atlantic salmon) and different times of year. Without that each claim remains hedged with a fair high level of uncertainty. Given the greater credibility of science-based observations, however, fishers' observations have been largely dismissed, however nicely phrased, as either error or exaggeration.

Case 2: Pollution from Cage Use

It is common for bottoms directly beneath floating cages to suffer various levels of perturbation. Unconsumed food pellets and waste, depending upon strength of currents and subsequent flushing, accumulates on the bottom. Over time, anoxic sludge-like residue can accumulate providing a low oxygen environment for a variety of sea worms and other low oxygen requiring organisms. For this reason, aquaculture regulations typically provide specific instructions for cage site selection, distance between cages, rotation and fallowing (Government of Newfoundland and Labrador 2005). Provincial government representatives tout current regulations governing these aspects of the industry as among the most rigorous and environmentally sensitive anywhere and to have formed the foundation of suggested international standards (NASCO 2003; Porter 2005).

From both government and industry perspectives, the above requirements are successful in minimizing marine pollution. While waste does accumulate immediately underneath cages, ocean currents ensure it is fairly minimal while limiting use of a particular site further prevents longer term accumulation and possible problems. As part of ongoing monitoring, underwater video surveillance and substrate sampling are carried out. The claimed result is that after a fallow period of a few years, the site area returns to nearly its original condition.

All the fishers challenged this view to varying extents. Their claims relate to both short term and longer term effects on the immediate area as well as zones extending up to two and three kilometers away. All but one recalled seeing discolored water around cages, particularly in the high growth months of summer. Several claimed that currents moved the creamy colored plume away from cages as well. This occurrence continued until heavier seas of Fall. Seventy percent (70%) of fishers claimed that, when winds blew towards the shore, the discolored water sometimes washed up on land and deposited a slimy residue on rocks.

A noticeable odor was also said to be sometimes present around the cages particularly downwind on warmer, summer days. One fisher described it in the following way:

> Oh, I guess I'd say it was something like rotting fish and rotting eggs mixed together or something like that. Whatever it is, it's something that's dead that's for sure ... It's really noticeable sometimes on hot days during the summer when there is a shoreward wind.

Several fishers claimed to have seen pictures of the ocean bottom underneath cages as well. Since local fishers generally have had cooperative relations with local fish farms, the pictures, both video and stills, came from the ongoing video monitoring fish farms to, among other things, monitor breaches in cages due to ocean storms, debris and marine mammals attacks from seals.

Though no one drew a direct causal link, a slight majority of fishers suggested aquaculture sites had effects beyond the immediate location of cages with potentially longer term consequences. The most noticeable effect was in the behavior of certain fish species. Herring, Mackerel and Capelin, all pelagic species traveling in large schools in the upper water column, were said to have altered their movements in the area in recent years remaining further away from aquaculture site areas than in the past. Since all three species were commercially important given the severe reduction in cod fishing, this alteration was noticed perhaps more readily than otherwise. The uncertainty associated with whether these changes might be due to aquaculture is clearly expressed in the words of this fisher who remained skeptical of the link:

> It's really hard to say (any connection). I've seen some of these changes before there was ever any fish farming here but some of the guys are convinced the two is related. It may be but I don't think anyone knows for sure one way or the other.

As with the discussion of parasite infestation, LEK and government science frequently do not address exactly the same dimensions of an issue. Fishers see aquaculture largely in terms of how it effects what they do and in more sensory terms (visual and smell). Government regulations and industry monitoring tend to be more site specific ensuring that fish farm operation does not create permanent damage primarily in the specific areas where cages are situated. In this important sense disagreements are not necessarily contradictory but may even be complimentary to the extent that important effects of aquaculture are 'exported' away from the immediate area of farm operation.

Case 3: Cage Site Selection and Local Access to Wild Fish Species

Many of the direct impacts of cage site selection on access to wild fish have been addressed with regulatory changes in anchor buoy visibility, removal of shore-fast morrings (cables to stabilize fish cages anchored directly to shore), sinking of mooring cables to minimal depths and participation in the Community-based Coastal Resource Inventory (CCRI) in which local fishers and communities compile, with assistance from the Canadian Department of Fisheries and Oceans, coastal maps of key resources and fishing locations (Government of Newfoundland and Labrador 1997). All informants indicated that they had been fully consulted by government and industry on fishing locations and were generally pleased with regulatory changes made to address their access concerns. Future expansion and several indirect effects were mentioned, however, as potential sources of friction.

With aquaculture expansion on the horizon, it is worth noting that every fisher mentioned one or more indirect effects on their fishing practices. Change in pelagic fish behavior was mentioned earlier. Effects on other species were also noted, especially scallop and lobster. Those fishers with lobster licenses noted a decline in areas 'down tide' from sites for a distance of a kilometer or more. A similar concern was expressed for scallop by those fishers harvesting bi-valves. Since lobster sites are typically not near the sandy, more open bottoms preferred for cage sites, effects most likely, if accurate, would be due to either/and effluent movement away from cages by tidal action or location of cages in lobster migration paths where bottom debris posed at least a partial barrier to migration. As one fisher indicated:

> "... lobsters always go up and down but I think there's no doubt that aquaculture has had some effect. Some of the better trap sites aren't very good anymore and most of them are downtide from a farm ... I'm not sure why it happens but there must be something the lobster don't like."

Since scallop fishing appears to occur closer to farm operations, the effects are also thought more direct. Fishers expressed a consensus that adjacent areas (from several hundred meters to a distance of a kilometer) had been affected.

As the primary concern of government and the industry is with the immediate site and the farmed fish, relatively little scientific research has been undertaken to assess fishers' concerns. Wild species harvest statistics have been reviewed for the region but data at this level cannot adequately evaluate changes in specific trap locations (lobster) or trawl drags (scallop). Since fishers appear to have made adjustments in where they fish, the issue is clearly not seen as immediate and pressing. Thus, conversations with government representatives and industry participants revealed little more than the occasional comment from fishers but no general sense of a problem.

Case 4: Impact of Escaped Farmed Fish

Some level of escapement from cages is an inevitable result of open water cage rearing. Given the magnitude of past escapements and the controversy surrounding their impact on the environment in Canadian and other jurisdictions such as genetically diluting wild species through inbreeding, strenuous efforts have been made in virtually all jurisdictions to build better cages, replace monofilament materials at earlier intervals, assess cage condition more often, update data on near-by seals and other marine mammals and install video monitoring equipment to continuously view underwater cage components. Newly introduced provincial regulations and improved industry compliance has dramatically reduced escapement levels from as many as several hundred thousand fish to as few as a few hundred on an annual basis according to government and industry sources. Fishers interviewed concurred that the problem of escaped fish has been reduced dramatically in recent years although large numbers of previous escapees may remain at large. Evidence of this escapee presence is the sizeable 'recreational fishery' for escaped steelheads pursued in the region with many fish exceeding fifteen pounds in weight.

Fishers' concerns regarding escaped fish are threefold. First, previously escaped numbers, particularly of wild rainbow trout create a potentially new source of predation on capelin and juvenile Atlantic salmon, in particular. Secondly, escaped fish may carry unwanted and potentially disastrous diseases to wild stocks. Finally, and of less concern in their mind, is possible movement to nearby rivers where they either inbreed with their wild brethren (Atlantic salmon) or displace wild salmon for preferred spawning and rearing habitat and thus establish a foreign presence (rainbow trout). (The risks of inbreeding and displacement of native populations are often distinguished as separate concerns – as assimilation is distinguishable from extinction.) As this fisher expressed it:

> There's thousands of them out there (in bay), mostly rainbow trout, and you know they're eating something! I've caught them spinning with the kids and when you open them up they're full of capelin, sandlance, sculpin and even food pellets from the fish farming. For all I know, they may be eating very small lobster since about the only place they could get sculpin is on the bottom. People says they eat the young salmon in the Spring too but I've

> never caught many of them then, maybe just the odd one in a bait net . . . Who knows what kinds of diseases they may be carrying too.

While many escaped fish are thought to remain adjacent to cages, fishers also believe that escapees may migrate considerable distances so that impacts may occur far beyond local sites.

Given the notoriety of fish escapement and possible consequences for wild fish populations, the provincial government has made some attempts to systematically understand the behavior of escaped trout and salmon with the objective of developing recapture strategies in event of a cage breach. To that end, at least one controlled escape has been undertaken with attempts to carefully monitor escaped fish behavior, in particular mobility and feeding patterns. While not a full-fledged experiment, two tentative conclusions emerged from that research. The first was that most, if not all, escaped fish appeared to remain very close to cages and that fish appeared to have some difficulty in adjusting to wild food sources. Stomach contents of several recaptured fish showed stones, debris and other objects similar in size and color to the pellets they had been fed in cages. Since most fish used in cage rearing were sterile as well, the overall conclusion was that any negative effect from escapees was most likely minimal, temporary and local.

With regard to the issue of entering rivers to either interbreed with native species or displace local stocks, Fisheries and Oceans Canada maintains a counting fence on a nearby major salmon river and records numbers of escaped rainbow trout and Atlantic salmon entering the river. Numbers have remained low during the operation of the counting fence (June to September) suggesting that if any impact does occur it would be by displacement rather than interbreeding since rainbow trout do not enter rivers until late fall or early winter for spring spawning. Discussions with Fisheries personnel suggest that while unlikely, displacement of salmon from spawning and rearing habitat is more likely than interbreeding (B. Dempson, Personal Communication).

Fishers view these claims with some skepticism. They or their friends encounter escapees at much greater distances than the controlled escapement suggests. Moreover, fish in the estuary are clearly growing larger in size in their view so that whatever initial confusion regarding food might exist for the fish is soon resolved. Until recently, the preponderance of evidence would seem to support government and industry. In the last two years, however, a more systematic controlled escape of farmed Atlantic salmon was made in New Brunswick's Bay of Fundy. The Bay of Fundy is home to the highest density of salmonid fish farming in the world (Porter 2005). Control of escapement was varied to distinguish light (day verses night escape) and tide (high verses low). Several fish were equipped with sonic transmitting devices to track movement (Whoriskey 2004). Several patterns of migration were detected with a significant number of fish moving out into the larger Bay and open ocean within a few days to a week. The dispersal pattern was substantially different than that reported in this region. The exact scientific meaning and implications for these differences are not clear without additional research in these two and preferably several other locations under varying controls and time frames.

Bridging the Gap(s): Lessons For Linking LEK and Fisheries Science

The primary tasks of this chapter have been to use aquaculture development as a context to present what has been termed fishers' Local Ecological Knowledge (LEK), compare and contrast LEK with the typical science underlying state resource management in a specific aquaculture setting and in so doing promote a number of strategies to link the respective approaches to understanding to sustainable and collaborative resource management. Five lessons appear particularly relevant.

Lesson 1: Conflicts Often Reflect Differences in Focus, Issue and/or Scale than Substance

The overwhelming majority of the apparent conflicts between fishers and government research noted in this chapter appear more parallel than oppositional. In other words, they reflect different, though related, issues and questions surrounding a particular topic drawn from different paradigms, articulated in different styles and stated at different temporal or spatial scales. As such, LEK and fisheries science would seem, regardless of their respective accuracy, to be more complimentary than oppositional in many instances—at least until each directly responds to the issues/problems of the other. Such complementarity should be encouraged if a more inclusive understanding is desired.

Lesson 2: Unwarranted and Invidious Comparisons Between LEK and Fisheries Science Must Be Addressed

Mention has been made of how fishers and scientists construct their respective understandings. Given the significantly higher status accorded the latter than the former, fishers are inevitably burdened with credibility and hence acceptability problems. This burden is significantly increased by two factors that further obfuscate any comparisons. These two are (a) the style and language with which each deals with uncertainty and (b) what might be called the 'complaint of consistency.'

Both fisheries science and LEK assign significant roles for 'uncertainty.' As we have seen, however, they do this in what might be termed 'active' and 'passive' ways with quite different public perceptions of control and confidence. These differences reinforce a wider view that science is professional, systematic and successful and LEK quaint, nonprofessional and anecdotal (in the pejorative sense).

Fishers readily acknowledge that there are many processes in the marine environment they can neither understand at a particular moment much less control. These are simply accepted and expressed in various passive stories and expressions. Fate is fate! Uncertainty quickly becomes enmeshed in superstition, avoidance and danger. Since LEK is dynamic and empirical, new experiences and the crafted understanding from it often supplant earlier ignorance but new forms almost as quickly arise. Fate reflects powerlessness and acceptance even if only temporarily.

Scientific presentation of uncertainty is dramatically different. Its reduction, and hopefully elimination, is arguably the primary objective driving the scientific paradigm. Stylistically, this activist quest to increase understanding and reduce ignorance is pursued with the most sophisticated statistical tools, utilizing a universal language and expressed in the most assertive if abstract of ways. One does not say we know very little about why such and such events occur (unless one is trying to get a grant to study it). Rather, results are communicated through statements on the amount of variance or variability explained by ones approach and statistical confidence limits indicating the range of values within which an answer likely falls. The message is clearly that we are making progress and reducing ignorance with full understanding possible at some future time even if we still have a great deal more to know!

Careful examination of many scientific reports suggest that, stripped of the presentational rhetoric, levels of uncertainty remain at least as high as in much LEK and, given the complexity of the subject, reduction not likely to occur any sooner than through the inductive, trial-and-error methodology of LEK. Nonetheless, the styles alone communicate profoundly different confidences in the strategies to wider audiences. Add to this a reticence fishers typically display in public and 'foreign' settings as well as the deference they, along with the public at large, demonstrate towards 'scientists' and it is hard to get the playing field (or consultative committee) level or equal for LEK. An important lesson is that presentation styles and approaches must not bury the important role uncertainty plays in each way of knowing nor the understanding that each, in their own quite different ways, continuously strives to convert it into understanding.

Increasing legitimacy for LEK is further hindered by what might be termed the complaint of consistency. Fishers appear not to agree on most issues. Leaving aside the obvious issues of local vernacular and species pursued (a bit like asking statistical modelers and fish physiologists to reach consensus), the comment reflects a kind of sampling difference when comparing a piece of fisheries research with fishers' views. An analogy might be in a juvenile court where the judge appoints one psychiatrist to prepare a report for the court on the juvenile's character and the defense solicits four people knowledgeable of the juvenile to testify about that same character. The four testimonies are not likely to be identical and, while perhaps sharing general conclusions, may dramatically differ in examples and presentation. In such a situation, both the higher status and the lack of diversity in the court's report will make it an influential document in subsequent decision-making.

Normal comparisons of fisheries science and LEK suffer a similar fate. Most fisheries research is undertaken as a single project. Often only one or two scientists are primarily involved along with technicians and assistants. The result is not unlike the psychiatrist's report above. Scientists often reply that whether one or many scientists undertake research is irrelevant due to their universal commitment to the scientific paradigm. Actual studies of routine science suggest the contrary (Latour 1999; Hutchings, Walters and Haedrich 1997; Finalyson 1991) with considerable debate and diversity of opinion regarding research design, sampling as well as interpretation normal rather than exceptional. Thus, it is not unreasonable to suspect that if twenty scientists each undertook a study of the same research problem at the same or an adjacent site, diversity might be as extensive as fishers. Without recognizing this, comparisons between LEK and scientists are likely to be flawed and often difficult to reconcile.

Where fisheries research occurs within a state bureaucracy, pressure towards scientific consensus is likely to be even greater and, as argued previously, the final product homogenized (see Hutchings, Walters and Haedrich 1997; Finlayson 1991 for examples with Canada's Ministry of Fisheries and Oceans). Managers require specific estimates on fish populations for allocation and other management purposes; politicians eschew controversy and disagreement, particularly within their own ministry. The result is a series of mechanisms and processes that encourage, if not outright demand, homogeneity and consensus. Only occasionally, as when scientists retire or citizens are recruited to sit on review committees does the diversity and disagreement that come to light even though, ironically, diversity and challenging are ideally at the core of the scientific paradigm.

An obvious lesson from this discussion is that context and how knowledge is created and communicated is nearly as important as the content itself when striving to elevate and legitimate the role of LEK in resource management.

Lesson 3: TEK Research Should be Promoted and Supported Independently

Before attempting to bridge TEK and fisheries science, it is desirable to promote each independently. Through such efforts, TEK can hopefully arise as a legitimate form of knowing in its own right. In some jurisdictions, particularly newly emergent political units associated with aboriginal land claim settlements, a movement in this direction is already apparent using the TEK associated with aboriginal residents. A recent report from Environment Canada on TEK and climate change in the Canadian arctic suggests something akin to this strategy (Environment Canada 2006). Once a body of TEK or LEK is assembled, the effort to bridge, potentially even integrate in some sense, is more likely to recognize and respect the validity and utility of the knowledge and thereby ensure it does not become too easily assimilated or co-opted into an ongoing scientific perspective. Our argument here is not that fisheries science deserves denigration but rather that LEK can make important contributions and should, as a result, be supported and encouraged in its own right. Ideally, for this to occur, fishers should be encouraged and supported in getting together to develop consistent views of marine issues and questions drawing on their extensive experience and knowledge. Fisheries scientists should be welcome and encouraged to support, and even assist, such efforts at the same time recognizing and respecting LEK for what it is. A initiative might be for agencies responsible for marine management to foster independent efforts to bring fishers and their understanding formally and comprehensively into the knowledge creating process guiding policy and management decisions.

Lesson 4: Mentoring as a Playing Field Leveler

Lesson four follows from the third. Potential conflicts and misunderstandings are likely to be reduced if scientists more intimately understand LEK and how it is created and tested. The suggestion flowing from our arguments here is that fishery science training should involve mentoring by commercial fishers. Presently, mentoring in the other direction is already well-advanced. It is common practice to encourage fishers to accompany scientists on vessels to sample stocks. While on board they often participate as quasi technicians. Though many scientists reciprocate through going out with fishers, there is great utility to building this into training. One might go further and suggest that would-be scientists, live with fishers for two or three months as a kind of residency requirement since much of LEK is shared, evaluated and modified on shore in conversations with other fishers. This collective, dynamic dimension is likely to be missed if mentoring is restricted to boat rides to set or haul gear. If combined with a modified curriculum in which LEK and TEK are presented as alternative and valid ways of knowing, greater appreciation and understanding might be acquired by scientists early in their careers.

Lesson 5: The Need for New Collaborative Structures

Our last lesson underscores the inadequacy of most previous attempts to cooperatively manage marine resources. If the changes enumerated are to emerge, new structures will be required. Power must be more meaningfully shared subject only to any Constitutional constraints flowing from legislation. Greater legitimacy for both fishers, and particularly their knowledge, must be reflected in the organization and operation of any new collaborative form. Foundations for such new ways of collaboration do exist. Rhetorically, many of the concerns voiced here about engaging fishers led to the creation of the Fisheries Resource Conservation Council (FRCC) following the groundfish collapse of 1992 (see Felt 2005). But more needs to be done. Promising beginnings are also occurring between government and First Nations in which latter with their extensive traditional knowledge (TEK) may assume critical roles in answering important questions and developing management structures. Structures suggested within a new land claims agreement between the Government of Canada, the Province of Newfoundland and Labrador and the Labrador Inuit Association also hold much promise. Arguably, recent experiences with resource failures open windows of opportunity along the lines suggested in this chapter. They should not be missed. A more systematic treatment of these challenges and opportunities as well as specific guidelines for addressing them can be found in Felt and Oliver (2006).

In fisheries management there is also likely much that can be learned from ongoing efforts to collect and utilize aboriginal Traditional Ecological Knowledge in resource management. While the specific forms aboriginal TEK takes vary generally as well as specifically from region to region and people to people, there may very well be important lessons to be learned from dialogues with them that are applicable to aquaculture in settings possessing long standing small scale fisheries. Recent land claims settlement and the constitutionally entrenched rights acquired in recent years have potentially provided a form of empowerment that may facilitate many of the lessons discussed above. Resource management of any type is essentially about finding ways to allow sustainable management to occur while protecting, even enhancing, the lifestyles and cultures of those who harvest and use the resources. Successful management cannot occur without the recognition that people are very much part of the process. Greater recognition of LEK, TEK and the resource users themselves is critical if any long term success in this larger sense is to be achieved.

Conclusion

Local Ecological Knowledge (LEK) represents a new and useful lens to in through which to 'see' and understand the marine environment. Sunglass lenses, like paradigms and viewpoints are neither true nor false. Rather, each accentuates certain features in greater detail and in so doing raises different ways of seeing things. Different perspectives can often lead to new and different understandings. While I do not know how far my fishing colleague meant to 'push' his lens analogy that sunny day, it is an exceedingly fitting metaphor from which to explore new ways of knowing and managing what should be renewable resources for the individuals, families and communities that depend upon them. Whether we all settle for the same lens, with it likely disastrous consequences, is dependent to some extent on the awareness that other lens do exist. To such a quest, this chapter is a modest contribution.

References

Agrawl, A. 1995. Dismantling the Divide between Indigenous and Scientific Knowledge. *Development and Change*. 26: 413–439.

Berkes, F. 1999. *Sacred Ecology: Traditional Ecological Knowledge and Resource Management*. Philadelphia: Taylor and Francis.

Calder, S.C. 1997. *Local Knowledge Matters: Knowledge, Technology and Power in Newfoundland Cod Farming*. Thesis prepared for MA, Department of Sociology, Memorial University of Newfoundland. Spring 1997.

Dempson, B. Personal Communication. June 22, 2005.

Felt, L. 1994. Two Tales of a Fish: The Social Construction of Knowledge Among Atlantic Canadian salmon Fishers. In C. Dyer and R. McGoodwin. 1994. *Folk Knowledge and the World's Fisheries*. Boulder: University of Colorado Press. 251–286.

Felt, L. 2004. Social and Demographic changes to the Baie D'espair region of South Coast Newfoundland ten years after the Moratorium. Paper presented to workshop on Decade of Change: Ten years after the Cod Moratorium. May 22, 2004. St. John's, p. 38.

Felt, L. 2005. You Can't Eat 'paper fish': Social Constructions of Fish in an Unexplained Crisis. In A. Herda-Rapp and T. Goedke, Eds. *Mad About Wildlife*. Volume 2, Human-Animal Studies. Briden:Brill, pp. 223–252.

Felt, L.F., Neis, B. and McKay, B. 1997. Comanagement. In J. Boreman, B.S. Nakashima, J.A. Wilson and R.L. Kendall, Eds. *Northwest Atlantic Groundfish: Perspectives on a Fishery in Crisis*. Bethesda, Md.: American Fisheries Society. 185–94.

Felt, L. and Oliver, S. 2006. Bridging Cultures and Negotiating Expectations: Challenges and Opportunities for Collaborative Watershed Stewardship in Canada. 3rd Annual Valuing Nature National Conference Theme Address. July 6, 2006. Corner Brook, NL. http://www.stewardship2006.ca. Accessed: July 18, 2007.

Finlayson, A.C. 1991. *The Social Construction and Reconstruction of a Scientific Crisis: A Sociological Analysis of Northern Cod Stock Assessments from 1977 to 1990*. Thesis prepared for MA, Department of Sociology, Memorial University of Newfoundland. Spring 1991.

Finalyson, A.C.1994. *Fishing for Truth: A Sociological Investigation of Northern Cod Stock Assessments*, 1977–1990. St. John's: ISER Books.

Fischer, J. 2000. Participatory research in ecological fieldwork: A Nicaraguan study In Neis and Felt, *Finding Our Sea Legs*. pp. 41–54.

Government of Newfoundland and Labrador. 1997. A Study of Conflict Management in the Newfoundland and Labrador Aquaculture Industry. St. John's: Department of Fisheries and Aquaculture, 23. http://www.fishaq.gov.nl.ca/publications/aqua_man_study.stm. Accessed: July 18, 2007.

Government of Newfoundland and Labrador. 2005. Regulations Pertaining to Aquaculture in Newfoundland and Labrador. St. John's: Department of Fisheries and Aquaculture. http://www.fishaq.gov.nl.ca. Accessed: July 18, 2007.

Government of Newfoundland and Labrador. 2000. Strategic Plan-Newfoundland and Labrador Aquaculture: Executive Summary and Listing of Recommendations. St. John's: Department of Fisheries and Aquaculture. http://www.fishaq.gov.nl.ca. Accessed: July 18, 2007.

Government of Newfoundland and Labrador, 2005. Department of Fisheries and Aquaculture. Year in Review. At URL http://www.fishaq.gov.nl.ca/publications/yir2005.pdf. Accessed: July 18, 2007.

Government of Newfoundland and Labrador.2007. Department of Fisheries and Aquaculture. Annual Report, 2004–2005. Found at http://www.fishaq.gov.nl.ca/publications/annual_report_05.stm. Accessed: July 18, 2007.

Government of Newfoundland and Labrador. 2007. Newfoundland Statistical Agency. Community Accounts. Found at http://www.communityaccounts.ca/CommunityAccounts/OnlineData/getdata.asp. Accessed: July 18, 2007.

Government of Canada, Department of Fisheries and Oceans (DFO). 2004. Estimates from data provided by Statistics Division, Policy and Economics Branch, Newfoundland Region.

Government of Canada, Department of Fisheries and Oceans (DFO). 1992. Overview of Fishery. Newfoundland Region. Available through Statistics Division, Policy and Economics Branxh, Newfoundland Region.

Government of Canada, Environment Canada. 2006. Ways of Knowing and Understanding: Towards the Convergence of Traditional and Scientific Knowledge of Climate Change in the Canadian North. http://www.msc-smc.ec.gc.ca/saib/atmosphere/atmch_e.html. Accessed: July 18, 2007.

Hacking, I. 1999. The Social Construction of What? Cambridge: Harvard University Press.

Hilborn, R. and Walters, C. 1982. *Quantitative Fisheries Stock Assessment*. New York: Chapman and Hall.

Hume, S., Morton, A., Keller, B.C., Leslie, R.M., Langer, O., and Stanford, D. 2004. *A Stain Upon the Sea: West Coast Salmon Farming*. Vancouver: Harbour Publishing. PP. 288.

Hutchings, J.A., Neis, B., and Ripley, P. 1996. The Nature of Cod (Gadus morhua): Perceptions of Stock Structure and Cod Behavior by Fishermen, 'Experts' and Scientists from the nineteenth Century to the Present. In D. Vickers, Ed. *Marine Resources and Human Societies in the North Atlantic Since 1500.* ISER Conference Papers #5. St. John's: ISER Publications.

Hutchings, J.A., Walters, C., and Haedrich, R. 1997. Is Scientific Inquiry Incompatible with Government Information Control?. *Can. J. Fish Aquat Sci.*51: 2363–78.

Johannes, R. 1981. *Words of the Lagoon:Fishing and Marine Lore in the Palau District of Micronesia.* Berkley: University of California Press.

Johannes, R. 1989. "Introduction" In Johannes, Ed. *Traditional Ecological Knowledge: A collection of Essays.* Gland, Switzerland and Cambridge

Kuhn, T. 1970. *The Structure of Scientific Revolutions.* University of Chicago Press.

Latour, B. 1999. *Pandora's Hope: Essays in the Reality of Science Studies.* Cambridge, Mass.: Harvard University Press.

North Atlantic Salmon Conservation Organization (NASCO). 2003. Report of the Sub Group on Salmon Co-operation (SalCo-op). January 2003.

Neis, B. and Felt, L.F. 1995. Fisheries Science and Local Ecological Knowledge in the Northwest Atlantic: Building Bridges. In A. Davis and C. Fawcett, Eds. *Japanese and Atlantic Canadian Fisheries Social Research: An International Seminar.* Antigonish, Nova Scotia: St. Francis Xavier University, pp. 6–25.

Neis, B. and Felt, L.F. 2000. *Finding Our Sea Legs: Linking Fishery People and their Knowledge with Science and Management.* St. John's: ISER Press.

Paquet, G. 2000. The new governance, subsidiarity and the strategic state. Paper delivered following the OECD Forum on 21st century governance: power in the global knowledge economy and society. Hanover, Germany, Spring 2000.

Porter, G. 2005. Protecting Wild Atlantic Salmon from Impacts of Salmon Aquaculture: A Country-by-Country Progress Report. Washington, D.C and St. Andrews, New Brunswick: World Wildlife Fund and Atlantic Salmon Federation, 56. http://www.asf.ca/docs/issues/impacts2005.pdf. Accessed: July 18, 2007.

Porter, T.R. Personal Communication. January 23, 2005.

Schering-Plough Animal Health. 2005. Product and Disease Directory. http://www.spaquaculture.com/default.aspx?pageid=536. Accessed: July 18, 2007.

Strum, S. and Fedigan, L.M. 2000. *Primate Encounters: Models of Science, Gender and Society.* Chicago: University of Chicago Press.

Whoriskey, F. 2004. Report of Atlantic Salmon Federation Research Initiatives, 2004. Atlantic Salmon Federation Annual General Meeting, Montreal 13, 2005.

Wright, M. 2001. *A Fishery for Modern Times.* Toronto: Oxford University Press, 196.

Chapter 11
"S'kuu See": Integrating Forms of Knowledge

Teresa Ryan

This chapter explores the political and practical context of knowledge claims about fisheries, and argues that integration of scientific, local ecological, and aboriginal knowledge is key to successful fisheries management now and in the future. The core elements of the current context can be seen in the views of just a few decades ago when society in general held the view that commercial fisheries industries were doomed because they were "managed" by the fishers themselves (Brown 2005; Royce 1985). The rapid decline of fish populations was certainly precipitated by increased mortality to fish, however, fishing is not the only cause of the decline of many species, particularly anadromous Pacific salmon (*Oncorhynchus sp.*). As state control over commerce took shape, it seemed appropriate that those with a direct interest in the commerce likely had a better grasp on the management of its activity. The fishers were somewhat included in early agency management yet over time they were marginalized away from decision-making. The early years of fisheries management by state agency included development of legislative instruments necessary for delivery of programs, budgets, and administration (Anderson 1998; Brown 2005; Harris D. 2001). Through trial and error, as these programs developed with constrained budgets and changing administration, the contribution of fisheries related science has slowly expanded. The evolution of fisheries science and its application in policy is in some ways analogous to modern medicine: both have evolved from an expanding knowledge base. The early years of fisheries science initially focused on enhancement activity for stocks in decline, Pacific salmon and halibut on the West Coast and Atlantic cod and haddock on the East Coast. International commissions were established by the 1930s on both sides of the North American continent for specific ocean areas and to conduct specific management activities including science research (Anderson 1998; Brown 2005). The integration of science into domestic fisheries management and policy is often constrained, such as the case of the predicament for management of Atlantic cod (*Gadus morhua*) fisheries (Harris M. 1998). As agency management evolved, local fisher participation

T.Ryan
Gitlan, Tsimshian , PhD Candidate, University of British Columbia, Institute of Resources, Environment and Sustainability
e-mail: ryan4@interchange.ubc.ca

K. Culver, D. Castle (eds.), *Aquaculture, Innovation and Social Transformation*,
© Springer Science+Business Media B.V. 2008

in decision-making was marginalized and in most cases outright removed. That was undeniably the intent towards Aboriginal fishers in Canada, especially with specific legislative instruments. As Felt (this volume) notes, the "palpable lesson ... is the necessity of meaningfully involving resource users in all elements of the management process...." and then provides a description of gathering knowledge from local fishers, typically meant to inform decisions, as opposed to actually engaging resource users in decision-making. The current decision-making process has repeatedly produced multiple discontented fishing sectors often conveyed through public media with acrimonious snippets. Although agency management is conducted by public servants, the current decision-making process does little to improve public confidence. The resulting fiscal-inertia is simultaneously compounded by the inability for institutional management to comprehensively gather and evaluate information relevant to requisite decisions. The resource users are critical to the decision making process, not just the information gathering elements of imposed management.

The transition from fisher participation to agency administration has created massive rifts between the agency and the fishers largely because the *administrators* of fisheries management are not involved in the day-to-day activities of fishing, nor do the benefits of fishing activity have any direct bearing on them. The administrators are part of an extensive bureaucracy intended to deliver information to the decision-makers, and in Canada under the present *Fisheries Act*, this is fundamentally one administrative position (Minister of Fisheries and Oceans). The bureaucrats rarely if ever get fish scales stuck to the backs of their hands. On the other hand, contemporary fishers are often uncomfortable sitting at a desk scrolling through spreadsheet data to calibrate stock assessment models. Some fishers, however, are leaping from the boats to the books and desks, largely as the result of depleted fishing opportunities.

For the purposes of this commentary, agency administration of fisheries management refers generally to government control (state) as an agent (agency) of legislative powers to achieve specific objectives (purpose and goals). The change from "fisher participation" to "state agency management" facilitated a top-down approach in constraining fishing and fishers, along with affecting dynamic free market trade. In the most recent past many fisheries management agencies have resolved to manage for sustainable fisheries in concert with some eco-thematic value, generally ecosystem management. This resolve may have emanated from the United Nations Food and Agriculture Organization (FAO) ministerial meeting that produced *The Rome Declaration on the Implementation of the Code of Conduct for Responsible Fisheries* (1999) and states in article "6. The Ministerial Meeting noted that greater consideration should be given to the development of more appropriate eco-system approaches to fisheries development and management." The Department of Fisheries & Oceans Canada (DFO) 2005–2010 Strategic Plan: *Our Waters, Our Future* (Strategic Plan) identifies changes to planning and reporting structures that includes "Healthy and Productive Aquatic Ecosystems." In the DFO Strategic Plan that provides context for department planning and for Science and Technology development it states, "... we will continue to move towards an ecosystem approach and take advantage of opportunities provided by new technologies ... " (DFO 2005). Yet even this thematic shift is not clearly for the final step toward socially, economically, and scientifically sound fisheries management. Not all elements within an ecosystem can be described with attributes that are applicable to resource management decision-making. Ecosystems are beautiful and they should be *conserved* at massive spatial and temporal scales primarily because of the services they provide to maintain the Earth's biosphere and also for the perpetuation of resource availability. How that can translate into the current resource management paradigm for a particular resource use is elusive. The science of ecosystem functions and services continues to evolve as knowledge expands; however, the linkages between species-species, species-habitats, and habitats-habitats are not easily defined or described with institutional knowledge for the expediency of policy engagement. The current resource management paradigm has an expectation of expediency in delivery of information pertaining to decision-making processes, making it virtually impossible to incorporate *"unknowns"* such as ecosystem linkages. The decision-making process itself must be changed to integrate holistic strategies that minimize risk to protect ecosystem functions and services.

The current hegemony between state agency and resource users has less to do with differences in forms of knowledge and more to do with legitimacy of the *rationale* leading to decisions. Often the rationale is precipitated by constrained budgets as well as capacity, *and* the inability to integrate knowledge from various forms.

Local Ecological Knowledge

Local fishers have been subject to the constraints imposed by state agency management for determining when, where, and how to fish, and how much fish to take. Many policy reactions to reduced fish populations have involved constraints costly to local fishers such as capture avoidance, live-tank revival systems, and, in some cases, improved monitoring. The costs of fishing have sky-rocketed in the last few decades largely as a result of increased costs of fuel and vessel purchase and maintenance, the imposition of increasingly complex rules and regulations, and also as a result of depressed fish prices. For the local fisher, times have been tough. After they finally get to leave the dock, grumbling along the way about the areas that are "open" and the allowed duration of anticipated fishing, they once again reinvigorate their sense of connection to the sea and the fish. Their sensory systems are heightened and their knowledge becomes their only reference to time (other than daylight); it is their time. It is a journey of survival, and with the right conditions, blessed with an abundance of fish.

The decisions that have directly impacted these fishers have often caused anxiety because of the *rationale* used to determine the decision. In many cases the decisions are based not on balanced knowledge but urgent *need* to make decisions that control exploitation in the absence of knowledge. The absence of adequate research and infrastructure funding to improve knowledge is another pervasive characteristic of the bureaucratic rationale, serving to further exclude fisher knowledge. The fishers have developed a repertoire of Local Ecological Knowledge (LEK) that could facilitate evaluation of alternatives to address the *need* perhaps more quickly than conducting assessments based on bureaucratic implementation of an exclusively empirical methodology.

Although Felt (this volume) states "... jurisdictions have formally incorporated the obligation of governments to utilize local understanding..." as a collaborative effort to improve knowledge, the obligation has not facilitated integration. The typical process in the recent past for Canada consists of the agency developing "draft" Integrated Fisheries Management Plans for the conduct of various multiple sector fisheries. These "draft" plans are reviewed in "consultation" sessions with various groups before being forwarded to the Minister for approval. The document captures decisions for exploitation rates, enhancement objectives, and enforcement and so on for the anticipated season. Those decisions were not made with local participation. Furthermore, the process is informal through the conduct of meetings after release of the draft without any opportunity to provide substantive formal comments/recommendations. The fishers want to have an active participatory role in the decision-making process to improve the *rationale* used to ascertain decisions. It is perhaps through the assertions of this desire that LEK and other forms of knowledge have actually become categorized differently than institutional knowledge. The incorporation of LEK to improve knowledge that informs decisions does not actually achieve the desired participation. In fact, as Felt (this volume) and other authors have noted, LEK as a form of knowledge is often marginalized as anecdotal and supplemental to institutionalized practice of gathering information. Even authors who intend to acknowledge the value (or higher value) of forms of knowledge other than science depict institutional knowledge as if it is superior by suggesting hierarchal integration of LEK and other forms of knowledge into the scientific form of knowledge (perhaps unknowingly; Campbell 1996; Neis et al. 1999; Stevenson 1996).

Forms of Knowledge

Differences in knowledge delivery may be difficult to recognize from outside the cultural experience from which it is derived, as knowledge is often constructed through specific language and a variety of cultural influences. As many scholars have argued, knowledge systems are derived from physical, psychological and social syntheses of cognitive interpretation about the experiential environment (Maturana and Varela 1980; Solomon and Higgins 1996; White 1967). Those experiences are often synthesized through social interaction that simultaneously validates, reinforces and expands the knowledge. Communication within the context of knowledge and exchanging knowledge is dependent entirely on shared language skills and cultural symbolism (Ryan, dissertation in progress).

The evolution of indigenous societies on coastal western North America was exceptionally successful in developing elaborate cultures with an abundance of wealth in diverse and complex ecosystems. It is no small coincidence the success can be attributed to the transfer of knowledge. It required extensive knowledge to survive and prosper in the maritime environment along with extensive networking throughout the coast and the interior continental margins. More than sixty distinct languages were spoken along the North American west coast (Helin 2006). In each of the coastal language groups, knowledge flowed through use of the language, stories, and law. These languages transcend the ancient uses of natural resources in highly complex ecosystems in a manner where specific words actually describe attributes of ecology. The local languages facilitated interconnectedness with the natural environment in understanding the natural rhythms and cycles of abundance. It is often difficult, if not impossible to translate the embedded values from one language to the English language. Shepert (this volume) also recognizes the difficulty in decisions for risk management as value-driven with differing values.

Aboriginal Knowledge

The cultural language-English translation expressions to describe profound knowledge-based environmental attributes have become known as "Traditional-, Indigenous-, Aboriginal- Ecological Knowledge" (T/I/AEK) or "Aboriginal Traditional Knowledge" (ATK). For the purposes of this discussion T/I/AEK and ATK are treated as Aboriginal Knowledge and will be referred to as AEK. The social construct of AEK and its derivation are distinctly different than LEK although many similar themes permeate both forms of knowledge. As Felt (this volume) acknowledges, some authors describe the main difference as the element of time giving AEK a longer life life-span. A common concept regarding AEK is that it is only elders that possess the knowledge form (bless each and every one of them). The source of AEK is embodied in the cultural context (e.g., social construct) it is expressed in with local "rules and regulations" spoken in a specific language. It is a knowledge form that is transmitted with cultural value for the learning experience, including natural resource use for users of all ages.

The "traditional" aspect of AEK is temporal and comes from generations of practices both learned from one generation and taught to another successively (the other terms also convey the element of generations). The "ecological" aspect is derived from the local area context of observations on multi-trophic scales. The "knowledge" aspect is systemic and resonates through cultural language with social values, structures, and human interactions. The resource *users* possess the knowledge and exercise its continuation, validation, reinforcement and expansion. Those resource users that have many years of experience often have a more richly characterized context of specific aspects of their knowledge, as is found in the knowledge of elders. The knowledge-base has also been recorded through elaborate symbolism (Ryan, dissertation in progress; Shepert, this volume).

It took many years for others to ascribe value or legitimacy to AEK as a knowledge source, partly because of the social constraints imposed on Aboriginal participation in economic activity. In fact, even well-intended recognition of AEK use has characterized AEK less in relation to economic activity and more in relation to "culture, lifestyle and identity" (Campbell 1996). The other side of recognizing AEK comes from the perspective that as Aboriginal knowledge has been conveyed to non-Aboriginals very little, if any, benefit has been reciprocated. The implications of using this knowledge-base have multiple dimensions and are considerably more important than superficial application reveals. Elements within the knowledge-base may describe specific attributes of ecosystems that if incorporated into projects and site specific purposes assist in understanding some of the ecosystem complexity (Stevenson 1996). A critical and often overlooked aspect of AEK is that it is *living* in the dimension of time. It is derived from activity that is integral to economy whether it is local or geographically extensive. The knowledge-base emanates from territorial jurisdiction of local and extensive geographic occupation and use (Weinstein 2000). The element of jurisdiction is not manifested in the form of knowledge referred to as LEK, making this distinction one of the more profound differences between the two knowledge sources. Clearly, the economic use of resources in Aboriginal territories is the center of conflict, not just simply remedied with an integration of knowledge.

Industrial Backwash

The marine environment on the Pacific coast is well known for its biological diversity and high abundance of species, aquatic and terrestrial. For thousands upon thousands of years these diverse ecosystems provided unprecedented opportunities for harvest. The actual harvest occurred in well-orchestrated activities that took considerable planning and understanding of the complexity of species population dynamics and habitat interactions. Massive fishing weirs were constructed to corral fish into enclosures and kept there for capture according to need (Harris, D. 2001; Newell 1993). That need included more than subsistence quantities because salmon was also a currency for acquisition of other products such as food, canoes, nets, and clothing. Most marine weirs were generally located in near shore areas for easy access, in areas where flows are slower than the open straits. Their use was so pervasive throughout the salmon waters that early European processors (canneries) perceived a threat to their existence and pressured government to ban them. Harvesting fish by other means was also common and once the weirs were banned, other gear was employed. In fact, processors/canneries were quick to employ Native fishers for their extensive *knowledge* of the resource. That worked well for providing employment and the opportunity to bring home fish, at least until the development of legislative instruments to curtail Native fishing entirely. Those legislative instruments not only attempted to prevent access in Aboriginal territories but many of them facilitated development of contemporary industries to enter these same areas. The Aboriginal people on the BC coast have witnessed destruction of ancient Aboriginal territories without their involvement. The activities were not really much different than those that supported elaborate societies however the impunity with which they occurred left much to be desired. The fact that these activities have occurred without engagement of the local groups is perhaps the most egregious offence. Some of the activities have had local and cumulative negative environmental impacts without any restoration mitigation leaving an industrial backwash their wake (i.e., extensive clear-cut logging, former mining, urbanization, etc.).

Enter at Your Own Risk

The BC finfish farming industry entered the coast amidst this backwash of previous activity. The industry entered the south coast and Broughton Archipelago at a time when environmental impacts started taking their toll on resource availability. These impacts may be from multiple sources such as logging, mining, fishing, urbanization and environmental impacts such as global climate change. The farming industry entered the area without participation of the local Aboriginal groups in *their* territories in most cases. In fairness, some operators indeed have partnered with local groups, and others may have offered. Regulations from the permitting agency regarding the industry were also not as comprehensively developed as they now stand. The fledgling entrepreneurs were a bit underdeveloped in their knowledge of local environments during the early years, including effective fish husbandry in these waters. The contemporary global industry is only about 40 years old.

In a recipe fashion, if you take the multiple and cumulative environmental effects from the industrial backwash, the inability to access resources in traditional territories, fuel it with environmental activism, shake with media sensationalism and *voila ici*, the results have been catastrophically generated *fear* and an inability to foster productive resolutions to either perceived or real concerns. Do salmon farms pose a significant risk in BC coastal waters? Robson (2006) has very eloquently answered that question because and space is limited, it will not be described here. The "pro/anti" approach is frustrating for resource managers to have to wade through rhetoric instead of spending precious resources on progressive innovation. Several of the items that Shepert (this volume) refers to as issues facing the industry have been addressed through improved feed conversion ratios and physical monitoring of feeding activity, daily inspections of double enclosed-net pens, and pathogen (disease and parasite) monitoring. Yet the "campaigns" regarding the industry have managed to distort information. For example, one of the disease pathogens, infectious haematopoietic necrosis (IHN) is an infectious disease of *wild* fish and the fear is actually based on production facilities coming into contact with infected *wild* fish. The juvenile fish planted at farm sites go through extensive sterilizing processes, much like hatchery fish, so that when they are placed, they have no diseases or other pathogens.

Science

The BC provincial government (2007) lifted the moratorium on fin fish farm expansion dating from September 12, 2002 based on recommendations from the Salmon Aquaculture Review 1997 and with caveats for additional research to help target answers for concerns expressed before the seven-year moratorium was implemented. Some scientists were conducting research at the time, and continue through research facilities such as the Centre for Aquaculture and Environmental Research, University of British Columbia, the BC Centre for Aquatic Health Sciences, the Pacific Biological Station and Bamfield Marine Sciences Centre, DFO. Additional research is being conducted through various organizations and other academic institutions as well. It takes years of research and volumes of data to begin understanding the dynamics of species, populations, and their habitats. The research activities are expensive and largely restricted to spring and summer on the rugged coast. The general nature of scientific methodology is to generate a refinement of question through multiple data collection cycles to advance and reject hypotheses to predict an inference.

For example, the present author conducted projects as Principle Investigator to collect baseline data of the Tsimshian marine environment (North Coast juvenile salmon habitat distribution and sea lice loading 2003 and 2004 projects) as an information set, something that was not done prior to siting of farms in the Broughton Archipelago. The data set collected from two years of projects does not contain enough information to formulate a null hypothesis, nor was that the purpose of the Tsimshian projects to date. These projects serve only as an information source as they were intended and also because of data collection biases (non-random, non-stratified, and observer). As Shepert (this volume) points out, "... science can only guide, not direct, difficult decisions in risk management ..." (see also Brown 2005).

Many industrial activities are anticipated on the North Coast of British Columbia, in an area that continues to contain high quality environmental conditions. The Tsimshian baseline projects also incorporated capacity building and skills development to prepare for potential industrial development. The remoteness of the coastal communities makes it difficult to access infrastructure services (health, education, fire/police, etc.). Most living requirements have to be obtained outside of the community because there is no economy to support infrastructure in the community, mostly comprised of Aboriginal people. In the last couple of decades some Department of Indian and Northern Affairs and other federal agency programs have developed to focus on "capacity building" although without strategic direction. One program clearly is an exception, the Fisheries and Oceans Canada Aboriginal Access to Resources and Oceans Management (AAROM; DFO 2007), although it too has challenges of jurisdiction.

Aboriginal capacity building is generally viewed as a one-way pragmatic direction of knowledge transfer *to* Native people. This is evident in that welfare-dependency capacity-building programs have no recognition of reciprocal knowledge exchange. Genuine capacity building should consist of more than just providing dollars to "fix the problems." The Aboriginal knowledge base is extensive yet remains *untapped* for the benefit of Aboriginals. The knowledge of resource use on the North American continent, including fishing, forestry, agriculture, and mining was learned *from* Aboriginal people. The coast of BC continues to have abundant opportunities although constraints to *access* limit the genuine capacity development through economic expansion for Aboriginal communities. Real capacity building will occur when the constraints are removed and capital-investment, operational-infrastructure and employment-generation are actively infused into building economy that actually contributes to GDP.

One of the most southern Tsimshian communities, Gitasoo (Kitasoo/Xais xias at Klemtu, BC; 2007), has successfully progressed in aquaculture development. Their unique story started with investing revenue that resulted from commercial fishing opportunities and has advanced to provide approximately 70 jobs in the community and a processing facility. Environmental monitoring continues on a regular basis at their open net-pen sites to ensure sustainable operations and to mitigate the *potential* risks associated with the environmental footprint of aquaculture operations. The progressive nature of economic development in this community is evident in the retention of residents. Out of the seven Tsimshian communities, this community retains nearly 70% living in the traditional territory where the other six communities are nearly the opposite with approximately 70% (in some cases up to 90%) living away from the community.

Own-Source Revenue

Before European contact on the western coast the terms "unemployment" and "welfare" did not exist as community members interacted collectively in the production of goods and services. The quality of life was considerably higher than compared to current general poverty conditions. Not only were individuals highly mobile but large groups of people traveled extensive distances for the purposes of commerce (and in some cases, war and raiding) and returned home because they had a stable, *and sustainable*, economy.

In many of these pre-contact coastal societies resource management was essentially a top-down state agency approach (Harris, D. 2001; Weinstein 2000). However the decision makers relied upon the extensive knowledge base of the resource users. In many cases, the decision makers participated in resource harvesting, received direct benefit of the activity and amassed substantial wealth. The knowledge base however kept the exploitation within manageable limits. Awareness of potentially severe consequences of error motivated decision-makers to integrate conservation and stewardship ethics to ensure successive generations (and years within generations) continue to receive the same benefits. By using access to natural capital to integrate knowledge into decision making, the historic harvesting activities did more than provide food. It was the skills of animal husbandry and the use of extensive weirs, also known as aquaculture (Harris, D. 2001), which supplied the natural capital for coastal communities. The revenues were derived from own-source production and created substantial wealth. Many communities diversified operations with various fin-fish and shellfish husbandry.

Environmental Integration

The Aboriginal knowledge of the environment is analogous to management of corporate operations. Without sound planning and efficiency, profit margins are elusive. A comprehensive understanding of how the operations work in tandem with providing production yields significant return on investment of *natural*, *human*, and *working* capital. The people living in coastal environments continue to rely on their knowledge base for their survival and their quality of life. The environmental and stewardship ethics that guided use of resources for thousands of years caused unprecedented success for development of a sustainable economy. The changes in the last 150 years have negatively affected sustainability. In fact, many of the non-aboriginal community members have started to feel the pinch of economic starvation resulting from external state agency resource management.

How are new economic opportunities developed with environmental integration? Another example is being set on the Tsimshian North Coast in a contemporary context. The Gitxaała have endeavored to enter into the aquaculture industry with applications for three operational sites in perhaps the most progressive approach to date in BC. Gitxaała have identified a method to integrate ancient environmental and stewardship ethics into decision-making to facilitate realizing new opportunities. By using contemporary applications and ancient wisdom, the Gitxaała Environmental Monitoring (GEM) program is capturing the wealth of information in AEK while simultaneously providing a tool useful for their decision making. Many activities are scheduled to develop in the north and Gitxaała is positioning itself to realize opportunities on their terms, in their territory without interference from external organizations (including *pseudo*-First Nations coalitions).

The GEM program is a combination of environmental modeling, training and skills development, and economic development driver. The modeling portion contains elements of mapping, however integrates mapping beyond a two dimensional representation into three (context depth) and four (time and movement) dimensions. The goal of model development is to provide a working scenario, rules-based methodology to examine alternatives for decision-making in Gitxaała terms. This particular model will capture AEK in an unprecedented manner.

The training and skills development portion consists of real time hands-on experience working in the environment with the credentials of accredited programs to advance skill levels and foster innovation. The model requires real data from baseline and monitoring data collection over time. The North Coast baseline data collection will continue through this program. The geographic area and its physical attributes are expansive and require specialized skill sets for vessel operation, data collection and analyses, and comprehensive understanding of the environmental attributes and amenities.

The economic driver portion of the program facilitates identification of production activities including location of fishing activities, sites for aquaculture facilities and operations, safe transportation corridors, mining possibilities, tourism amenities, and power generation, just to name a few. These sectors also require the sophistication of specialized skill sets, including business operations. The results will have profound benefits on the ability for Gitxaała to be economically self-sufficient (again) with an improved caliber of skill sets that effectively perpetuate employment generation. The program is a compilation of *knowledge* for Gitxaała to use on their terms through their optic lens to determine decisions of resource use.

Conclusion – "S'kuu See"

A fundamental difference with Felt's commentary from this perspective is, "If legitimacy and hence increased rule compliance is to occur," *decision-makers* must understand the resource and the knowledge of the resource users to effect sound resource management, conservation and sustainability. Aboriginal groups' knowledge of the aquatic resources and environment is based on complex associations of geographic occupation, stringent Aboriginal law, consistent resource use and distinct languages to communicate generations of knowledge. The change from fisher and Aboriginal participation to *state agency* management facilitated a top-down approach in constraining fishing *without* using the knowledge of the resource users. The challenge for "co-management" and "collaborative governance" between resource users and state agency is to align the management of the resource with local context. Although most scientists may be empathetic regarding the use of local knowledge, they generally are providing information to others for decision making that is constrained by capacity and the inability to conduct much needed research to support information leading to decisions. In this sense, science could be considered supplemental to other knowledge forms, particularly compared to AEK that spans thousands of years. The accommodation of resource users must be more than an exercise of gathering key elements of knowledge because their use of the resource is directly linked to sustaining future conditions. The decisions that affect their use of the resource in the contemporary paradigm emanate from universal policy designed for efficiency of administrative program delivery and not for specific local conditions.

Interestingly, Robson (2006) devotes a chapter to "Policy and Politics," a fully vested component of the contemporary top-down state agency approach to resource management. Pre-historic resource management was more inclusive of knowledge in decision-making. It fostered sustainability with conservation and stewardship as the primary driving tenets. The ancient salmon weirs that were prolifically used throughout the Pacific salmon watershed were of concern only to those with an interest in generating profits from the salmon. Banning the salmon weirs undeniably altered the Aboriginal economy. The canneries have come and gone, with remnant companies remaining as processors/buyers of wild product where supply is controlled by agency. The communities on the BC coast north of the Sunshine Coast are remote and isolated. Most of these communities have high numbers if not entirely Aboriginal populations. They are situated as "islands" (metaphorically and physically) amidst an abundance of resources that could drive GDP yet they are unable to realize economic potential.

The access to aquatic and terrestrial resources was the economic driver for elaborate commerce; the buying and selling of fish by Native people was an economy not just minimal subsisting. It was a flourishing economy that humbles any commerce in human memory. That economy was based on the knowledge of the local environment and interactions that allowed sustainable use of highly abundant resources. Clearly, the economic use of resources in Aboriginal territories is the center of conflict.

The use of the *knowledge* regarding the aquatic environment takes many shapes. The integration of knowledge in the last couple hundred years has been perhaps the most challenging because of the evolution of bureaucratic decision-making. The contemporary top-down state agency approach for resource management has not been successful integrating the triple bottom line; environmental (*natural*), social (*human*) and economic (*working*) attributes for sustainable use. It has however facilitated divisive factions that may or may not have valid assertions within each of the attributes of resource management. Whether or not these assertions are valid by logic, reason and/or data is uncertain on many levels.

The pre-historic top-down approach was interactive with the knowledge of the resource users integrated into decision-making. The evolution of current resource management has brought us unrelenting arguments and fear of the unknown. Is that fear *real*? Yes, undoubtedly it is real but is it necessary? No. Knowledge is powerful and an opportunity to use that knowledge, in all of its forms, will advance the necessity of obtaining food, conserving the environment, and maintaining economic infrastructure based on environmental conservation and stewardship. The *integration* of scientific knowledge, LEK, and AEK is the transformative language for development of cooperative resource management. The "integration" implies a working relationship of communication. It does not mean that all knowledge must be exchanged to affect that relationship with surrender of other values. Science alone is not the appropriate antidote to provide "all of the answers" either and in fact, it actually should generate more questions. A commitment to funding the necessary elements to effective resource management and actually conducting the activity that is desperately needed may prevent a repeat of the Atlantic Cod story for Pacific salmon.

Finally, reflection on the term "S'kuu see" can provide insight into the difficulty of communication with and between different languages about the same subject. The very word continues to bring laughter to many Tsimshian people of all generations. The humor is undoubtedly profound. At the time of European contact, many activities and mannerisms were "introduced" to Aboriginal communities, or intended to be introduced. One of the great marvels of these introductions was the "potato" (likely a Russet). The idea was to entice an interest in farming on the land, in the dirt (this being for marine people) and producing potatoes. It is said that a man was standing in front of a group of Tsimshian and was holding a potato out in front of him trying to relay a descriptor of the object, "S'kuu see, S'kuu see" which the Tsimshian had never seen before. A Tsimshian repeated what he was hearing, "S'kuu see, S'kuu see." After the language barriers were somewhat dissipated, the topic was revisited. The good man was trying to convey the object as, "It's good seed, it's good seed" although not knowing English the Tsimshian heard "S'kuu see, S'kuu see." The North Coast lands are not conducive to agriculture food production, as the good man eventually learned. However the Tsimshian still refer to potatoes as "S'kuu see" more than 150 years later. Farming fish, well that could be a different story.

References

Anderson, E. D. 1998. The History of Fisheries Management and Scientific Advice – the ICNAF/NAFO History from the End of World War II to the Present. *J. Northw. Atl. Fish. Sci.* 23: 75–94.

British Columbia Government – Ministry of Agriculture and Lands (BC Government). 2007. Available online: http://www.al.gov.bc.ca/fisheries/faq.htm#moratorium; http://www.al.gov.bc.ca/fisheries/reports/aqua_backgrounder.pdf. Accessed: July 18, 2007.

Brown, D. 2005. *Salmon Wars: The Battle for the West Coast Salmon Fishery*. Harbour Publishing Co. Ltd., Madeira Park, BC.

Campbell, T. 1996. Co-management of Aboriginal Resources. *Information North* 22 (1).

Fisheries and Oceans Canada (DFO). 2005. Fisheries and Oceans Canada 2005-2010 Strategic Plan: *Our Waters, Our Future*. http://www.dfo-mpo.gc.ca/dfo-mpo/plan/plan_e.pdf. Accessed: July 18, 2007.

Fisheries and Oceans Canada (DFO). 2007. Aborignal Access to Resources and Oceans Management (AAROM). http://www.pac.dfo-mpo.gc.ca/tapd/aarom_e.htm. Accessed: July 18, 2007.

Fisheries Act, R.S.C. 1970, c. F-14, ss. 34, 61 (1).

Gitasoo. 2007. Available online: http://www.kitasoo.org. Accessed: July 18, 2007.

Harris, D. C. 2001. *Fish, Law and Colonialism: The Legal Capture of Salmon in British Columbia*. University of Toronto Press. Toronto.

Harris, M. 1998. *Lament for an Ocean: The Collapse of the Atlantic Cod Fishery: A True Crime Story*. Toronto: McClelland & Stewart Inc.

Helin, C. 2006. *Dances with Dependency: Indigenous Success Through Self-Reliance*. Orca Spirit Publishing & Communications, Inc. Vancouver, BC.

Maturana, H. R. and Varela, F. J. 1980. *Autopoiesis and Cognition*. Boston: D. Reidel Publishing Company.

Neis, B., Felt, L. F., Haedrich, R. L. and Schneider, D. C. 1999. An Interdisciplinary Method for Collecting and Integrating Fishers; Ecological Knowledge into Resource Management. In D. Newell & R. E. Ommer, eds. *Fishing Places, Fishing People: Traditions and Issues in Canadian Small-Scale Fisheries*. Toronto: University of Toronto Press.

Newell, D. 1993. *Tangled Webs of History: Indians and the Law in Canada's Pacific Coast Fisheries*. Toronto: University of Toronto Press.

R. v. Gladstone (1996) 2 S.C.R. 723–725.

Robson, P. A. 2006. *Salmon Farming: the Whole Story*. Surrey, BC: Heritage House Publishing company, Ltd.

Royce, W. F. 1985. The Historical Development of Fisheries Science and Management (Taken from a lecture given at the Fisheries Centennial Celebration (1985) by William F. Royce. Available online: http://www.nefsc.noaa.gov/history/stories/fsh_sci_history1.html. Accessed: July 18, 2007.

Solomon, R. C. and Higgins, K. M. 1996. *A Short History of Philosophy*. New York: Oxford University Press.

Stevenson, M. G. 1996. Indigenous Knowledge in Environmental Assessment. *Arctic* 49 (3): 278–291.

United Nations Food and Agriculture Organization (FAO). 1999. The Rome Declaration on the Implementation of the Code of Conduct for Responsible Fisheries. Adopted by the FAO Ministerial Meeting on Fisheries Rome, 10–11 March 1999. Available online: http://www.fao.org/DOCREP/005/X2220E/X2220E00.htm. Accessed: July 18, 2007. Accessed: July 18, 2007.

Weinstein, M. 2000. Pieces of the puzzle: Solutions for community-based fisheries management from Native Canadians, Japanese cooperatives and common property researchers. *The Georgetown International Environmental Law Review* 12(2): 375–412.
White, L. 1967. The historical roots of our ecological crisis. *Science* 155: 1203–1207.

Chapter 12
Oral History and Traditional Ecological Knowledge

Marcel Shepert

Before European contact, First Nations in British Columbia along the north, central, and south coasts had developed societies based in large part on the anadromous salmon species found throughout the entire region. For thousands of years prior to contact with Europeans, the First Nations Peoples enjoyed fishing activities, especially those related to the local region, whether coastal or inland. Harvesting for food and trade have been the foundation of First Nations' economic, cultural, and social lifestyles along the major rivers and waterways of this land. All groups had access to a variety of fish. In British Columbia the distribution and population densities of the People are directly linked to their access to this resource. Large populations could only be maintained with sufficient food resources; in this case salmon that was caught and smoke-dried. Pre-contact, the ability to process and preserve fish influenced the amount of fishing that would take place. The outcome for many First Nations people was a culture and economy heavily intertwined with the fisheries resource, and the development of an extremely efficient and productive fishery. Because of this rich and diverse history, passions tend to run high – what more than salmon evokes the quintessence of what it means to be an Aboriginal in British Columbia? Since first contact with European settlers and the cannery developments of the 1870s, the sport/recreational fishery, and now aquaculture, First Nations have always felt themselves to be, at the very least, the stewards, and in some cases the owners, of these animals.

Evidence exists of an ancient practice of catching fish using traps and weirs in shallow waters, generally at the mouths of tributaries. Fish caught were not only used for food but also as a valuable trade item that the Nation exchanged with neighbouring First Nations, particularly where fish were not as abundant. Fishing sites were traditionally regarded as a form of property. This remains the case today with certain families retaining a degree of exclusive fishing rights at specific places. It was the responsibility of chiefs to allocate individual fishing sites along the weirs, and the times and the order in which clan members were allowed to fish. Indian Agent, R.E. Loring, noted in 1905 that there was a high degree of organization to

M. Shepert
Fraser River Aboriginal Fisheries Secretariat, Prince George, British Columbia
e-mail: mars_shepert@shaw.ca

K. Culver, D. Castle (eds.), *Aquaculture, Innovation and Social Transformation*,
© Springer Science+Business Media B.V. 2008

such fisheries, and that fishing activities were carefully timed, an indication that the First Nation fished in a manner that ensured large numbers of fish were caught and that sufficient numbers were able to spawn (Harris 2001).

Aboriginal people since time immemorial have spoken for the fish with which they have had such a long and complex history. Within the songs dances, names, and stories of the Aboriginal peoples of British Columbia there is a rich and detailed oral history relating to these species. "Our oral history tells the story of Syalustsa, one of the first people to fall from the sky. He sacrificed his baby girl so that the salmon would spawn in the Cowichan River and provide our people with food forever" – Chief Harvey Alphonse, Cowichan First Nation "All Cowichan are descendants of Syalustsa and the other eleven people who fell from the sky, and because of this we have a right to the salmon." This manner of passing information along from generation to generation is not directly associated with the "European/academic" scientific model of investigation.

There is an interesting point of intersection with Felt's paper insofar as there is an underestimated, or at least little recognised, similarity between two kinds of traditional community and their response to new technologies. While First Nations have authority to demand repatriation of land, Newfoundland fishers do not. Yet the two communities share similar positions insofar as both have local or traditional knowledge which tends not to be valued by bureaucratic science in federal and provincial ministries, or at least not when it comes to natural resource management issues. While there have been many papers published and there have been many discussions about the use of oral histories and traditional ecological knowledge, recognition of First Nations interests and inputs into the planning and management of aquaculture has been minimal at best. The experience in British Columbia of First Nations people is that they are generally not involved in the process of approving or disapproving of planned fish farms in their respective areas. The Broughton Archipelago First Nation, Namgis, is a case in point.

Oral traditions are continuous, can be both public and private, intertwined and interrelated, and may be recorded on physical objects and held by a nation, or group of nations, communities, families, and individuals. Oral knowledge may by communicated with gestures and performance and shared at gatherings, feasts, ceremonies or as part of everyday living and activities. From the storyteller's perspective, the story will communicate aspects of the culture and of the social structure and tradition which relates to the use and beliefs of the resource in question. The story can validate hereditary claims to the authority, prestige, lands and resources, while providing information as to the practices and activities associated with a given resource. The Gitxsan people of north western British Columbia have well documented their use and occupation of the land. Evidence of the Gitxsan's occupation is extensive and beginning to be well documented on maps. More than 400 days of oral testimony on the Gitxsan's court 1997 battle in *Delgamuukw* litigation led to the *Delgamuukw* decision. This historic court ruling, *Delgamuukw* (named after one of many Gitxsan and Wet'suwer'en chiefs who first brought the case to court) is a precedent-setting decision that may forever change aboriginal entitlement to land and resources-in particular, forests.

In order to research oral history and mine the information for management purposes, it is important to be strategic and respectful and remember that this is about relationships and in the long run may extend the boundaries of existing knowledge for both science and First Nations. In the long run, we may see the end of what appears now to be a gap between scientists and First Nations' knowledge, and ways of building knowledge. Science may come to confirm First Nations' knowledge, or First Nations may choose to integrate more scientific approaches into their understanding of aquaculture and the environment. The interactions between people, cultures, industry, and biodiversity are difficult and full of uncertainty, and we are some distance from achieving a shared understanding of these interactions. Testimony of traditional knowledge holders, for example, may be questioned as to the reliability of their knowledge and to explain the transmission of traditions and history. And, as Felt points out, this goes both ways – neither science nor TEK has a unique claim to possession of reliable, relevant knowledge. They might overlap, and the ingredients of each may change over time – First Nations members study science, and scientists are at least more aware now than in the past that TEK may contribute to science. It may be appropriate to question the admissibility and then to assign a weight to the knowledge to be used for management. In some cases Bands or Tribal Councils are making use of "elders committees" to discuss TEK, copyright, and other intellectual property rights. The Skeena River First Nations from upper lakes region, Lake Babine First Nations to the estuary, and Tsimsian Nation, have formed an organization called, the Skeena Fisheries Commission (SFC). It is based on traditional law, which is TEK and is drawn from any number elders or elder societies (www.skeenafisheries.ca).

For First Nations, the issue of trust is still an overshadowing concern and will need to be rectified at some point before TEK can be fully utilized. In British Columbia, Aboriginal people have experienced cultural domination including physical violence, residential schools, territorial occupation through the reserve allocation system and many elements of legislative and political oppression. More importantly are aspects of psychological oppression; when different systems of thought meet and are contested – but on an unequal basis. These issues will make trusting difficult unless a reconciliation process is agreed to by the parties.

For many First Nations, protection of information is vital, particularly with so many of the knowledge holders entering their twilight years. Many communities, at a time when western scholars and researchers are looking to mine the rich and diverse knowledge base within the First Nations communities, are confounded by the problem of what to do and how to access this information, and even whether or not to share the information.

Legal Rights and Inclusion of First Nations' in the Legal Fabric of Canada

However First Nations' traditional ecological knowledge is used, it is one part of the complex relation of First Nations to the Government of Canada and provincial and territorial governments. The structure of First Nations' relations to governments has changed very significantly in the last thirty years since the *Constitution Act*, 1982, for reasons including a series of Supreme Court rulings setting out the reach of aboriginal rights.

The foundation of the change in First Nations' position in Canada is found in Section 35 of the *Constitution Act* 1982, which affirmed that there are existing aboriginal and treaty rights of the aboriginal peoples of Canada. In 1990 the Supreme Court's decision in *Sparrow* (*R. v. Sparrow*) confirmed Aboriginal fishing rights including the controversial *priority access clause*, which places First Nations second on the priority list only after conservation. The Court states this priority ordering in plain and unambiguous terms: "Any allocation of priorities after valid conservation measures have been implemented must give top priority to Indian food fishing (*R. v. Sparrow*)."

Sparrow was, however, just a first step in the Court's development of what Lamer, C.J. (as he then was) later characterised in *Van Der Peet* (*R. v. Van der Peet*) as "the framework for analyzing s. 35(1) claims." In *Van der Peet* Court clarified the nature of aboriginal title and aboriginal rights. In *Delgammukw (Delgamuukw v. British Columbia)* writing for the majority Chief Justice Lamer wrote: "In *Adams*, and in the companion decision in *Côté*, I considered and rejected the proposition that claims to aboriginal rights must also be grounded in an underlying claim to aboriginal title. But I held, nevertheless, that aboriginal title was a distinct species of aboriginal right that was recognized and affirmed by s. 35(1)." Throughout its reasoning, the Court emphasised the fiduciary nature of the Crown's relation to First Nations, and emphasised that any ambiguities in law should be interpreted in favour of First Nations. As Lamer, C.J. wrote in *Van der Peet*: "This judgment will thus, after outlining the context and background of the appeal, articulate a test for identifying aboriginal rights which reflects the purposes underlying s. 35(1), and the interests which that constitutional provision is intended to protect (*R. v. Van der Peet*)." The Court said that in order to be an aboriginal right an activity must be "an element of a practice, custom or tradition integral to the distinctive culture of the aboriginal group claiming the right (*R. v. Van der Peet*)." As Lamer, C.J. explains by reference to legal scholarship, this is not an easy task: "*The challenge of defining aboriginal rights stems from the fact that they are rights peculiar to the meeting of two vastly dissimilar legal cultures*; consequently there will always be a question about which legal culture is to provide the vantage point from which rights are to be defined .... a morally and politically defensible conception of aboriginal rights will incorporate both legal perspectives (*R. v. Van der Peet*)." Incorporation of both perspectives is admittedly difficult, as seen in the 1997 Supreme Court decision in *Delgamuukw* considered *Van Der Peet* in affirming and clarifying the fiduciary obligation to incorporate First Nations perspectives into decisions regarding infringements on aboriginal rights and aboriginal title. In *Delgamuukw* the Court ordered a new trial for reasons including the trial judge's failure to take sufficiently seriously the weight of oral histories of First Nations. Lamer, C.J. wrote of the trial judge that:

> Although he framed his ruling on weight in terms of the specific oral histories before him, in my respectful opinion, the trial judge in reality based his decision on some general concerns with the use of oral histories as evidence in aboriginal rights cases. In summary, the trial judge gave no independent weight to these special oral histories because they did not accurately convey historical truth, because knowledge about those oral histories was confined to the communities whose histories they were and because those oral histories were insufficiently detailed. However, as I mentioned earlier, these are features, to a greater or lesser extent, of all oral histories, not just the adaawk and kungax. The implication of the trial judge's reasoning is that oral histories should never be given any independent weight and are only useful as confirmatory evidence in aboriginal rights litigation. I fear that if this reasoning were followed, the oral histories of aboriginal peoples would be consistently and systematically undervalued by the Canadian legal system, in contradiction of the express instruction to the contrary in *Van der Peet* that trial courts interpret the evidence of aboriginal peoples in light of the difficulties inherent in adjudicating aboriginal claims (*Delgamuukw v. British Columbia*).

Here the Court clearly requires future decisions – and the new trial judge – to respect the evidence of First Nations even when it does not arrive in a familiar package. In the Court's conclusion to its ruling in *Delgamuukw*, the Court stated again the importance of giving due respect to First Nations' oral histories, recognising that without this respect, the fiduciary obligations of the Government of Canada cannot be achieved. Lamer, C.J. wrote:

> The trial judge's treatment of the various kinds of oral histories did not satisfy the principles I laid down in *Van der Peet*. These errors are particularly worrisome because oral histories were of critical importance to the appellants' case. They used those histories in an attempt to establish their occupation and use of the disputed territory, an essential requirement for aboriginal title. The trial judge, after refusing to admit, or giving no independent weight to these oral histories, reached the conclusion that the appellants had not demonstrated the requisite degree of occupation for "ownership." Had the trial judge assessed the oral histories correctly, his conclusions on these issues of fact might have been very different (*Delgamuukw v. British Columbia*).

In all of these remarks we can see that the Courts have made progress toward understanding and giving the knowledge of First Nations its place in negotiation.

Rights and title may be established through oral history evidence and courts must come to terms with the oral histories of aboriginal societies. Courts now recognise that the perspective of the aboriginal people must therefore be taken into account. Further, the laws of evidence, in order for the oral historic facts to be accommodated and placed on an equal footing with the types of historical evidence the courts are familiar with must be adapted. This is the key task and challenge for holders of traditional ecological knowledge and those who operate under the European-derived scientific paradigm. In law then, according to the decision in *Sparrow*, both government and industry have a duty to provide full information to the First Nations regarding proposed activity, development or decision regarding use of resources to which First Nations have a claim; they must then seek information from First Nations groups about the impact a decision or action may have on the rights or interests of a First Nations. It is precisely at this point that the two cultures merge in ways discussed in cases subsequent to *Sparrow*. Both government and industry have a duty to fully inform themselves regarding the practices of the First Nations, and to avoid or minimize infringements. In order to do this it is crucial that governments and industry understand the systematic nature of the traditional use and knowledge the First Nations carry, and understand how First Nations rely upon the resource and management structures used for sustenance and economic well being. Most importantly, this understanding needs to be incorporated into modern management structures.

Even with the progress made in relations between First Nations and the Government of Canada, much remains to be done and there are many complicating factors including the diversity of perspectives and priorities with First Nations. This point can be put more strongly: The current system is in direct opposition to the more holistic approach many First Nations have in common, which is to say that there is a clash between one system operating on economic imperative and another system operating by connection to the natural world. Environmental degradation and habitat destruction, for example, are pivotal issues that many First Nations see as constantly eroding their ability to earn a living. However, it is important to understand that many First Nations have similar attitudes about the environment and business as many European settlers do, which suggests that the environmental morality espoused by the "noble savage" worldview is misguided, simplistic, and Eurocentric. Incorporation of First Nations' perspectives into resource use is not, therefore, a simple matter of plugging in a set of First Nations views. Those views vary, just as views on resource use vary among other citizens of Canada. First Nations' perspectives on aquaculture are a part of this complex interaction between greater legal respect for First Nations, and differences within First Nations regarding the balance of priorities between economic well being and environmental conservation.

Finfish Aquaculture in British Columbia

Finfish aquaculture has steadily increased in British Columbia since the early 1980s and created both problems and opportunities for British Columbia First Nations. Canada is the fourth largest salmon producer in the world, and of the entire salmon farming industry in Canada, 65% is farmed in British Columbia. Aquaculture in British Columbia is conducted on over 120 site tenures, most on the central or south coast (see Map 1). A key issue for many coastal First Nations is the lack of consultation by industry and government over a wide range of issues associated with these tenures. Contrary to popular belief, not all First Nations are averse to aquaculture in British Columbia. It is a fact that some coastal First Nations favour aquaculture because it creates much needed jobs and builds capacity and self reliance. For example, the partnership between Creative Salmon and Tla-o-qui-aht First Nation was featured in hearings of the British Columbia Legislative Assembly Special Committee on Sustainable Aquaculture in 2006 (Legislative 2006), however on the flip side many First Nations have witnessed the habitat destruction that net pen aquaculture brings via the disappearance of once-abundant species like clams, sea cucumbers, crabs and oysters. In this respect, there is an interesting parallel here

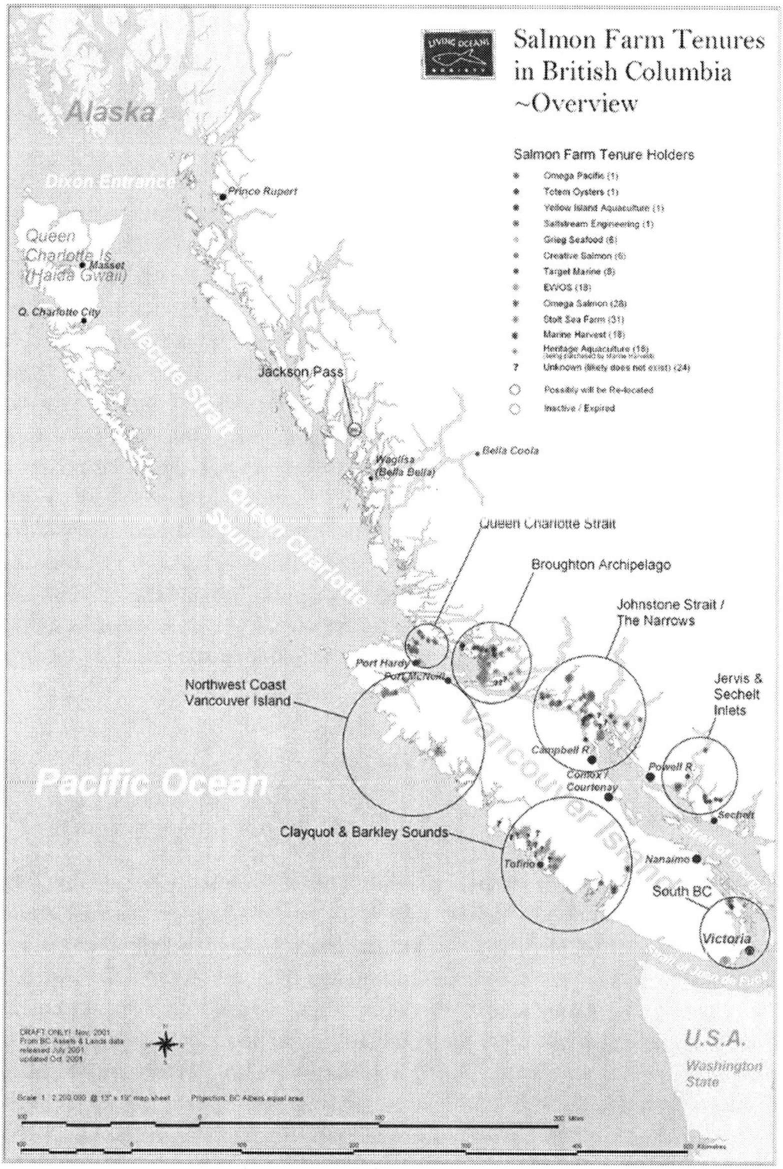

Map 1 This map was the property of the British Columbia Aboriginal Fisheries Commission (BCAFC) which is no longer a viable organization, it is officially defunct

between these observations from First Nations and Felt's recording Newfoundland fishers reporting similar changes to the aquatic environment after the introduction of aquaculture. Some First Nations have begun to realize that employment creation has not been as promising as once thought due to automation and technological breakthroughs. What is more, First Nations are beginning to understand that science can only guide, not direct, difficult decisions in risk management, because these decisions are, for the most part, value-driven and many of the values are not in line with First Nations values. It may be necessary to think of creative convergences of values, as Creative Salmon and the Tla-o-qui-aht First Nation appeared to have done, to guide the use of science.

The issues facing the industry and First Nations are universal and not specific to First Nations. First Nations are not just faced with the question of how best to oppose aquaculture. They are instead faced with a much more complex set of tasks, both evaluating whether aquaculture makes sense for them as coastal communities – as Felt points out Newfoundlanders must also do – *and* remaking themselves as communities in light of the new possibilities opened by the series of Supreme Court decisions from *Sparrow*. At the practical level, the issues confronting First Nations include a range of well-known issues: Understanding and mitigating the negative impact of fish faeces and feed as waste material, escaped Atlantic salmon from net pens, disease transfer, density, parasites, and food chain contaminations. Salmon aquaculture is not only an ecosystem concern, it is social, and encompasses values, traditions, and ancient laws. The applicable laws are not only provincial, or federal, but are also aboriginal and balancing these three levels of law is difficult. As Map 1 shows, aquaculture sites are located in many different locations which vary in their proximity to human settlements and other activities, presenting location-specific issues and opportunities which must be understood with respect to their individual characteristics.

Inter-tribal Consultation

The diversity of British Columbia First Nations is nowhere more obvious than on the Fraser River which is home to ninety six bands from the headwaters down to the confluence. Just as diverse is the geography that lies between the mountains and estuary. The anadromous salmon literally sew British Columbia First Nations together in a long and continuous stitch. It has long since been recognised that, when it comes to salmon, including farmed salmon, inter-tribal consultation is essential and even vital. First Nations of the Fraser River speak of the former relationships held amongst the Nations and the variety of agreements between Nations who individually governed their Nation's use of resources according to their own laws. Non-Natives may not understand, or may simply dismiss, these past governance practices. In a modern context all legal systems need to be recognized, but representation of the diversity of First Nations law is beyond the purview of this chapter. It is possible, however, to express a general Aboriginal Legal perspective.

There are core values which underlie Aboriginal systems, including the laws, customs, traditions and ways of life of First Nations (Gaertner et al. 2004). A few of these values could be understood as those of *respect, integration, and stewardship*, including:

1. Respect for all parts of creation, which includes a responsibility for past, present, and future generations;
2. Recognition that all components of life and culture are related and integral to the whole; and,
3. Commitment to stewardship sustaining the land and resources for future generations, which requires a respectful relationship with the land, resources, and people under one's direct stewardship, and in the past included those resources and people that the proposed actions would directly and indirectly affect. Stewardship implies, for example, *fishery methods which both ensure sufficient fish for the villages, and which allow fish to pass by for those upriver.*

What exactly does that mean and what, if anything, has this to do with aquaculture? Aquaculture is not exclusively a marine issue affecting only those First Nations communities in close proximity to the net pen sites. If you consider the migratory routes of Fraser, or Skeena River bounds Salmon, you will see that they come in close contact with net pens. There is a potential for aquaculture to have a negative affect on migratory salmon stocks, and consequently for First Nations far inland yet dependent on salmon in rivers. To the extent that First Nations people and the wild salmon on which they have depended for generations are endangered by aquaculture, the First Nations fishery is constitutionally protected under section 35 of the *Constitution Act*, 1982 and subsequent interpretation of the Constitution by the Supreme Court of Canada.

There are reasons to believe that reliance on legal safeguards for the First Nations fishery may need to be put into play, as there are reasons to believe that aquaculture may threaten that fishery. As mentioned above, while there is a potential for disease transfer to adult salmon from net cage fish, this life stage is not as critical as the juvenile stage. It is the juvenile salmon that is at most risk from net pens, and the reason is simple: when juvenile fish enter the marine environment they enter from rivers, and almost all net pens are located at or near the confluence of rivers, which gives a better flushing effect for the waste generated by the net pens. It has been argued that sea lice can have an affect on juvenile salmon and therefore can affect the wild salmon. The Gixan have demonstrated that sea lice transferred to juveniles can affect the ability of Interior First Nations to harvest fish in the future, which puts the federal management agency in a conflicting position. On the one hand Fisheries and Oceans Canada openly promotes aquaculture, on the other hand has the fiduciary responsibility to protect wild salmon for First Nations. This argument has created a problem for Fisheries and Oceans Canada, particularly since the release of the Wild Salmon Policy (WSP). The problem gains further complexity one it is recognized that in British Columbia the anadromous fish migrate past many First Nations communities creating complex dynamics which make Inter-tribal relationships difficult as demonstrated. Map 2 shows the migratory routes and the First Nations communities they come in contact with.

It is worth emphasising again here that the complexity of First Nations' relation to aquaculture is not simply a complex series of rejections. Just as the diversity of

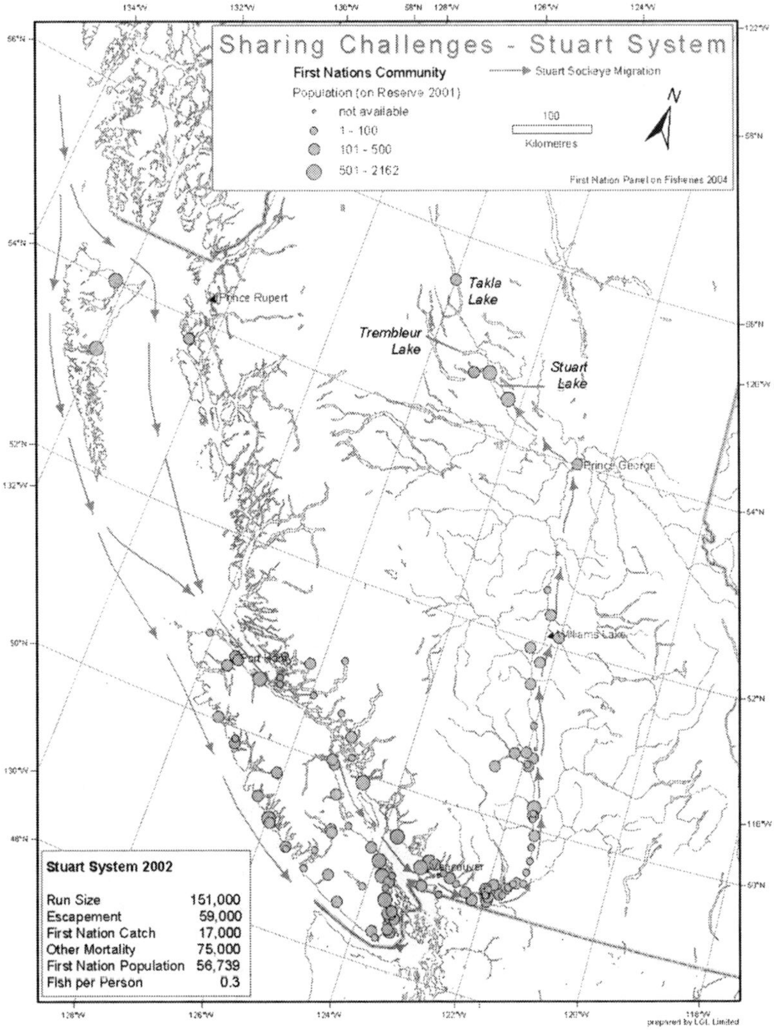

Map 2 This map is from the First Nations Panel report, the inter-tribal map was commissioned by the First Nations summit in BC, a public document

First Nations communities in British Columbia is a reality, so too is the diversity in opinion over aquaculture. The range of opinion is broad, not different than the non-native community in British Columbia. First Nations and non-native communities share a range of common concerns including: sustainability of the environment, continued use of local resources, social stability, economic well being, and biodiversity.

Recent Developments

"Sustainability" in natural resource management is a buzz word that evokes a lot of passion in people, particularly First Nations. For many tribes in British Columbia, mapping has proven to be an effective tool for First Nations efforts in developing and communicating their understanding and views regarding the natural resources under their stewardship. Communities have begun to use maps to determine resource values, traditional uses, and ancient sites to help in the pre-planning stage of development. This method has been useful to evaluate potential impacts and explore alternatives by many British Columbia First Nations and in contexts other than fisheries and aquaculture such as the Carrier Sekani Tribal Council's dealings with the oil and gas Enbridge Gateway Project (Enbridge 2006).

Mapping of resources and taking inventory has given many First Nations the feeling of empowerment and is being used in local land and marine use planning. The future challenge is integrating these data with other data sets and synthesizing the information so communities can understand what is at stake and make better informed decisions. There are gaps between traditional and scientific knowledge, and so it has become clear to First Nations that members need access to learning about both traditional knowledge and the scientific paradigm. The value of mapping goes beyond what I have just described, a process of what may seem similar to accounting. The concept of mapping is of interest to First Nations because it encompasses traditional ecological knowledge *and* current practices and is representative of those living closest to the resource. All levels of government including First Nations governments recognise the potential for better outcomes as governance includes both First Nations' traditional ecological knowledge and knowledge derived from the scientific paradigm.

Much has been done to enable First Nations and the Government of Canada to understand each other's perspective. Now the challenge is to go beyond understanding to positive action. Inclusive governance processes need to be developed in order for sustainability to be achieved. The word co-management is often overused, and more needs to be done to put substance to this often elusive jargon. Through a deliberate program of capacity building First Nations are becoming increasingly prepared to take on a more central role in resource management including aquaculture. In order to build a better management framework which would include structures to incorporate traditional ecological knowledge, these structures will need to come up with common bases for interaction and this will vary region by region. The structures will need to find common ground on what the Aboriginal interest is from a historical use, practical and working perspective. Once these issues have been worked out implementation plans will need to be developed and tracked for effectiveness. An effective co-management structure should have the following characteristics:

- Integration of traditional ecological knowledge
- Respect for Aboriginal interests at all levels of decision making: Local, district, regional, industry, and both federal and provincial headquarters

- Formalized organizational structures that would be respected and appear within existing structures
- Priority focus on implementation at the local level
- Consultative structures that recognize the current legal landscape, as discussed above in the train of decisions following *Sparrow*
- Deliberate programs of capacity development
- Greater understanding of ecosystem values from a First Nations perspective
- Inclusion of culturally sensitive sites and perspectives
- Community benefits including access to decision making, economic development leading to self reliance, and opportunities to fulfill supply and services contracts

Through the local and regional structures and through regular face to face meetings, an integrated network of Aboriginal co-managers could develop to support a sustainable shared used of aquatic resources including the traditional capture fishery, and aquaculture.

References

Delgamuukw v. British Columbia, [1997] 3 S.C.R. 1010; http://scc.lexum.umontreal.ca/en/1997/1997rcs3-1010/1997rcs3-1010.html; accessed September 10, 2007.

Enbridge Gateway Pipelines. 2006. http://www.enbridge.com/gateway/; accessed August 1, 2007

Gaertner, B., Danesh, R. 2005. Inter-Tribal Consultation on the Fraser Watershed. Fraser River Aboriginal Fisheries Secretariat (FRAFS). www.frafs.ca; accessed September 10, 2007.

Harris, D.C. 2001. *Fish Law and Colonialism*. Toronto: University of Toronto Press.

Legislative Assembly- Province of British Columbia. 2006. Special Committee on Sustainable Aquaculture (minutes). http://www.leg.bc.ca/cmt/38thparl/session-2/aquaculture/hansard/W60606p.htm; accessed August 1, 2007.

R. v. Sparrow [1990] 1 S.C.R. 1075; http://scc.lexum.umontreal.ca/en/1990/1990rcs1-1075/1990rcs1-1075.html; accessed September 10, 2007.

R. v. Van der Peet [1996] 2 S.C.R. 507, http://scc.lexum.umontreal.ca/en/1996/1996rcs2-507/1996rcs2-507.html; accessed September 10, 2007.

Part V
Messages, Consumers and Aquaculture: New Products, New Worries

Introduction

In typical developed countries, less than 5% of the population is involved in primary agriculture, fisheries or aquaculture. Urbanization has distanced populations from the physical sources of food, and conceptions of the ties between farms, processors and supermarkets are deliberately manipulated by advertisers and retailers. Many supermarkets decorate product displays to evoke associations with farmer's markets, often using visual cues of a simpler era of food production and marketing such as straw bales, wooden barrels and wheelbarrows. Supermarket murals and labels of packaged foods reinforce attempts to associate food with a nostalgic sense of natural purity, using backdrops of rural and coastal scenes inhabited by iconic fisher and farmer figures – including, for example, Captain Highliner and the Jolly Green Giant.

Consumers' lack of knowledge about the origins of their food is not, however, evidence of a lack of concern about their food. This paradoxical situation is particularly striking in situations where public awareness of a production system is low, yet public concern is high when food safety appears to have been compromised. Consider, for example, media attention to the purported presence of the antifungal agenent malachite green in British Columbia aquaculture salmon (CTV 2005; Creative Salmon 2005; AMB 2007), and worries about carcinogenic chemicals such as chloramphenicol and nitrofuran in farmed shrimp (Dobbs 2007). Equally, consumers register concern over novel products and production techniques. One case in point is the undying controversy about genetically modified food, seen in specific concerns as worries about the use of genetically modified feed limiting African beef exports to Europe and the potential for cloned animal products in the food system (Rudenko and Matheson 2007).

Where does aquaculture fit in this spectrum of concerns? As we have suggested in the general introduction to this volume, aquaculture is in itself alarming to those who are uneasy about the employment of science and technology in intensive food production. Romantic misconceptions about fisheries no doubt help many to look past technology-intensive harvest practices. Just the same, even fish originating from hatcheries are nearly the most purely wild food we consume, as company testimonials suggest (VitalChoice 2007; GoldSeal 2007). Aquaculture products, by contrast, are the direct result of human intervention and and continued application of technology throughout the aquaculture product's life cycle, and cannot claim so plausibly the virtues of untouched 'naturalness' or 'purity.' Organic aquaculture might be the next best thing to wild fish, an appealing idea were a suitable standard of 'organic' widely available for all kinds of aquaculture.

Yet the very idea of organic aquaculture brings us around to the problem once again: whether the complaint against aquaculture is a worry about the health benefits and safety of the products resulting from aquaculture, or whether aquaculture itself is objectionable as a kind of human meddling with a natural food which ought to be left alone in its natural life cycle until it is harvested in a traditional manner. As you read our authors' exploration of consumers' attitudes toward aquaculture as a technology, and aquaculture food products, it is perhaps worth bearing in mind that the stark division between aquaculture and capture fisheries might well be a false dichotomy in many places. The recreational and commercial fishers in British Columbia who are attraced by the ideal of the wild salmon caught in the pristine waters of its natural habitat may sometimes overlook an inconvenient fact. As we noted above in contrasting the 'naturalness' associated with the capture fishery against the 'artificiality' of aquaculture, many salmon brought to consumers by fishers are the products of human interventions in the form of salmon hatcheries releasing juveniles into the wild. Hatcheries in British Columbia have operated (somewhat sporadically) since 1894 in service of a range of goals (MacKinlay et al. 2004), and by 2000 Fisheries and Oceans Canada reported that the Canadian Salmon Enhancement Project "release about 600 million juvenile salmon into streams each year, accounting for about 10 to 20 per cent of the Canadian salmon catch, depending on the abundance of wild fish in any year (DFO 2007)."

Castle, Finlay and Clark face the issues occupying the most controversial part of the debate in their study of the potential attitudes and purchase intent toward genetically modified fish. At the time of writing, one company in North America (Aqua Bounty Technologies 2007) is seeking approval to market for human consumption its genetically modified, growth-enhanced Atlantic salmon. While the technology is a reality, the commercial application is not, leaving a rare opportunity to engage consumers regarding real product development *prior* to the confounding influence of the market. Based on a study of Canadians' attitudes and purchase intent toward genetically modified salmon, Castle et al. found that potential purchasers were initially uncomforable with the product and did not intend to purchase it. The provision of balanced information, however, changed these attitudes somewhat, suggesting that information and time to reflect on it helps people to overcome the shock of the new. Far from market analysis, this study shows that if there is ever to be a market for genetically modified aquaculture products, providing timely, balanced and meaningful information to potential consumers will be of paramount importance.

How and when to provide information to consumers is a difficult matter to gauge. Frode Nilssen considers consumer attitudes toward farmed fish, and salmon especially, in two dimensions: the product itself, and the processes by which the product is created. Consumers want their food to be safe, nutritious, pleasing, and a good value. But beyond that, given increasing awareness of the prevalence of farmed fish, consumers are also becoming aquainted with the underlying technology of aquaculture – at least to the point of understanding the meaning of advertising which distinguishes between captured and cultured fish. Nilssen points out that while the aquaculture industry might be focussed on addressing consumer interests in the qualitative and safety aspects of farmed products, issues about the way fish are farmed must also be communicated to consumers. The future of aquaculture may well turn on whether the aquaculture industry can explain its processes in ways which convince consumers of both the honesty of producers and the merits of their industrial processes. If this hurdle cannot be overcome, efforts to develop genetically modified fish for human consumption may be in vain.

The food industry is notoriously secretive, limiting consumer access to basic information about their food. Without the efforts of people such as Marion Nestle and Michael Pollan, for example (Nestle 2003; Pollan 2006), public access to knowledge about food production systems might remain between the industry and regulators. Interest in aquaculture's sustainability, safety, and ethical standards follows unsurprisingly from prior concerns about the food industry. As prior experience shows, those concerns are likely to grow along with aquaculture's growth and attempts to fill the looming gap in seafood production left by diminished capture fisheries. The question now is how industry balance its opportunities with good public relations. Aquaculture journalist Kenny McCaffrey observes that the aquaculture industry has realized its position and is willing and able to communicate about its products and and processes. Admittedly, like any other industry, the aquaculture industry will attempt to put itself in the best light, but McCaffrey's considered view is that the aquaculture industry has generally succeeded worldwide in developing good relations with the media and generating informative message to the public.

References

Aqua Bounty Technologies. 2007. http://www.aquabounty.com/. Accessed September 13, 2007.

Department of Fisheries and Oceans Canada (DFO). 2000. The Future of the Salmonid Enhancement Program. http://www-comm.pac.dfo-mpo.gc.ca/pages/release/bckgrnd/2000/bg0008_e.htm. Accessed September 13, 2007.

Dobbs D. 2007. Pass the cocktail sauce! *Audubon Magazine*. http://audubonmagazine.org/audubonathome/audubonathome0701.html. Accessed September 13, 2007.

Creative Salmon. 2005. Update Malachite Green – Creative Salmon was given the green light from the CFIA to resume harvesting. http://www.creativesalmon.com/creative_salmon_news_2005.htm. Accessed September 12, 2007.

CTV. 2005. Cancer-causing chemical found in B.C. salmon farm. http://www.ctv.ca/servlet/ArticleNews/story/CTVNews/20050625/cancer_salmonfarm_050625?s_name=&no_ads=. Accessed September 12, 2007.

GoldSeal. 2007. Wild Salmon Spotlight. http://www.goldseal.ca/wildsalmon/. Accessed September 13, 2007.

MacKinlay D., Lehmann S., Bateman J., and Cook R. 2004. Pacific salmon hatcheries in British Columbia. *American Fisheries Society Symposium*. 44:57–75.

Nestle M. 2003. *Food Politics: How the Food Industry Influences Nutrition and Health*. Berkely: University of California Press.

Pollan M. 2006. *The Omnivore's Dilemma: A Natural History of Four Meals*. New York: Penguin.

Rudenko L. and Matheson J.C. 2007. The US FDA and animal cloning: Risk and regulatory approach. *Theriogenology*. 67:198–206.
VitalChoice. 2007. Purity. http://www.vitalchoice.com/purity.cfm. Accessed September 13, 2007.
West Coast Vancouver Aquatic Management Board (AMB). 2007. Tofino salmon farm wins defamation case. http://www.westcoastaquatic.ca/article_Tofino_salmon0107.htm. Accessed September 12, 2007.

Chapter 13
Public Engagement Regarding Aquaculture Products Produced Through Biotechnology

David Castle, Karen Finlay and Steve Clark

Introduction

Aquaculture of fish is a centuries old practice that can provide a stable supply of dietary protein. In the contemporary context, expanded use of traditional forms of aquaculture, coupled with technological innovation in culturing practices, has led to the "Blue Revolution" in aquaculture. Amidst rapid decline of the remaining oceans stocks that can be sustainably harvested, aquaculture production of food fish contributes one third of all fish consumed (FAO 2002). The increase in aquaculture productivity on a global scale has been rapid, including in developing countries where population growth puts renewed pressure on aquacultured fish. The most rapid growth in aquaculture has been in industrialized countries, however, which, having had little traditional aquaculture, were almost exclusively reliant on the capture fishery and imports.

It has been observed that, like the green revolution that preceded it, the blue revolution is becoming an industrial mode of food production. Like terrestrial forms of industrial food production, the effects of scaling up are not always benign (White, O'Neill and Tzankova 2004). Apart from the issues of escapes, disease and waste, aquaculture in industrialized countries tends to focus on high commercial value fish like salmon and trout. While aquaculture generally may appear to be a sustainable alternative to the capture fishery, the high protein feed required for salmonids is acquired by overfishing species generally considered unfit for human consumption. The resulting ecological pressure is now considered one of the major problems associated with salmon aquaculture.

A situation in which overfishing of feedfish undermines industry claims of environmentally sustainability may strike many as odd. It is perhaps even more peculiar given that salmon is not among the species expected to feed the starving, and, meanwhile, consumption in industrialized countries has plateaued and prices are in decline. Yet from the industry perspective, salmon aquaculture is a highly

D. Castle
Canada Research Chair in Science and Society, and Director Institute for Science, Society and Policy, University of Ottawa, Ottawa, Ontario
e-mail: dcastle@uottawa.ca

© Springer Science+Business Media B.V. 2008

competitive business in search of incremental marginal gains now that the gold rush that early salmon aquaculture operations experienced is over. In this context, finding non-fish sources of feed is one option for reducing the high cost of feed which is estimated at 75% of the operating costs of salmon aquaculture. Another approach is to find fish which convert feed more efficiently, or which can grow in environments which are not tolerated by other breed lines. Another competitive advantage can be achieved by filling in the troughs of the supply cycle which peaks twice yearly from production north and south of the equator.

Research conducted at Memorial University of Newfoundland in the 1970s identified antifreeze proteins and their genes and their promoters that could be used to develop an array of gene constructs (Fletcher, Goddard and Wu 1999). Aqua Bounty Technologies, located in Canada and the United States, exploited these technologies to produce transgenic salmon under the AquAdvantage Bred name. This salmon uses the cold-tolerance gene from the ocean pout as a ubiquitous promoter of its own growth hormone. As a result of the expression of the salmon's own growth hormone in all of its tissues, transgenic salmon grow four to six times faster than their conventional counterparts and reach market size in 18 months, about half the normal time. Because they spend less time maintaining basic metabolism and swimming, the transgenic fish are better feed converters – roughly 5% better – which is not a negligible result for a competitive industry. An unanticipated, but welcome by-product of this modification is that the transgenic salmon can grow in colder than normal conditions, which in the Canadian context means that previously undesirable sites can be used for salmon aquaculture, and the general cold coastal waters of Canada, can stay in competition with Norway and Scotland's slightly warmer waters. The AquAdvantage salmon is before regulators in Canada and the United States, and if it were to be approved, it could have significant impact on the production of aquacultured salmon. In Canada, salmon represent 15% of the total fish and seafood produced by (Agriculture and Agrifood Canada 2003) at 132,000 tonnes or roughly $520 million annually (Department of Fisheries and Oceans 2003). The question is, however, whether anyone will buy it.

Public Engagement on Transgenic Salmon

In no area of product technological innovation have consumer values been more ignored, and industry interests been elevated, than in the pursuit of products based on genetic modification. Not surprisingly, a lack of public consultation has been associated with the market failure of products produced through biotechnology by researchers in the United Kingdom. (Frewer, Howard and Shepherd 1995; Frewer, Howard, Hedderley and Shepherd 1997). The deficit of information about the existence of these products and their potential risks and benefits, together with a lack of consideration of consumer value creation around food products are implicitly understood as the cause of outright public opposition to the technology in that country. Given that new product introductions are costly and risky, it may strike one as counter intuitive that market failure rates averaging as much as 75% are not being addressed by industries which remain largely incapable of reliably satisfying consumer needs and achieving new product success (Joshi and Sharma 2004). Perhaps the antidote is more thorough public engagement to find out what motivates consumers in the marketplace, and how their values will determine their interactions with new products like transgenic fish (Prahalad and Ramaswamy 2003; Urban and Hauser 2004).

There are some recent studies which have investigated consumer information needs and response to information exchange concerning biotechnology. Lambraki (2002) reported that Canadian consumers react with surprise when advised that products produced through biotechnology are available on grocery store shelves without public disclosure. A combination of self-administered questionnaires (n=58) and follow-up in-depth interviews (n=35) gauged consumers' perceptions of risks and benefits associated with genetically modified foods. A host of issues entered into the decision process for consumers relating to biotechnology, including effects on the environment, ethical issues, long-term health effects, and inadequate scientific testing. Respondents felt it was their right to choose what to eat and to be informed regarding a food's processing technique and makeup through labelling and/or other forms of communication.

In a study using an experimental methodology, Porter, Sheeshka, Finlay and Marmurek (2005), examined consumer reaction to different types of labels and label content for genetically modified plant-derived foods. Two hundred and forty respondents judged the informativeness and appeal of labels being considered by the Canadian General Standards Board's committee to develop a voluntary standard for the labelling of genetically modified foods. Some respondents were exposed to label information regarding the reasons for the genetic modification, while other respondents saw a label that simply told them that the product was 'genetically modified.' In 50% of each of these two conditions, a prominent symbol on the front of the package indicated that there was "GM" content in the food, creating four possible label conditions: front panel symbol with nutrition label announcing benefits of the genetic modification; front panel symbol with nutrition label containing no benefit information; no front panel symbol with nutrition label announcing benefits of the genetic modification; and no front panel symbol with nutrition label containing no benefit information. After viewing mocked-up breakfast cereal, soy oil and potato packaging for one of these conditions, respondents were asked to evaluate the labels overall, how informative they were to their choice of product, and to provide open-ended rationale for their ratings. Respondents felt most informed and rated the label highest overall when information was provided to indicate the reason for the modification, (e.g. reduced use of pesticides) regardless of whether the front panel of the package included a "GM" symbol or not. Canadian consumers want disclosure; they want to be informed about the content of products they are eating and the production processes they have undergone. Labelling is one important mechanism for exposure but broader communication programs would provide a network of communication that would be valued within the consumer context.

We engaged a representative sample of the Canadian public about the potential for transgenic salmon to enter into the marketplace. The approach was based on four main considerations. First, it was decided that assessing potential consumer reaction prior to any actual product release, or discussion in the media, would provide a valuable baseline of information about attitudes toward and intent to purchase transgenic salmon. Second, we wished to test a hypothesis that progressive release of information would allow people to consider information about the technology without out rightly rejecting it. The effects of progressive release of information could therefore be tested. Third, and related to the second consideration, we wanted to gain insight into consumers' reactions to disclosure of information about products in a climate in which industry has proved hesitant to inform the public about the processes by which products are made if those processes include genetically modification. Fourth, and finally, we wanted to learn what the potential role of public engagement could be in managing the success or failures of new food biotechnologies.

The over-arching assumption is that no matter how one might wish to engage the public about genetically modified food such as salmon, communication planning begins with an assessment of the existing beliefs and knowledge structure held by a target group concerning the topic of the communication. In our study, we started the discussion about transgenic salmon by first soliciting beliefs and opinions about the potential products on an open-ended basis to ensure the enquiry is tapping into pre-existing awareness and knowledge levels of consumers. This approach to getting attitudes disclosed also supports subsequent measurements of individuals' prospective intentions to purchase transgenic salmon. Individual beliefs about a concept like transgenic salmon are associated with beliefs about the concepts importance once it is considered in the decision making process. Once subjects become aware of their attitudes toward transgenic salmon, they will incorporate that knowledge into the weighting of this and other concepts in their overall purchase decision.

In our study, quantified evaluations of the concept of transgenic salmon are taken after consideration of new information, after which respondents indicate the extent to which they would be likely to purchase transgenic salmon. More information is then provided about the transgenic salmon, allowing respondents to again react on an open-ended basis to determine whether additional beliefs have been consolidated in memory as a function of interim communications. They are again asked to give quantified evaluations and rate their purchase intent. Since new beliefs arising from new information about the products may influence evaluations of the concept and purchase intent, these are re-gauged on a scaled basis. The entire procedure is repeated after a third and final wave of information dissemination.

A total of 1365 subjects in eight Canadian cities (Vancouver, Kelowna, Calgary, Guelph, Toronto, Montreal, Fredericton and Saint John) participated. The eight cities selected permitted representation of diverse consumers across Canada. In addition, two cities on each coast were visited, one larger city and one smaller coastal city where fishing is an important economic and cultural construct. For maximum control, a single research assistant oversaw data collection in all eight cities. Respondents were recruited through a professional recruiting firm who used a broad adult sampling frame to reflect the diverse potential purchasers or purchase influencers for the pork and salmon products.

Qualitative data were collected using individual booklets where respondents wrote answers to open-ended questions, later transcribed to text and theme analysed using NVivo software. Open-ended responses were coded on two bases. First, the substantive ideas contained in the protocols were categorized as to their thematic content. Second, each idea was assessed as positively or negatively valenced depending upon whether an individual's response on that theme was favourable or unfavourable. A second judge coded 25% of the open-ended data with inter-judge reliability on that portion high at 82%.

Quantitative data were collected using individual wireless handheld units and Resolver Ballot software, previously applied to management decision-making and educational settings and adapted for consumer use. Resolver software permitted research sessions to include up to 30 respondents in a theatre-style setting. All survey questions were exposed to participants in a uniform manner on a large viewing screen. Maximum experimenter control was thus realized and data collection was efficient since quantitative responses were automatically coded to a dataset using the wireless technology.

Respondents were exposed to the first question of the survey on a projector screen at the front of the room that asked them to "please take a moment and describe everything that comes to mind when you hear the term transgenic salmon." Respondents indicated their free association responses in an individual booklet. The next set of questions that appeared on the screen asked respondents to rate transgenic pigs on four, 7-point attitudinal scales anchored by "bad ... good", "not interesting ... interesting", "not important ... important", and "not acceptable ... acceptable." To each of these enquiries, subjects indicated their response using the wireless hand-held units with corresponding numbers from 1 to 7 indicating levels on the scale. Following this first set of responses to the term "transgenic salmon", subjects were provided with additional information about the concept that appeared on the screen before them. They were told that:

> One significant type of transgenic salmon is modified to have increased the production of growth hormone, and grow faster and reach maturity more quickly then unmodified salmon.

Having received more information about the transgenic salmon, subjects were again asked to use their booklets to indicate everything that came to mind when they thought about the concept. They were then cued by the instructions on the projection screen to respond using their hand-held units to the same 4 attitudinal questions.

A second information intervention revealed more information about the benefits and risks of the product to determine any changes to respondents' attitudes and purchase intent as a function of more knowledge. Specifically, the following was presented:

> Growth enhancement in salmon is promoted by adding a gene from another fish, the ocean pout, combined with the salmon's own gene for its growth hormone. Since the salmon would be ready for human consumption at an earlier age, many benefits could result, including economic spin-offs in coastal towns, a more stable supply of salmon throughout the year,

global price continuity, and a greater supply of protein for developing nations. Although there are many benefits that result from the introduction of transgenic salmon, there may be a concern of the long term effect that this introduction could cause.

The content of information interventions was determined in pre-tests. Interventions were chosen which provided consumers with levels perceived understandable, while not overly taxing in terms of information load. Both economic and food safety risks were referenced, the latter set of risks being unknown in more specific terms. Respondents were prompted with the same free association and 4 attitudinal after this new information was presented. They were then debriefed concerning the purpose of the study, and paid $20 for their participation.

Results

The 1365 respondent national sample was comprised of 48% females and 37% males; 37% were single, 34% married, and 11% divorced. In the case of income; 53% made less than $40k, 19% made $41–$79k, and 15% indicated earning more that $80k. Percentages were in line with the Canadian population average, with a few exceptions: a slight under-representation of males, 15% fewer married people and some over-representation of individuals in lower income brackets.

Overall, consumers felt most positively informed about transgenic salmon at the final information intervention. Initial mean attitude (purchase intent) to the concept was 4.19 (2.80) on a 7-point scale. Once subjects learned a little about the concept (modified to increase growth hormone; reach maturity faster), responses dipped to 4.08 (2.67), returning to an attitude (purchase intent) level of 4.24 (2.76) following more information about both the benefits and risks of transgenic salmon. The quadratic trend for both attitudes and purchase intent were significant at $F(2,1344) = 30.80, p<.00)$ and $F(1,1238) = 13.89, p<.00)$ respectively. Providing sufficient information about the modification to transgenic salmon products does not negatively impact consumers' intent to sample the technology as industry feared.

Additional analysis of the sample of respondents revealed that consumers differentiate themselves in at least three different ways when it comes to their reactions to transgenic salmon. Amongst the sample overall, across the baseline and two information interventions, mean purchase intent for transgenic salmon was 2.74. One group of consumers emerged who would never knowingly buy products produced through biotechnology (mean purchase intent 1.0 at the absolute bottom end of the scale). This group represented 33.2% of the total sample. A second group remain unconvinced but open-minded concerning the new technology (a high 49.8% of the total sample), with mean purchase intent registering 2.86. The final group (17% of the sample) already see benefits of trangenic salmon or they are willing to compromise levels of one attribute delivered by transgenic salmon (e.g. taste) for gains on another attribute (e.g. lower price), with purchase intent averaging 5.81 (Table 13.1). Reactions, demographic make-up and information needs of each of the three groups are discussed below.

Table 13.1 Responses and Demographic Composition by Categories of Consumer Acceptance of Transgenic Salmon

		Categories of Consumer Acceptance (% of sample)		
		Never Endorse (33.2%)	Open-Minded/ Questioning (49.8%)	Converted (17.0)%
Mean Purchase Intent (S.D.*) (across three interventions)		1.00 (0.00)	2.86 (1.02)	5.81 (0.69)
Purchase Intent by intervention				
1st		1.00 (0.00)	3.02 (1.48)	5.68 (1.20)
2nd		1.00 (0.00)	2.69 (1.43)	5.89 (0.98)
3rd		1.00 (0.00)	2.88 (1.52)	5.84 (1.01)
Demographics (%'s within each category of consumer acceptance**)		%	%	%
Gender	M	35.1	39.8	54.8
	F	60.0	54.6	33.7
	Education < High School	5.2	7.5	12.4
	High School	29.2	30.5	31.1
	College	25.6	21.9	26.3
	University Degree	22.1	25.1	17.2
	Graduate Degree	17.9	15.1	12.9
Age	18–24	14.3	18.8	20.6
	25–24	24.8	20.1	9.6
	35–44	18.2	16.6	15.3
	45–54	18.2	13.7	11.0
	55–64	11.1	9.7	14.4
	65+	4.4	9.6	13.4

* S.D. = standard deviation.

** differences versus a 100% total are due to non-responses by demographic.

Consumers Never Endorsing Products from Biotechnology (33.2% of the Sample)

One third of the sample indicated that they would not buy transgenic salmon products (indicated a "1" on the 1–7 purchase likelihood scale). These "non transgenic salmon buyers" appear to be so adversely inclined toward GM food modifications that no level of information about benefits or risks could induce them to purchase. Not one of the 669 respondents in this group deviated from a "1" purchase intent rating for any of the information phases (baseline reaction to transgenic salmon, neither first nor second information intervention). It appears there is nothing to say or do, based on their reaction to the information about transgenic salmon, which would convince this group to abandon their convictions about transgenic salmon.

Sixty percent of the non-buyer group were females, versus the 48% that females represented in the sample overall. This over-representation of females in the non-buyer subgroup was significant. Cross-tabulations between gender and the three consumer groups generated a significant X^2 (42.18, p<.00). This group was disproportionately more likely to have a university or a graduate degree and less likely to have not finished high school ($X^2 = 18.80$, $p<.02$). In terms of age, the non-endorsing group was less likely to be 18–24 years or 65+ years; on the other hand, more like to fall into a middle (25–54) age group, $X^2 = 49.95$, $p<.00$. Consistent with the fact that this group tends to be more likely to have 1 or 2 household members working full-time ($X^2 = 21.20$, $p<.01$), it also tends to be earning a higher income in the range of $25M–$95M ($X^2 = 37.50$, $p<.01$). Finally, the non-buyer group disproportionately tends to be single ($X^2 = 15.62$, $p<.05$). Table 13.1 provides percentages for the three consumer groups across relevant demographics.

This group indicated far-reaching opinions regarding the technology. The largest category of negative open-ended responses sourcing to the non-buying group related to the intertwining of negative "genetically modified," "environment," and "hormone" related mentions including:

> "It might alter taste and health in me. Young children and older adults may be more susceptible to picking up sickness or changes in it. Alter the environment from its natural process. The world would change."
>
> "We are not gods therefore what makes us think that we can tamper with nature in such a way? We have destroyed nature enough, we must find better ways to repair the damage we have caused"
>
> "I don't think messing with genes and increased hormones are a great thing. There are probably side effects and other things that have not been discovered yet. I wouldn't buy them."
>
> "Environment could endanger other species that salmon may prey on and playing God and not letting natural selection to occur."
>
> "In time, this genetically modified fish could mutate and wipe out complete species of other fish and cause more damage than good. Tampering with God's creation. Send this idea to starving countries."

A small percentage (3.4%) of comments related to concern over the generation of cancer in humans or the perceived interaction between hormones and cancer. Over 90% of comments of this nature in the study sourced to the "never endorsing" consumer group, including:

> ". . . it causes cancer, disease to human beings. Should be banned"
>
> "In time we will get cancer from the different hormones"
>
> "Who cares if something is cheaper to produce and saves the environment if we are going to drop dead from cancer in eight years."
>
> "Growth hormones, not naturally, artificial. Moral issues about messing with nature, owning a genetic code for a living thing, potential cancer risk"
>
> "Why change what God gave us. Man shouldn't tamper with these things. That is why we have such an amount of disease in this world, cancer etc. From the foods and the air etc. that man has tampered with"
>
> "It is time to stop putting hormones into the things we eat. I believe children are especially vulnerable. In the future we will probably find that this is one reason so many children have A.D.D."
>
> "Growth hormones being transferred to people, not a good thing. Effects on people not tested, report it to the public so they are competent to make intelligent, informed choices."

Despite potential misconceptions about transgenic salmon, the processes by which they are made and the attributes of the resulting product, this group had hardened attitudes and demonstrated utterly no intention to purchase genetically modified salmon.

Consumers Committed and/or Willing to Make Benefit/Risk Tradeoffs (17% of the Sample)

This group scored 5 or higher (5.81 mean) across the three information interventions. Individuals are disproportionately male, the least educated in terms of a college/university degree and less likely to have finished high school. They are more likely to be 18–24 (students or new to the workforce) or 55+ years of age (retirees). Consistent with this profile, consumers in the group are more likely to have no household member working full-time, and tend to be in a lower income bracket (especially $15,000–45,000). For this group, purchase intent was initially high, and it rose after the first detailed information was exposed (Table 13.1). At the time of the last information intervention, attitudes toward transgenic salmon remained constant.

Many of the committed individuals registered unqualified positive thoughts about the new technology, whether those thoughts related to price discounts, protein or other benefits:

> "Faster growing fish means more food sooner. Larger supply should mean lower prices."
>
> "Mostly all positives in the food (healthy) of quality protein would be readily available, economical and healthy."

Others in the committed group understood that benefits of biotechnology might need to be traded off against risks or discounts on other attributes. A meta-analysis of the responses across the three information interventions indicated a significant interaction between negative taste thoughts and positive thoughts around a lower cost explains attitudinal ratings. This model explained 42% of judgement formation across subjects. Figure 13.1 portrays the interaction of individuals who indicated few versus many thoughts about potential negative taste against those with positive thoughts relating to the anticipated price reduction involving transgenic salmon. Individuals who indicate positive thoughts relating to anticipated price are equally split between having and not having concerns about taste. Consequently, a lower price compensates, at least in part, for possible poorer taste. Furthermore, the number of positive price thoughts in absolute terms was 32% higher than negative taste thoughts:

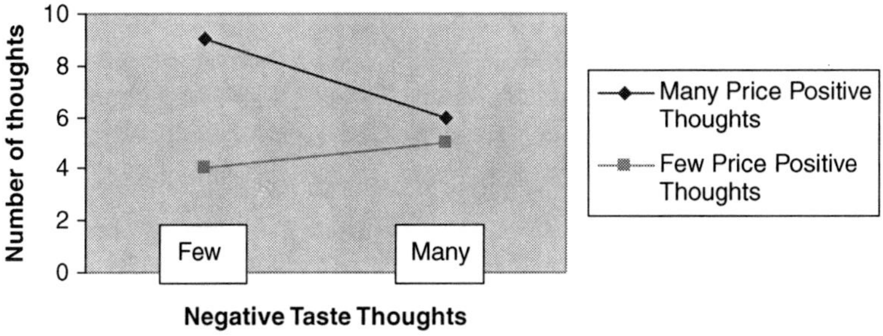

Fig. 13.1 Interaction between positive price thoughts and negative taste thoughts

"Safe, better for environment, cheaper, some taste issues."

"Flavor not as good as fresh water. Probably adds up to the fact that we will get it at a better price."

"I don't think the flavour and taste would be as good as a salmon left to grow on its own . . . .but if it were more affordable .."

"Higher fat content, more mush than unmodified salmon, lower flavor taste, higher odors, cheaper"

Individuals in the converted group also appear willing to trade off a perceived lower taste of transgenic salmon for the benefit of eating more protein. This model of attitude formation explained 30% of the variance in the judgement data. Individuals interested in consuming higher levels of protein were equally likely to think such consumption might be associated with poorer or better taste. Evidence of benefit and risk compromises for this group include the following open-ended responses:

". . . consumption of more "fish" food which may have no use anywhere else, edible protein, possibly more healthy than "meat" i.e. red meat, but not the same taste as in the wild."

"Taste may be a problem . . . certainly good for more protein for my family."

". . . perfecting transgenic salmon so there are no health risks, society accepting it in terms of ethics and taste appeal to public of transgenic salmon."

"Omega 7, good and healthy food. Great flavor. Little difficult to eat, high protein less cholesterol."

The converted group of consumers appears committed to transgenic salmon in absolute terms or in acknowledgement of acceptable benefit and risk tradeoffs.

Consumers with Questions and Who Are Open-Minded (49.8% of Consumers)

The largest group of consumers in the study were those registering in the mid-range of the purchase intent scale (2–5 range in score with a mean of 2.86). These individuals were more likely to have a university or a college degree but not a graduate degree. They were proportionately well split among all age groups and categories of marital status, but more likely to be earning <$55M or $95M+. Mean purchase intent for this group declined when they first learned some information about transgenic salmon, but rose again once more information about benefits and risks was conveyed (Table 13.1). Information therefore helped this group allay fears that arose when first exposed to the concept. Purchase intent with the amount of information communicated in this study registered 2.86, indicating there is still potential for this group to become more accepting in the future. Verbatims tended to contain questions and represented a broad range of issues that this group would want addressed in future communication programs:

"Sounds like it has something to do with 'changing genes' of the salmon, which I expect would mean altering features for the fish for such things as higher yield, better taste, etc."

" Something work looking at , to study for health risks. Cheaper price for salmon, quality of the salmon, by that I mean, nutrients . . . it is as good as 'natural' salmon?"

"It could be cheaper to buy. More food for the global world."

"If a study of long term affects could be done and no side effects are present, I would accept this. But it's kind of playing God with genetic codes."

"Economy price more affordable if this what they're doing it good for sure I would try.

"A cloned salmon, does it taste the same?"

"Will it alter the taste of the salmon? What if allergies occur? Will it be more expensive or cheaper?"

Discussion

In this study, public reactions to transgenic salmon were gauged using innovative methodology as respondents progressively became better informed about the biotechnology. A combined open- and closed-ended survey technique was used to identify potential consumer reactions to progressively more information concerning the benefits and risks of transgenic salmon. Against the overall sample, opinions about transgenic salmon were maintained as more information about the use of transgene biotechnology was communicated. One of the most important initial conclusions one can draw from this study is that even prior to product release, or widespread public discussion of the prospect of product release, the Canadian publics' response might be stratified into three groups. This result confirms our first motivation for engaging the public early, namely that it is useful to have a baseline of information prior to any actual product release, or discussion in the media, in order to understand the potential influence of subsequent information, or the influence of certain kinds of information sources. As we shall discuss below, this study suggests that at least for some potential consumers, information might be a determinant of their attitude and purchase intent.

One third of consumers staunchly opposed the concept of transgenic salmon both on ethical and environmental grounds. Not one of the 453 respondents opposed to the technology deviated from their position during the course of the study. Verbatim protocols contained negative attitudes and misconceptions held with conviction. Individual importance of negatively-valenced beliefs is clearly strong. All efforts to reduce the importance of these beliefs and/or convert negative beliefs to positive ones are judged futile.

Another 17% of consumers are the converted and indicate a commitment to purchasing transgenic salmon. This group is the least educated, more likely to be younger or older, not working and with a moderate income. Presumably many of these individuals are students or retired adults. Attitudes towards transgenic salmon increased for this group across the interventions, indicating that more information only increased embedded feelings. Future communication efforts targeting this group should reinforce positive beliefs and further establish beliefs that biotechnology offers trade-offs that this group already judges acceptable.

The most fruitful opportunity for future consultation and communication efforts with consumers would appear to be amongst the half of the sample who are uncommitted and, in many instances, open-minded. These individuals tend to be better educated and proportionately well dispersed across gender, age, marital status and income categories. Attitude and purchase intent levels induced by the first detailed information about the concept of transgenic salmon were lower for this group than the baseline rating. As more of the questions on the minds of these individuals began to be answered in the second detailed communication, judgements rose again to their baseline level. Existing beliefs were not overly negatively-valenced, neither were these beliefs judged extremely important to decisions regarding the purchase of transgenic salmon. Given the size of this group, its reactions drove the overall attitude and purchase intent results of the study. Several issues are on the minds of these individuals including testing procedures, health effects, environmental effects and animal welfare. Where these issues are positively addressed in the case of transgenic salmon, this should be reinforced. Positively-valenced issues should further be positioned in the minds of consumers as being important or relevant to their purchase decision regarding this biotechnology. The open-minded group is searching for disclosure and reassurance.

One significant result from this study pertains to the study design. Because respondents had roughly half an hour to consider the idea of transgenic salmon, and to weigh increasingly detailed information into their evaluation and intention to purchase the product, we achieved some measure of deliberation in our respondents, and avoided knee-jerk reactions associated with phone polling techniques. This confirms our hypothesis that progressive release of information, conducted in accordance with the study design we outline above, is a useful method for allowing the public to consider complex information before rendering definitive judgments. In this respect, we have combined the virtues of large scale, demographically representative polling with traditional focus group methodology and deliberative democracy methodologies.

In this study we also wished to gain insight into consumers' reactions to disclosure of information about products about which they know very little. The current approach endorsed by the Canadian government is to insist that product novelty is the only trigger for safety evaluation. While information about product manufacturing processes is generally publicly available, unless protected by trade secret or patent, the trend is to downplay biotechnology as a natural progression from conventional breeding. As our study indicates, not only is product safety a concern when considering transgenic salmon, but so too is the underlying technology used in the production of the product. The public's concern about the naturalness of their food, and the potential for harm arising in new forms of product development, elsewhere confirmed in the case of genetically modified crops, applies to genetically modified salmon as well.

Fourth, and finally, we wanted to learn what the potential role of public engagement could be in managing the success or failures of new food biotechnologies. As we have learned in this study, a third of the Canadian public find the idea of transgenic salmon abhorrent, and it appears that no amount of information provided to them will change their attitude or intention to purchase. Considering that the information came from university researchers with no overt bias for or against the product, it may be the case that less trusted sources of information would deepen the resistance to transgenic salmon. The second smaller group, about 17% of our sample, is willing to make trade-offs about transgenic salmon, but are generally less receptive to the idea than the single largest group, almost half, which was open to the idea of transgenic salmon and would generally be willing to purchase it once they have more information. The size of this group, and their willingness to consider more information about the product suggests that there is a fairly large group of Canadians who would be willing to take time to make considered judgments about the technology and to incorporate new information into their decision making process. For those entertaining the prospect of approving, regulating or marketing a product like transgenic salmon, this kind of information may make or break entrance into the market.

As we mentioned at the outset, this study takes advantage of the opportunity to proactively engage the public and understand consumer attitudes prior to market introduction of a new technology. Even under conditions like those in Canada where trust in regulators is relatively high, consumers are interested in knowing more about novel products and the processes by which they are derived (Lambraki 2002). We learned that while consumers are the first to admit that they do not always have the background knowledge or expertise to assess all new information they learn about transgenic salmon, they are nevertheless willing to consider new information and integrate it into an evaluation of the product and make a judgment about their intent to purchase the product. We found that information will have differential effects on consumers, ranging from reinforcing commitments to avoid food biotechnology, to open-mindedness about wholly new products of food, such as genetically modified, aquacultured fish.

In light of these findings, we hold our commitments to the basic tenets of communication theory renewed – that testing unveils current knowledge levels of the target audience as well as information required to make an informed decision. If the public is given an opportunity, free of the coercive effects of marketing or the sense of the inevitable consumption of some genetically modified food, many will integrate new information into their initial beliefs and weigh their options to embrace or avoid the product. Without this baseline knowledge about potential consumers' response to the technology, it is inconceivable that anyone could give reliable odds about the likelihood of success of a genetically modified salmon product launch. As our study suggests, on balance providing information about genetically modified salmon to the public does not appear to decrease public attitudes toward, or intent to purchase, genetically modified salmon. It will remain the task of those wishing to sell transgenic salmon to find those early adopters.

References

Agriculture and Agrifood Canada. 2003. "Canada's Fish and Seafood Industry," http://ats.agr.ca/seafood/industry-e.htm. Accessed: July 2, 2007.

Concerted Action Group. 1997. "Europe ambivalent on biotechnology." Nature 387: 845–7.

Department of Fisheries and Oceans. 2003. "Aquaculture" http://www.dfo-mpo.gc.ca/communic/statistics/aqua/aqua02_e.htm. Accessed: July 2, 2007.

Einsiedel, E.F. 2000. Consumers and GM food labels: providing information or sowing confusion? *AgBioForum* 3(4): 231–235.

Einsiedel, E.F., Finlay, K., and Arko, J.C. 2000 "Meeting the Public's Needs for Information on Biotechnology." Technical paper prepared for the Canadian Biotechnology Advisory Committee.

Fletcher, G.L., Goddard, S.V., Wu, Y. 1999. Antifreeze proteins and their genes: From basic research to business opportunity. *Chemical Technology* 30: 17–28.

FAO. 2002. The state of the world's fisheries and aquaculture. Rome: FOA

Frewer, L.J., Howard, C., Hedderley, D., and Shepherd, R. 1997. Consumer attitudes towards different food-processing technologies used in cheese production: The influence of consumer benefit. *Food Quality and Preference* 8: 1–10.

Frewer, L., Howard, C., and Shepherd, R. 1995. Genetic engineering and food: What determines consumer acceptance? *British Food Journal* 97: 31–6.

Joshi, A.W and Sharma, S. 2004. Customer Knowledge Development: Antecedents and Impact on New Product Performance. *Journal of Marketing* 68(4): 47–59.

Lambraki, I.A., 2002. An exploratory qualitative and quantitative study on consumers' attitudes towards genetically modified foods. Dissertation Abstracts International- MAI 40/06, p. 1504, Dec. 2002, UMI#AAT MQ67357.

Mallin, M.A. 2000. Impacts of industrial animal production on rivers and estuaries. *American Scientist* 88: 26–37.

Miles, S, and Frewer, L.J. 2001. Investigating specific concerns about different food hazards. *Food, Quality and Preference* 12(1): 47–61.

Porter, M.V., Sheeshka, J., Finlay, K., and Marmurek, H. 2005. Judgment of Labels for Genetically Modified Foods among Canadian Consumers. (unpublished)

Prahalad, C.K, and Ramaswamy, V. 2003. The New Frontier of Experience Innovation. *Sloan Management Review* 44(4): 12–18.

Urban, G.L and Hauser, J.R. 2004. 'Listening In' to Find and Explore New Combinations of Customer Needs. *Journal of Marketing* 68 (April): 72–87.

White, K. O'Neill, B, and Tzankova, Z. 2004. "At a Crossroads: Will Aquaculture Fulfill the

Promise of the Blue Revolution?" A SeaWeb Aquaculture Clearinghouse (http://www.seaweb.org/resources/documents/reports_crossroads.pdf). Accessed: July 2, 2007.

Chapter 14
Consumers and Aquaculture, New Products – New Worries

Frode Nilssen

The overall aim of this chapter is to discuss some critical factors related to how people perceive aquaculture technology and farmed fish as food. Public perception of farmed fish begins with consumer awareness of the existence and availability of farmed fish only; there are numerous factors that have influence on whether or not consumers actually would choose or even actively choose not to purchase farmed fish. The aim of this chapter is to shed light on some of the factors and how they might affect consumers' perception – and consumption, of farmed fish.

In order to discuss the issue of public perception and consumption of farmed fish, it is appropriate to focus the discussion on farmed salmon. Farming of Atlantic salmon is one of the more successful large scale marine aquaculture development initiatives that have taken place in recent history. The annual production has grown tremendously both in volume and value since the infant industry stage in the 1970s. The industry has also created many new jobs and provided badly needed economic activity in sparsely populated areas in many remote parts of the northern hemisphere. Further, through this period, problems related to spawning, feeding, and breeding of Atlantic salmon have to a large extent been resolved. At the same time, some of the main fish diseases that created significant worries in the earlier phase of the development are no longer a major concern due to increased knowledge about amongst others fish behaviour and the reason for and treatment of various fish diseases, together with provision of effective vaccines. There are of course still challenges around these issues, but many of the problems from the early days have been solved. So, by and large there are due grounds to regard salmon aquaculture as a success. The achievements leading to the positive development are perhaps one of the main explanations for the interest for further expansion of salmon farming as well as the interest and effort to develop aquaculture based on other marine species such as cod, spotted wolf fish, and halibut.

But alas, along with success has also come potential obstacles for a continuation of the development over the past decades. These obstacles are both of practical and

F. Nilssen
Department for economics and marketing research,
Norwegian Institute for Fisheries and Aquaculture Research
e-mail: frode.nilssen@nofima.no

© Springer Science+Business Media B.V. 2008

technical nature, but can also be of a more conceptual kind. Here, the notion of conceptual refers to the underlying factors that influence the perception individuals or groups hold towards farmed fish and/or the technology itself. It is these aspects that will be addressed here seen from a market or marketing perspective. The discussion may also be seen as a part of the more generic public concern that has evolved lately related to food safety issues and environmental concern – in a broad sense – that tend to influence consumers' food choice.

Food Choice, Social Considerations and Farmed Salmon

Food choice is a complex function of sensory (taste, odour, texture) and non-sensory factors, for example attitudes, health concern (Bech-Larsen and Grunert 2003; Miles and Frewer 2001), price (Ritson and Petrovici 2001), personal values (Furst et al. 1996; Honkanen and Verplanken, 2004) social influence (Shepherd 1989), availability (Marshall 2001), perceived risk (Frewer et al. 1996), convenience (Jaeger and Meiselman 2004), and self-identity (Charng et al. 1988). Various models have been introduced to explain food choice. Most of them include aspects of food, individual factors and the environment as determinants for food choice. In a review paper Shepherd (1989) brings forward a selection of conceptual models that visualise the issue. Recently, issues like animal welfare and ethical production have received increased attention (Honkanen and Verplanken 2004; Lindeman and Väänänen 2000). The fact that food consumption behaviour could be habitual has also gained attention recently (Honkanen et al. 2005; Verbeke and Vackier 2005). Although much of the food choice research is carried out on a conceptual level (Leek et al. 2000; Sparks and Shepard 1994) there seems to be a general acceptance that dietary patterns are relatively solid and stable over time, and that that the sensory characteristics of the food are important (Southgate 1996; Warde 1997). It is also important to notice that food choice may be seen as an aspect of culturally supported learned behaviour (acquired taste), moderated by sensory properties of the product and influence of external stimuli such as commercials, new product displays, and new life style – or value – aspects. Altogether the many factors represent a delicate mixture of central constructs; values, attitudes, perceptions and social considerations. Although it is possible to identify relationships between individual factors and food choice based on quantitative research and analytical techniques, it is notwithstanding the sum of all the internal and external influence that constitutes the underlying motivation behind consumers' food choice. Considering the many factors, it seems reasonable that the average consumer needs to simplify the decision process related to the daily food purchases. This simplification is often referred to as heuristics – which basically means that consumers develop simple decision rules for solving frequently raised decision problems. The choice of food is one of the many every-day decision problems that involves a series of considerations, and hence is subject to simplification (Köster 2003; Miles et al. 2004). This simplification process leads to the development of product cues that the individual purchasers use when carrying out the daily/weekly routine shopping. It is however important to stress that multi-factor considerations still lie behind, and motivate the choice.

But, in normal, everyday life, purchase of food products may yet be seen as a decision where a few simple criteria are used to reach a decision. Some of the factors that seem to define the main consideration set are price, convenience, taste and some nutritional criteria. When the main shopper in the family buys today's dinner in the shop, the decision process is likely to be severely simplified where only a few decision rules come into force as the main premises for the decision are set already. This is mainly explained by the generally low level of involvement in food shopping, characterised by amongst others negligible information seeking behaviour and little deliberation in brand choice (Beharrell and Denison 1995). An example of this is that the decision criteria used to make a purchase decision may not be well formulated (Bettman and Sujan 1987). The shopping list (if available) may for instance say "dinner" instead of "fish", "salmon", "farmed vs. wild salmon", or even "organically farmed salmon." Judging from a pure visual inspection of the available products in a randomly picked supermarket in Europe or North America, it seems that convenience and price dominate over other considerations for the vast majority of the consumers.

In some instances, however, a seemingly ordinary product may change its position from a low involvement to a "political" item, which in turn has the potential to raise the level of involvement significantly. As a result the sales of the product may be substantially reduced or increased. This is an additional aspect of buying behaviour that complicates the study of consumers' decision making related to the choice of food. Some of these aspects that may influence consumer choice can be defined as social and ethical concern and risk perception. These factors do not necessarily have anything to do with the actual product per se, but may notwithstanding have some impact on the decision of whether to purchase or not. One example of this is when the French Government detonated a nuclear bomb under the Mururoa Atoll in 1995. This governmental decision upset many European consumers, and one means of showing public protest to the government was to stop buying French wine. This action or campaign was initiated by several international Non Governmental Organisations (NGOs) and gained significant support throughout Europe. As a result sales of French wine dropped markedly. Although the actual figures varied somewhat the sales in Sweden, for example, were down by 50 percent and in Germany by about 30 percent the same year. Another and slightly different example related to another dimension of social and ethical concerns is that consumers in some Scandinavian countries stopped buying cod after Scandinavian media published a story on the potential threat for the sustainability of the local cod stock in the Baltic sea due to over-fishing in the area. While the consumer restraint to buy cod only lasted for a very limited time and in a very limited geographical area, the consumer "boycott" of French wine had far reaching consequences for the industry, both cases illustrate the impact social and ethical concerns can play on food choice – provided that good alternatives or substitutes exists.

During the past decade, the salmon farming industry has drawn some negative attention to the industry. The bulk of these voices are articulated from environmentalists, who are questioning the possible negative impact that fish farming in general, and salmon farming in particular, play on the biological environment (Tidwell and Allan 2001). Another dimension of this issue that is also pointed out by some NGOs is the concern for human or animal welfare. In sum, the arguments are basically built around three aspects of possible negative biological impact: "genetic contamination" from escaped farmed salmon on wild species as well as bacterial cross contamination across species, and general pollution caused by the large volumes of fish gathered in a relatively small area. Yet another, more recent issue that has been raised is the question of possible food safety risks related to consuming farmed salmon. This latter aspect received substantial media coverage recently when the journal *Science* reported that farmed Atlantic salmon contains pollutants that may pose a health threat if the diet contains large amounts of farmed salmon (VOA 2004). (The fact that a diet consisting of eating farmed salmon as a main course every fourth day would be safe was under-communicated. In spite of these only minor temporary drops in consumption of farmed salmon occurred in a few markets.) The effect of this particular case was dramatic – albeit very short in duration. As opposed to what happened with the French wine industry, but perhaps more similar to the "cod concern" in Sweden, the environmental discourse does not seem to have had a strong longer-term negative impact on the consumption of farmed Atlantic salmon among the average consumer.

This may be partly explained by the fact that the negative rhetoric on salmon farming basically has been based on arguments that create uncertainties of to what extent the technology represents environmental threats, and not so much related to the qualities of the fish itself as food. Hence, the potential reluctance to salmon farming may be seen as a different matter than the propensity to consume salmon, as a general environmental concern among consumer groups is not necessarily directly associated with the food product Atlantic salmon. To our knowledge such association still remains to be identified.

Yet another adjacent question then, is whether farmed salmon is one of the many low-involvement food items where the consumer choice to purchase is mainly based on automatic, habitual, and subconscious decisions, or a product that entails a higher degree of involvement. Moving from a habitual mode in shopping behaviour towards a more complex judgement or decision-making process inevitably leads to an increased level of information seeking and processing. One reason for pursuing a more complex decision process may be may be perceived uncertainty, or perceived new information related to the product— as the cases of the Mururoa Atoll and the local cod stocks along the southern Swedish skerries as described above. Another factor that may trigger a more complex decision process is if a consumer is about to purchase some food for a special occasion or is looking for some specific qualities, then the search for and use of information increases.

In October 2003 the journal, American Demographics, published a brief piece on what they call a "snapshot of today's consumer" (American Demographics 2003). The article reports on survey findings that Americans claim they care about protecting the planet and are concerned about the environment while their actual buying behaviour reflects something else. In practical terms, the finding from the survey suggests that environmental concern in general does not necessarily have influence on food choice – or choice of other products. It would then be interesting to raise the question of, to what extent – if at all – information about whether the fish offered is farmed or wild caught would influence on their choice in any direction – positive, negative or neutral, and further, if aquaculture technology would raise concerns for consumers? So far this distinction does not seem to represent a particularly big issue for the consumers in an average purchase situation.

In the discussion above some general aspects of food choice, applied to farmed Atlantic salmon, have been discussed. There are of course a wide range of factors that influence consumers' food choice. In the above discussion four major dimensions related to modern food production that influence on people's choice of seafood have been discussed. These four dimensions are: environmental and amenity concerns, ethical concerns, food safety issues, and sensory quality aspects of the products. Some of the above (environmental and amenity, and ethical concerns) are typical "political" aspect that may elicit emotional reactions against purchase of a product as mentioned above. Food safety issues are yet another matter and refer to other dimensions of involvement related to food purchase decisions that are mainly related to risk perception and personal hazards. So far, research has consistently demonstrated that hazards dreaded or worried about, or products perceived to be unnatural or potentially catastrophic, elicit public concern. In general consumers tend to distinguish between technological hazards related to food (food irradiation and genetic modification), and life style hazards (healthiness and hygiene issues) (Slovic 2000, Sparks and Shepard 1994).

Genetic Engineering, Food Production, and Consumer Concern

One particular aspect of food technology that has proved to elicit uncertainty among consumers is the issue of genetic modification of organisms (GMO) in food production. Research on consumer attitudes to genetic modification in seafood (production) has been relatively sparse. One explanation that has been put forward to explain this phenomenon is that there are few genetically modified food products available in the market place and the consumers have therefore had very little knowledge about and experience with such products and their qualities. To illustrate this one may refer to a recent USDA-funded national survey examining public awareness of, knowledge about, and feelings toward genetically modified foods in the United States. The main finding in this study is that Americans generally are unaware of GM foods, both of its presence and application in foods production. Although they generally express a desire for more information and labelling of GM foods, when asked directly about this, they do not seek information about the subject (Hallman et al. 2004). This lack of awareness is comparable in Canada, although the findings of Castle and Finlay (this volume) suggest that Canadians might be more inclined to seek and act upon information about genetically modified salmon than their American counterparts.

It should be noted, though, that European research results indicate that consumers tend to find the concept of genetically modified (GM) plant-derived foods more acceptable than GM animal-derived foods (Kuznesof and Ritson 1998). In their article Kuznesof and Ritson draw on data of explorative nature from a focus group discussion. Through the discussion process they revealed quite a bit of scepticism to GM food products. While only one of the participants (out of 62) took a definite adverse attitude to the application of genetic modification, several of the discussants expressed some concern about the healthiness, safety and moral aspects of using gene technology in foods production.

As indicated, the body of research on consumer reaction to GM food where seafood production is included is relatively modest. One explanation for this may be that there are very few food products commercially available where GM technology has been applied. Interestingly, farmed salmon has been included in several of the European empirical studies on consumer reactions to GM food products (Bredahl 2001; Kuznesof and Ritson 1998; Magnusson and Hursti 2002). The typical statements from the respondents involve negative associations related to the "tampering with nature," "acting God/ruling over nature," "threat to the environment," and "threat to human health." The consumers tend to be significantly less negative to GM in plants and for use in medicines as to GE of animals and fish (Atlantic salmon). For example, Magnusson and Hursti (2002) found that consumers perceived that application of GM to animals (including farmed salmon) gave no/very little benefit, was not at all necessary, and not at all healthy at the one hand. On the other hand they also reported a strong reluctance to purchase such products, were negative to the tampering with nature, and that GM to a large extent was used for profit alone. In general, most of the GM food products were perceived as negative by the majority of the respondents. The fast growing salmon and lean pork meat were given the most negative ratings, while rice and wheat the most positive ones (Magnusson and Hursti 2002). It may also be added that those who scored high on health aspects of food were more negative to genetic engineering applied in production of food. This is in line with earlier studies that have shown that GM products are perceived as less natural than conventional products (Frewer et al. 1996).

In a cross-national survey in three of the single largest markets for farmed salmon in Europe (Germany, Spain and France) the food industry and trade also expressed a strong reluctance to introduce GM salmon to their consumers (Nilssen and Monfort 2000; Johnsen and Nilssen 2001; Hansen et al. 2002). In general the GMO issue was seen as a dimension of the food safety discussion in all three European markets. From a strictly technical/technological perspective from the producer's perspective, the use of GMO mainly represents a means of enhancing the productivity of food production. The worry from the industry was that consumers would hold a different perspective, namely that the main concern is related closely to emotional aspects and perceived uncertainty related to extensive use of GMO at the one hand, and a fear for potential irrevocable large-scale damage to living species as a result unsuccessful manipulation with the genetic coding at the other hand. The majority of the respondents believed that supplying genetically modified salmon to the market would have a negative impact on sales. Likewise most of the respondents across the three countries reported that use of GMO-based fodder would have negative impact on future acceptance. The retail/wholesalers in both France and Germany unanimously explicitly stated that they would not introduce GM animals/salmon in their range of products as long as they perceive the strong reluctance of consumers, while in Spain it was reported that this basically was not seen as an issue at present. The reasons for the reluctance to accept genetically modified fish products are related to ethical/emotional aspects pertaining to genetic engineering and the tampering with nature as well as the perceived potential risks on human health. This is perhaps also further enhanced by the very negatively charged discourse promoted by some NGOs.

Since the application of genetic modification in food production is a relatively new phenomenon, consumers have had little or no opportunity to actually purchase and try such products. Likewise, the information about GM technology in general and the impact of GM technology on the fish/animals have been sparse. Considering this, it is reasonably to assume that the average consumer is reluctant to accept fish and animals-based foods where GM technologies have been applied. One could hypothesize that the main reason for reluctance is related to perceived uncertainty of the consequences of extended use of gene technology on animals – both on the environment and for humans consuming the GM food. Empirical research on perceived risk related to food indicate that knowledge of the technology, and knowledge over potential hazards seem to reduce uncertainties and negatively charged emotional problems related to food production (Sparks and Shepard 1994; Miles et al. 2004).

A supplementary explanation may be that consumers' attitudes toward, for example, GM farmed salmon are influenced by their overall attitude to the use of genetic modification in general. Amongst others, Bredahl (2001) argue for this view, and refer to previous research showing that attitudes towards genetic modification in food production are directly linked to a number of more general attitude domains such as nature and technology. In the discussion Bredahl argues that these perceptions perhaps are linked to general consumer values which in turn constitute the frame of which the GM food purchase/accept decision is made – the perspective through which a decision maker views the alternatives in a decision problem. While discussing the effect of decision problems Kahneman and Tversky (1979) argue for decision frames to be reliable predictors of choice in situations involving risk. Honkanen and Verplanken (2004) investigated the role of values and attitude towards GM foods among European consumers. Their research results clearly indicate that the attitude towards GM food is a strong predictor of intention to buy GM food. An important aspect in their discussion is that attitudes towards eating genetically modified food will be strong and resistant to change even if the technology proves to be harmless, provided that the attitudes are based on values. But, if the attitudes are weak, consumers may be susceptive to change all the same. It is interesting to note that a key finding in their research is that there is variation in the strength of negative attitudes towards GM foods. This basically means those holding a weaker, albeit negative, attitude to GM food may change towards a more positive attitude provided certain conditions. One such condition may be provision of credible information about the advantages of applying GM technology in producing farmed salmon or safeness of eating such GM food. This latter point is also in line with Magnusson and Hursti (2002) as they argue that knowledge is an important determinant of support for science and technology; "the more informed the public are, the more likely they are to be supportive." Other research does, however, suggest that information at best plays a modest part in determining attitudes towards science and technology.

One of the main results of the research carried out by Castle (this volume) reveals much of the same scepticism toward GM of animals for enhancing food production. Contrary to other research carried out within this stream of research Castle exposed the consumers participating in their research project to various aspects of transgenic salmon – that is a salmon which is genetically modified—and probed on their perception of the product as they were gradually supplied more information about the technology. Although only minor alterations were made of the biological properties of the salmon, it was notwithstanding, genetically modified, and this technological operation was initially perceived as negative by a majority of the respondents. After gradually having received more information about the technology applied in this concrete case, some respondents took a more positive stand to the so-called transgenic salmon.

It is not easy to draw solid conclusions based on the discussion above. Several interesting aspects nonetheless stand out as areas of research that call for more in-depth investigation. Some problems seem to be related to ethical, environmental and amenity concern among some consumer groups and the emphasis on some of these aspects varies across nations and cultures. In Canada First Nations issues seem to be important while Germans are more concerned with environmental aspects of aquaculture. In the woodland of research problems related to the theme the GM issue stands out as a major problem for the future development of the aquaculture production and aquaculture products marketing. The body of research along this theme does so far give some indications that attitude strength may be a key factor to investigate and measure along with the relationship between values and motivation (motivated behaviour).

Some Concluding Remarks and Implications for the Industry

During the period from the introduction of farmed Atlantic salmon to the French market in the late 1970s and untill today there has been a tremendous development throughout Europe. At the one side there have been great changes in the intermediate market with a focus on a much wider use of farmed salmon. At the other side the product itself has seen a widespread consumer acceptance in the market. Perhaps due to this effort, or at least partly, the salmon have seen a significant market penetration with a wide use across many market segments. This development may of course also be attributed to the improvements in effectiveness in production – which in turn has enabled the suppliers to reduce the price and at the same time deliver a relatively consistent technical quality. This has again moved Atlantic salmon from being a (very) expensive product to a relatively affordable and accessible food product for the average European consumer. Some would also claim that there has been a corresponding move for farmed salmon on the downward slope on the "vanity scale" from being an exclusive product for special occasions to a more common item if not an ordinary everyday product today. This apparent move on the "vanity scale" toward a broader use and a wider potential customer base has opened a whole array of market possibilities for the industry.

The growing concern about food safety, increasing demand for quality guaranties, and high consumer awareness on these issues spurs the industry to develop a consistent production and processing strategy that meets future consumer demands. Together with the environmental and welfare issue the GMO questions represents the main "joker" in the development of the future market for farmed salmon. All-in-all one impression is that there will be an increasing demand for increased communication attached to the tangible product (the fish) as well as the intangible aspects related to technology.

References

American Demographics. 2003. Indicators: Snapshots of todays consumer: Eco-friend or foe? October 2003, pp. 12–13.

Bech-Larsen, T. and Grunert, K.G. 2003. The perceived healthiness of functional foods: A conjoint study of Danish, Finnish and American consumers' perception of functional foods. *Appetite* 40:9–14.

Beharrell, B. and Denison, T.J. 1995. Involvement in a routine food shopping context. *British Food Journal* 97(4):24–30.

Bettman, J.R. and Sujan, M. 1987. Effects of framing on evaluation of comparable and non-comparable alternatives by expert and novice consumers. *Journal of Consumer Research*, 14:141–154.

Bredahl, L. 2001. Determinants of consumer attitudes and purchase intentions with regard to genetically modified foods – results from a cross-national study. *Journal of Consumer Policy* 24(1):23 61.

Charng, H.-W., Piliavin, J.A., and Callero, P.R. 1988. Role identity and reasoned action in the prediction of repeated behaviour. *Social Psychology Quarterly* 51:303–317.

Frewer, L., Howards, J., and Sheperd, R. 1996. The influence of realistic product exposure towards genetic engineering of food. *Food Quality and Preferences* 7:61–67.

Furst, T., Connors, M., Bisogni, C.A., Sobal, J., and Falk, L.W. 1996. Food Choice: A conceptual model of the process. *Appetite* 26:247–266.

Hallman, W.K., Hebden, W.C., Cuite, C.L., Aquino, H.L., and Lang, L.T. 2004. Americans and GM Food: Knowledge, Opinion and Interest in 2004. Publ. no, RR-1104-007, Food policy Institute, Cook College, Rutgers State University, New Jersey

Hansen, K., Fruensgaard, L., Nilssen, F., and Johnsen, O. 2002. The market for farmed salmon in Spain and Germany – A study of supermarket and wholesale buyers, MAPP Research Centre, Aarhus, Denmark.

Honkanen, P. and Verplanken, B. 2004. Understanding attitudes towards genetically modified food: The role of values and attitude strength. *Journal of Consumer Policy* 27:401–420.

Honkanen, P., Olsen, S.O., and Verplanken, B. 2005. Intention to consume seafood – the importance of habit. *Appetite* 95:161–168.

Jaeger, S.R. and Meiselman, H.L. 2004. Perceptions of meal convenience: the case of at-home evening meals. *Appetite* 42:317–325.

Johnsen, O. and Nilssen, F. 2001. The future for salmon in Germany, Research report 10/2001 Fiskeriforskning, Tromsø.

Kahneman, D. and Tversky, A. 1979. Prospect theory: An analysis of decision under risk. *Econometrica* 47:263–279

Köster, E.P. 2003. The psychology of food choice: some often encountered fallacies. *Food Quality and Preferences* 14: 359–373.

Kuznesof, S. and Ritson, C. 1998. Consumer acceptability of genetically modified foods with special reference to farmed salmon. *British Food Journal* 4/5:39–47.

Leek, S., Maddock, S., and Foxall, G. 2002. Situational determinants of fish consumption. *British Food Journal* 102:18–34.

Lindeman, M. and M. Väänänen. 2000. Measurement of ethical food choice motives. *Appetite* 34:55–59.

Magnusson, M.K. and Hursti, U.-K.K. 2002. Consumer attitudes towards genetically modified foods. *Appetite* 39:9–24.

Marshall, D. 2001. Food availability and the European consumer. In *Food, People and Society* Frewer, L., E. Risvik & H. Schifferstein (Ed.). Springer, Berlin, pp. 317–338.

Miles, S. and Frewer, L.J. 2001. Investigating specific concerns about different food hazards. *Food Quality and Preference* 12:47–61.

Miles, S., Brennan, M., Kuznesof, S., Ness, M., Ritson, C., and Frewer, L.J. 2004. Public worry about specific food safety issues. *British Food Journal* 196(1):9–22.

Nilssen, F. and Monfort, M.C. 2000. The future for salmon in France., Research report 8/2000 Fiskeriforskning, Tromsø.

Ritson, C. and Petrovici, D. 2001. The Economics of Food Choice: Is Price Important? In *Food People and Society*, Frewer, L., E. Risvik and H. Schifferstein (Ed.) Springer, Berlin, pp. 339–363.

Shepherd, R. 1989. Factors influencing food preferences and choice. In *Handbook of the Psychysiology of Human Eating*, Shepherd R. (Ed.) John Wiley & Sons Ltd., pp. 3–23.

Slovic, P. 2000. (ed). *The Perception of Risk*. Earthscan Publications Ltd, London.

Southgate, D.T. 1996. Dietary Change: changing patterns of eating. In *Food Choice Acceptance and Consumption*, Meiselman, H.L. and H.J.H. MacFie (Eds.), Nlackie Academic & Professional, London, pp. 365–391.

Sparks, S. and Shepard, R. 1994. Public perception of the potential hazards associated with food production and food consumption: an empirical study. *Risk Analysis* 14(5):799–806.

Tidwell, J.H. and Allan, G.L. 2001. Fish as food: aquaculture's contribution. EMBRO reports 2(11):968–963.

Verbeke, W. and Vackier, I. 2005. Individual determinants of fish consumption: application of the theory of planned behaviour. *Appetite* 44:67–82.

VOA News. 2004. Science in the News—Pollutants in Salmon. http://www.voanews.com/specialenglish/Archive/a-2004-02-03-1-1.cfm. Accessed: July 2, 2007.

Warde, A. 1997. *Consumption, Food, and Taste*. Sage Publications, London.

Chapter 15
Aquaculture Innovation and the Role of Popular and Trade Media

Kenny McCaffrey

Aquaculture has existed for thousands of years, yet as the capture fisheries decline precipitously aquaculture has recently gained fresh prominence, and intensive aquaculture now represents an innovative departure from our past practices. The technological advances propelling innovative aquaculture are sometimes complementary to integration with the capture fisheries, and at other times disruptive. Together or separately, these two aspects of aquaculture innovation – and other, less obvious effects such as disruption of viewscapes or recreational areas – affect coastal communities and everyone in the aquaculture value chain. Yet to portray aquaculture innovation and social transformation in such sweeping terms obscures the day-to-day, pragmatic business of being in the aquaculture industry, and how the activities of the industry, including the adoption of new technologies, are discussed and reported within the industry and to the public.

Kenny McCaffrey, as the long-standing editor of the trade publication, *Fish Farming International* (FFI), has a unique perspective on the development of the global aquaculture industry, changes in the industry's self-conception, and the content and purpose of media stories about aquaculture. Popular media coverage of aquaculture often consists of sensational stories about genetically modified fish, the use of other biotechnology, the food safety of farmed fish, or stories about environmental damage. The writers of these stories often have little familiarity with the aquaculture industry, and may be accused of unreflective use of controversial evidence. Industry reporting, by contrast, typically addresses issues bearing on industry practices, markets, regulation and related matters of interest to business-people. Aquaculture trade publications often cover stories arising from the aquaculture industry itself, and sometimes take up stories derived from parallel industries such as soybean agriculture, of interest to the aquaculture industry because it has a bearing

K. McCaffrey
Editor, Fish Farming International
e-mail: kenny.mccaffrey@fisfarminginternational.com

Editors' note: Mr. Kenny McCaffrey, Editor, *Fish Farming International*, was interviewed at the offices of *Fish Farming International*, Albert House, Singer Street, London, March 9, 2007 by Dr. Keith Culver. This chapter is an edited version of that interview, prepared by Drs. Castle and Culver. Mr. McCaffrey approved the final version of the chapter.

K. Culver, D. Castle (eds.), *Aquaculture, Innovation and Social Transformation*,

© Springer Science+Business Media B.V. 2008

on fish feed. As McCaffrey reports his early experience in aquaculture journalism, he initially conceived of his role as limited to reporting exclusively on aquaculture and related industry activities, but he has since come to believe that *FFI* can also address the gap between the conventional roles of trade publications and popular media. Sometimes this means setting the record straight about actual industry practices and their effects in cases where sensationalistic reporting misrepresents aquaculture. As McCaffrey says, this is bringing about a change in the way that the aquaculture industry views itself. "What I find is that fish farmers are beginning to realize they are part of the food chain, realize they are part of the food industry, and that they are not a special case, and they really have to make themselves heard to get their story across. And they realize how important the story is now. I think formerly they didn't. So I think the fish farming industry is beginning to be less scared, say, of addressing the public – speaking straight to the public."

There are many technologies in the aquaculture industry that could be of general interest to in the public media and to readers of *FFI*. For example, net pen technology is sometimes portrayed as inadequate in popular media stories reporting escapes of farmed fish (Macaskill 2004). Similarly, stories regarding genetically modified fish appear in the popular media (Black 2004). McCaffrey, however, considers the aquaculture industry to be technologically advanced rather than a victim or proponent of inadequate technology, to the point where general public perception lags far behind the facts in their understanding of the effectiveness of aquaculture technologies. Technology has, for example, enabled aquaculturists to better mimic nature, to feed fish remotely and with Doppler feeding systems that automatically adjust feed amounts to fish demand. McCaffrey believes that consumers are "ready to be impressed" by the technology and that it is "up to the fish farming industry to realize that." The public is asking more sophisticated questions about the aquaculture industry and the technology it uses, which means that the curiosity and background knowledge of the public, particularly consumers, is on the rise.

Net pen technology is an interesting case in point because of reports about massive fish escapes, for example, in the Shetlands. People are often distressed by these escapes because of the large numbers of escaping fish, and net pen technology is held responsible in the public's eye. McCaffrey points out that reports usually do not delve into the issue of whether escapes are in fact harmful to the surrounding environment. "They don't breed, and somebody from an NGO reporting on wild salmon interests admitted to me that there's really no chance of these fish making any impact on the local fish that have taken, say 10,000 years to become established. So over the course of the short term they actually die off quite quickly as they haven't a clue what to do, and are without food. That's not really much of a problem." As McCaffrey points out, these fish are easy to catch and return to pens because they tend to school around the pen from which they escaped. The fish farming industry would nonetheless do well to address these issues about net pen technology, since the public is concerned by media reports that the technology is failing and causing environmental damage.

In the face of sometimes hostile relations between proponents and opponents of aquaculture, it is worth considering more deeply the possibility that more accurate and thorough reporting, supported by the aquaculture industry and *FFI*, could lead to improved public understanding of the technologies used in aquaculture. A shift of this kind seems most likely in the case of net pens and escapes, yet public acceptance of other aquaculture technologies might not be so quickly or easily won. McCaffrey observes that use of drugs in fish farming, presence of colourants in feed, and capture of wild fish to make fish meal are problems in the public's eye that "can't be gainsaid but you can certainly tell the story properly, in such a way that people would be impressed about doing that" – understanding that whatever else might be said about the aquaculture industry, it faces its issues squarely and recognizes that misinformation is ultimately harmful to the industry. McCaffrey explained further that: "Somebody told me from a very, very big, well known German chemical company - who had a big but hitherto not well-publicized position in the industry – that they are beginning to be more publicized in the industry." This change in policy is "purely because they see that consumers, the consumers of today and of tomorrow, basically from undergraduate age on downwards, are very, very keen on finding out what the information is on what they're eating." Implied in what McCaffrey says is the idea that general education, and specifically some basic scientific knowledge is a major determinant in the acceptance of new technologies. "People are much more knowledgeable about what it is in the food industry and what is in the environment and they're able to make choices based on that."

The role of *Fish Farming International* in the coverage of science and technology now and in the future relates to the industry's self-conception and its interface with the public. According to McCaffrey, *FFI* has been devoted almost exclusively to new technology in response to the high rate of technological change in the industry. While the overall rate of technological change has slowed somewhat, the reported technology domain has broadened to include related issues such as the integration of fish life cycles into aquaculture, vaccines, broodstock development, use of GIS and the development of polyculture. McCaffrey observed of polyculture, for example, that "aquaponics is a new sector that's building up that comes a lot from urban farming." Land-based fish farming is another important area of technological development, as is its opposite, offshore aquaculture in high-energy sites. Technological expansion is also evident in development of alternative feeds. McCaffrey notes that "This is potentially great news for the marketers, because people are beginning to find ways of, in the case of polycheat worms for example, farming the feed; and on the other hand the big, big companies are finding ways of using natural products to replace other ingredients in feed that we possibly didn't really publicize in the first place." There is work to be done, however, in producing fish feed that has the appropriate nutritional balance while maintain desirable physical properties such as whether it sinks or floats. While consumers are probably not so concerned about binders and pellet size, they are concerned about what goes into fish feed, and whether healthy fish foods will make for healthy fish products. This point raises the issue of whether the aquaculture industry should respond to concerns about the healthfulness of their foods by switching to organic feed and culturing systems. McCaffrey has reservations regarding the organic movement, which he describes as "more a marketing term than anything," a term which does not really allay concerns about food quality and safety since it is not defined consistently and clearly in different countries, especially for the seafood industry. "Until that's done there's a chance that people will jump on something about organics that is not quite organic" and in that way devalue aquaculture producers' attempts to respond to consumers' concerns.

A related issue associated with the healthfulness of consuming farmed fish is the current interest in the levels of Omega 3 fatty acids in farmed fish. McCaffrey described a recent situation where he was invited to join a radio program aimed at consumers, specifically women who do the bulk of household food purchasing. He was told the discussion would be about Omega 3 fatty acids in farmed fish, and that the discussion would revolve around whether Omega 3 was a finite resource because the wild fish stocks used to produce fish feed were declining. "They wondered whether the growth in fish farming, given that it uses fish as feed, would cause us to run out of capture stocks." The shift of the topic from the presence of healthy compounds in farmed fish to the notion of a managed fishery signaled to McCaffrey that there is still significant work to be done in communicating about the work done by the international fish meal and fish oil organizations to manage the stocks that they have. "The fact is that they are catching wild fish. So it's a case of finding out ways of making the best of what we've got. I think a lot of companies are doing very good work on that, and there is a lot of stuff that can go into the public domain."

In other areas of agriculture there is interest in enhancing the nutritional composition of the food, such as in the case of Omega 3 fatty acids in chicken eggs, and there has been similar interest in increasing Omega 3 fatty acids in farmed fish (Purdue 2003). There has not, however, been a concerted effort in the aquaculture industry to use genomics or genetics to achieve these goals. Instead, McCaffrey pointed out that significant effort has gone into using science and technology to "fingerprint" fish to tailor them to specific markets. McCaffrey considers this one of the "great biotechnology advances of the last ten years of fish farming. You can actually secure your future supplies, in business terms, you can get to this position where you've got a type of strain of a farmed fish that fits a particular market and this is all to do with genetics, but instead of manipulating genetics, it is identifying genetics." The Norwegians, in particular, have done much of this research with government support for the industry, a situation which has given the Norwegians a significant competitive edge in aquaculture research. "As far as other genetic work goes," McCaffrey observed, "the industry has been very loath to associate itself with modifying genetics, and understandably so." Following long and acrimonious disputes about the use of genetic technologies in crop biotechnology, many people in the industry simply do not want to talk about the genetic modification of fish. McCaffrey predicts, however, that companies holding biotechnology patents, such as Aqua Bounty (Hew 1996), might one day find themselves in a favorable regulatory and consumer environment. Aqua Bounty is an interesting case, since their AquAdvantage growth enhanced fish have satisfied early stages of the process required for approval from the United States Food and Drug Administration

(ABT 2006; Weintraub 2006). "I think they're sitting on something that could make them a lot of money if the wind blows their way."

Some technology is finding its way into the aquaculture industry because the industry is responding to consumer concerns at a time when inexpensive technological solutions are coming available. For example there is recent work on fish genetics that makes traceability of fish like farmed tilapia possible. As McCaffrey says, one can now trace "from the egg right through to the market, all contained on a chip that is in a gill tag, which is absolutely amazing to think that it can be done economically." The impact that technology innovation has on the development of the industry is considerable: "When I first became editor ten years ago it wouldn't have even been countenanced because that type of thing was too expensive." Now, consumer and industry interest in traceability is technologically feasible, and it is a "good news story" in the aquaculture industry waiting to be told.

McCaffrey argues that there are many good news stories to be told. "The technology for vaccinating each 5-10 gram fish in a batch of 10,000 – it's that type of technology that's really come along. Maybe nanotechnology is the next thing that's going change biotechnology in fish farming." Traditionally, however, the industry has fought a rear-guard battle against media mostly interested in looking for environmental problems, the need for aquaculture versus capture fishery, and the "visual pollution" of near-shore aquaculture sites (see Chopin in this volume). But to get the media on board with the beneficial aspects of aquaculture will require considerable education efforts. "Once they've realized what the industry's doing and once they're comfortable with it, they'll starting telling good stories like what we've been telling, that's their job. There are a lot of great stories to tell that come from the industry, and I think the industry is just beginning to realize that. I think the industry might need one big PR company."

McCaffrey admits that the industry is only gradually coming to grips with the need to engage the public, partly motivated by their realization that the consuming public is often alarmed by the appearance that a food industry is rapidly adopting unfamiliar new technologies. The main driver for the turn to public engagement, however, is a change in attitude, as aquaculture firms retreat from understanding their business as production-driven, and begin to understand aquaculture as a market-driven industry. "I think that when it was production driven, everybody was almost like in the old Soviet Union, they used to say never mind the quality, fuel the works sort of thing. And if they could say 'we produced a million tons of salmon' they were doing well, but a million tons of salmon at sort of dumping level prices, it's not really much use to anybody." Consumer demand for safe and healthy aquaculture products has instigated the change to a market-driven industry. "I think that is a good thing in any industry, that your customers tell you what to do, and that you're responding to your customer [...] I think the fish farming industry is by and large producing products that they are being asked for."

Despite the fact that the aquaculture industry responds to consumer demands with selective use of new technology, interest in aquaculture innovation is localized. Although *FFI* is based in London, McCaffrey hears very little about aquaculture locally. Yet, he contends, if he were in Iceland or Norway, technological innovation would be front page news. "In Norway they're very proud of biotechnology – there's a lot of amazing stuff that comes out of there. Stuff that's actually really interesting that you can make [TV] programs from, in the quest to change how they produce their food. The example I gave of [gill tags in Tilapia developed in] Singapore, that was actually on their version of a business news program." At *FFI*, McCaffrey and colleagues "live and breathe new technologies, new products, new moves" and so it was heartening to see that a positive story about traceability technology could make its way into the mainstream media. "It's an old journalistic saying that dog bites man is not a story, but man bites dog is. In general, newspapers writing about what happens in fish farming aren't really interested in the good news, the good news comes after the bad news I think, and there's a bad news fatigue. When there have been some scare stories that hit the media about fish farming, you can guarantee that within a couple of months, some of the news outlets will be looking for the opposite story because they want to be different. So, in a back-handed way you eventually get your good story out."

Even as consumer concerns are handled and sometimes mishandled, advanced technology such as biotechnology and genomics are slowly making their way into the aquaculture industry. One particularly prominent example can be found in genomics-enabled marker assisted selection of broodstock exhibiting preferred characteristics. Yet the quite limited use of this technology, and the lack of public awareness that it may become mainstream in the industry, raises questions about the preparedness of the aquaculture industry for media interest in genomics and biotechnology in aquaculture. McCaffrey thinks that some larger companies, particularly those with established public relations departments, will manage media interest in genomics and biotechnology without difficulty. So too will seasoned companies like Aqua Bounty, developers of growth-enhanced, genetically modified salmon. "I know from my days, even before I became editor of *Fish Farming International*, I knew that the guys who were running Aqua Bounty were really having to be sort of guarded at certain exhibitions and conferences around the world because of the threat from NGOs and protesters against transgenic salmon. After several years of that you get pretty adept at handling the media industry, and more importantly, what you say and how you say it."

This is not to suggest that the aquaculture industry should manipulate the media and the public with its news releases and responses to reporters. About biotechnology and genomics in particular, McCaffrey counsels "honesty is always the best policy" even if the industry finds it beneficial to put a positive spin on a story about fish farming. "I think one of the benefits of the fact that it's new, and it's technology-based, and it's transparent... [W]hen these new consumers come along, thirsty for knowledge, they'll find more knowledge from fish farming than any other food source or protein production source. I think you'll find that maybe the beef industry, the lamb industry, the poultry industry will have to open up more than they have other done, whereas the fish farming industry started from being open, and they're maybe now starting to control what information they're getting over." This comparison is anchored in what McCaffrey considers to be a general pride of those in the industry, which he suspects might have something to do with the fact that aquaculture is rural. "In general, a fish farmer is very proud of what he does and that comes across [. . .] maybe it's another thing of town folk and rural people. Even if people come from the town, most fish farming is done in rural communities, and there must be an effect on how they put themselves across to the wide world."

When questioned about how industry might foster appropriate relations with media to convey aquaculture's story to the general public, McCaffrey takes a positive attitude about what must be done: "[I]nvite them in, really. It's always said that you shouldn't shun the media; if you shun the media they'll definitely look for bad stories." He imagines a scenario in which a giant public relations firm manages all of the media encounters of the industry. Newspapers would catch the front news stories, but also feature news about aquaculture. There could also be opportunities for documentaries, some of which could be done by the main stream media. While McCaffrey doubts that many people "would actually go out of their way to watch fish farming television news," there may be multi-media uptake of the work of the media company Intrafish, which has started a web TV station (NHST 2006). "With multimedia it's not just newspaper, websites, and all the rest of it, but things coming through wi-technology – all of the ways people are taking technology into the home, these are things which will affect how our industry tells its story."

Turning to the future of aquaculture, and the innovations in biotechnology and genomics, McCaffrey anticipates that the media and communications will play a major role in the social acceptance or rejection of the technology. The fish farming industry has been built on biotechnology and will continue to rely on technology-intensive methods. "I think the growth or development or changes made in media and how media are delivered will have wide-ranging effects on the fish farming industry. I think the fish farming industry is in a very good position to take advantage of that because most fish farmers are already experts on latest communications technology, which is actually used to grow their fish. So they're in a good position, if you think about it, and it's a way of getting them all together to use their strengths to get their message across. I think this will happen in the future. I think the future's quite bright for fish farming."

References

Aqua Bounty Technologies (ABT). 2006. Aqua Bounty successfully completes key study for AquAdvantage Salmon; FDA accelerates review process. Press release, December 4, 2006. http://www.aquabounty.com/media/AquAdvantagecompletionofstudiesfinal.html. Accessed: August 6, 2007.

Black, R. 2004. Push to have GM salmon approved. *BBC News*, March 24, 2004. http://news.bbc.co.uk/2/hi/science/nature/3565041.stm. Accessed: August 6, 2007

Hew, C. L., Fletcher, G. L. U.S. Patent 5,545,808, 1996. Transgenic salmonid fish expressing exogenous salmonid growth hormone. Date: August 13, 1996.

Macaskill M. 2004. McConnell's brother faces salmon inquiry. *The Times Online*, September 19, 2004. http://www.timesonline.co.uk/tol/newspapers/sunday_times/scotland/article484552.ece. Accessed: August 6, 2007

NHST Media Group. 2006. DN-TV starting up. Oslo, September 13, 2006. http://www.nhst.no/reports/press/article2369.ece. Accessed: August 6, 2007.

Purdue University. 2003. Purdue University Design Fishes' Diets to make you Healthier. *Science Daily*, January 1, 2003. http://www.sciencedaily.com/releases/2003/01/030101223108.htm. Accessed: August 6, 2007.

Weintraub, A. 2006. Salmon that grow up fast. *BusinessWeek*. January 16, 2006. http://www.businessweek.com/magazine/content/06_03/b3967111.htm. Accessed: August 6, 2007.

Part VI
The Final Frontier: Integrated Coastal Zone Management

Introduction

Most of the world's population lives within a 2 km of a body of water. This fact presents a problem for aquaculture: new aquaculture sites must join a set of established uses of land and water, and existing aquaculture sites can face pressures from established resource users whose evolving use of land and water is incompatible with aquaculture. Given that technologies for high-energy offshore sites are in their infancy compared with low-energy sites in coastal zones, aquaculture sites have often been sought in near shore coastal zones. Aquaculture has not been always been a welcome newcomer to coastal zones, particularly in industrialized countries where few people work as fishers or aquaculturists, and otherwise have limited daily interaction with water for travel, trade and food. Coastal areas are more typically associated with nostalgic images of rugged, plain-speaking and plain-living fishermen, sunsets over bucolic sea-side cottages, and long strolls on unperturbed beaches.

Consider, for example, the growing controversy in the tiny town of Vinalhaven, Maine regarding a 500 acre inlet on a 23-square mile island. As the New York Times reports (NY Times 2007), the State of Maine granted a 10 year permit to a local developer for an oyster farm. The farm has been met with undiluted criticism, and worse, vandalism. It mattered neither that the developer is a local resident known for good aquaculture management, nor that he posted a bond with the State to indicate his commitment to sound environmental practices. The issue, as the director of the Maine Aquaculture Association put it, is that "wealthy summer folk want to use the waterfront for recreational purposes, not for commercial purposes. We have that conflict up and down the coast of Maine." Indeed, the owner of a local kayak company compared the siting the oyster farm in the Basin, as it is known locally, to defacing a priceless work of art. This sentiment is echoed by another resident: "I think oyster farming is great. But not in that spot. The natural value of the Basin is far greater than its commercial value."

Disputes over siting of aquaculture are similarly problematic on Atlantic and Pacific shores of Canada. A moratorium on new aquaculture sites in British Columbia, originally imposed to quell disputes, has only recently been lifted. How can similar conflicts, like the one erupting in Vinalhaven, be resolved? To begin with, it helps to recall the distinction between *technological* innovation and ensuing change in aquaculture, and *social* innovation and ensuing change in local, social contexts. Given that commercial operators must have viable sites for their enterprise, disputes about aquaculture siting are not really about correct siting and its use of technological innovation. Instead what makes siting contestable is the use of new and old technology in places where it is no longer wanted or not wanted in the future. The core problem is change, in the sense attributed to Burke in the introduction. Aquaculture sites threaten to change actual coastal zone use, or they threaten ideals about what coastal zone use ought to be. The act of siting an aquaculture operation is therefore a potent source of social transformation.

In the first chapter of this section, Dan Lane and colleagues try to defuse siting disputes by containing them in an ostensibly objective and neutral decision framework. The integrated management tool they developed attempts to collect, rank, and decide upon the uses, values, and the strength of participants' commitment to particular outcomes. The framework begins with a multi-criteria evalution of the commercial viability of marine aquaculture sites, then adds an overlay of participant evaluation of the sites based on non-commercial criteria. The resulting decision support model is intended to provide decision makers with comparative assessments of sites, and an evaluation procedure to make final coastal zone governance decisions. It is this last phase of the model that worries Jamey Smith, a long time aquaculture manager from New Brunswick. Smith lauds the multi-criteria approach for its ability to draw out and keep track of different positions on aquaculture sites. His concern is that the decision model does not make explicit whose values are ultimately going to do the heavy work of arriving at a siting decision. If the siting decisions simply go back to the regulators, does Lane's approach really improve upon less formal methods? Smith's criticism is expanded upon by Thierry Chopin, the second commentator in this section. Lane's model presumes tension beween aquaculture and other uses of coastal areas, leaving Chopin to wonder if presumed complementarity of uses would not be a better starting point, one which would lead to more socially harmonious outcomes. Chopin's second major concern centres on the decision model's realism – whether it captures all facts salient to a sound siting decision. While this issue lies more on the side of wise deployment of aquaculture science and technology, it is relevant because a decision process for resolving social tensions about siting needs to start with a list of truly viable sites. Like Smith, Chopin is fundamentally unpersuaded that the messy process by which disputes are normally resolved can be smoothed with a process that promises formal objectivity. The decision model, Chopin argues, is only as good as its inputs; a formal process does not prevent lack of transparency or accountability.

References

Berstein, F.A. 2007. When conflict washes up on a quiet Maine island. *New York Times*, Sunday August 12. National Perspectives p. 16.

Chapter 16
Integrated Systems Analysis for Marine Site Evaluations and Multicriteria Decision Support for Coastal Aquaculture

Daniel Lane, Wojtek Michalowski, Robert Stephenson and Fred Page

Introduction

Traditional fisheries are based on wild, spatially uncontrolled stocks whose habitats are the coastal and marine environments. Aquaculture attempts to control and manage the supply of products from the sea by establishing designated marine farms. In recent years, the aquaculture industry is growing more rapidly than all other animal food producing sectors (Allen et al. 1992; Fletcher and Neyrey 2003). However, aquaculture has also been challenged by complex interactions between marine resources, ecosystems and multiple resource users. An integrated valuation approach that incorporates the ecosystem, aquaculture, and other marine activities is needed to promote sustainable development in the coastal zone and to find feasible solutions to resolve conflicts (Nath et al. 2000).

While the benefits of coastal aquaculture have been widely recognized (CAIA 2003), Barg (1992) noted "... aquaculture may increasingly be subject to a range of environmental, resource and market constraints. Aquaculture is competing for land and water resources, which in some cases resulted in conflicts with other resource users. Also, there is growing concern about the environmental implications of aquaculture development, comprising the adverse effects of aquaculture operations on the environment as well as the consequences of increasing aquatic pollution affecting feasibility and sustainable development of aquaculture." Similarly, Brindley (1991) pointed out the necessity to evaluate coastal aquaculture from different points of view and that a multi-disciplinary team, including specialists involved in decision making, should consider the economic, social, cultural and environmental aspects to assess all possible impacts of coastal aquaculture development (Lane and Stephenson 1998). At the Department of Fisheries and Oceans, the Oceans Action Plan featured in the recent 2005 federal budget (Department of Finance Canada 2005, p. 194) acknowledge that improved integration of marine impacts is an important matter for management of our ocean ecosystems.

D. Lane
School of Management, University of Ottawa, Ontario, Canada
e-mail: dlane@uottawa.ca

K. Culver, D. Castle (eds.), *Aquaculture, Innovation and Social Transformation*,
© Springer Science+Business Media B.V. 2008

This discussion paper focuses on coastal aquaculture developments around Grand Manan Island, New Brunswick, in the Bay of Fundy region of Atlantic Canada. The objective of the paper is to develop a framework for evaluating the integrated ecosystem and the socioeconomic impacts of alternative uses of the coastal zone for aquaculture as well as more traditional uses. Overall, this research project seeks to develop an integrated, multicriteria decision support system for the evaluation of marine sites by multiple participants in support of marine governance and decision-making processes in the coastal zone.

Methodology

There are three methodological structures used in the framework model development exercise. These are depicted in Fig. 16.1 and are described as follows:

1. *Marine site components* – describes the geophysical state of the ecosystem area of interest using spatial and temporal GIS indicators;
2. *Selected site specific valuation* – assigns areal values to the inventories of the principle primary resources recorded in the database that are present in selected marine sites; and,
3. *Site comparison and multicriteria group ranking* – ranks alternative marine sites according to the preference tradeoff information about the ecosystem inventory provided by the various participants in the marine governance system using multicriteria decision analysis.

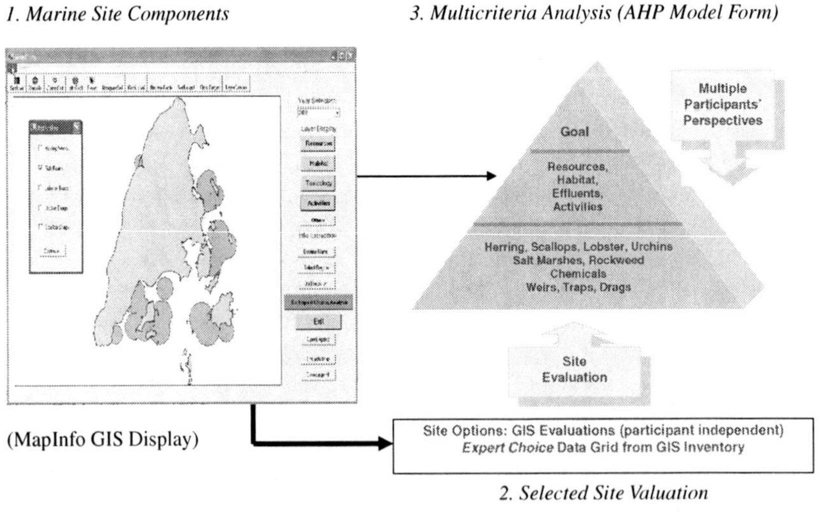

Fig. 16.1 Main methodological components for the coastal aquaculture problem

Marine Site Components. The Grand Manan Island ecosystem is described by four key spatially defined components:

(1) Biological – the dynamic spatial distributions of natural resources of several select species, including subcomponents for herring schools staging areas and inshore feeding areas distributions, lobster molting and feeding area distributions, scallops and urchins resident population area distributions;
(2) Habitat – the spatial distributions of natural habitats of the area including the subcomponents rockweed, salt marshes, benthic community and bathymetry descriptions, and current flow areas;
(3) Effluents – the pollution of potentially harmful effluents arising naturally from the ecosystem and due to the fish farm wastes; and
(4) Activity – the areal distribution of human commercial activities including the subcomponents of traditional herring weirs, lobster traps, scallop and urchin dragging gear, recreational activities, designations for reserved aboriginal activities, as well as fish farms sites.

GIS area data for Grand Manan sources for each of the components and their subcomponents described above are compiled into a databased inventory of the status of marine and coastal sites around the Island. Figure 16.2 illustrates the complete set of 9 overlapping subcomponents currently in use for this study area.

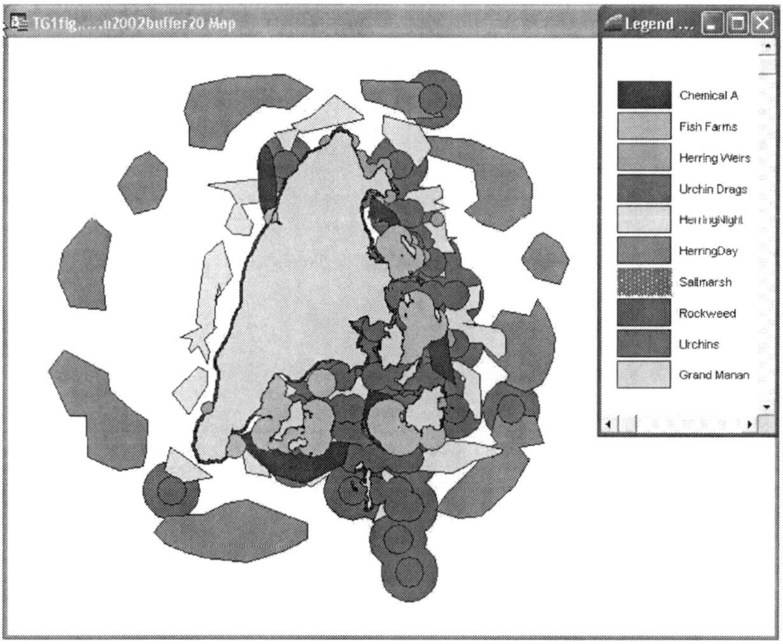

Fig. 16.2 MapInfo overlay information for the whole study region of Grand Manan Island

Selected Site Specific Valuation. For each of the subcomponents of the natural ecosystem a unit area value denoting "yield" is assigned. Full valuation of a selected marine site takes into account the presence of singular and overlapping layers of resource, habitat, effluents and activities. If layers do not overlap, the Yield values are a simple function of the area of the non-overlapping layers for the components and the unit area Yield values provided as input for each Grand Manan subcomponent. For overlapping subcomponents, "overlap rules" are determined to deal with these more complex valuation cases (Table 16.1). For a selected site, Yields from all subcomponent items including overlapped and non-overlapped areas form a vector of values for the site that characterize its important ecosystem inventory. This vector is provided as input to the decision making participants who are charged with evaluating and ranking the site against other site options, for example, different site comparisons, or the same physical area with or without selected commercial activities, and so on.

Site Comparison and Multicriteria Group Decision Ranking. A comprehensive evaluation of ecosystem components and their subcomponents are provided from the attributions of importance assigned by different participants in the decision making problem. The Analytical Hierarchy Process (AHP) is adopted to evaluate different participants' perspectives on the ecosystem characterizations in a common hierarchy model built for all participants (Saaty 1980). The hierarchy specifies the Goal, Components, and Subcomponents levels of the Grand Manan ecosystem as presented in Fig. 16.3 below for the "Local Communities" group.

Weights are assigned to each level of the hierarchy from the participants' feedback in the form of pairwise comparisons among the pairs of the elements in each level of the hierarchy. The Analytic Hierarchy Process determines the scaled weights applied across the elements by soliciting feedback on the ordinal comparison between each pair of elements. For example, suppose a level of the hierarchy contains elements A, B, C. If the ordinal comparison of the importance of A is judged to be 2 times more important than B, and the importance of A is 4 times more important than C, and B is 2 times more important than C, then A has twice the weight of B and B has twice the weight of C, or, the effective relative scaled weights for A, B, C are (4/7, 2/7, 1/7).

Table 16.1 Summary of rules for single and multiple overlapping layers in pairs

	Yield affecting components			
Yield affected components	Resources	Habitat	Effluents	Activities
Resources	+	+	−	−
Habitat	+	+	−	−
Effluents	+	+	+	+
Activities	+	+	−	−

Note: Plus signs (+) denote that the overlapping effect on yields for the layer indicated at the left hand side is positive; minus signs (−) denote negative yields. Also, this is not a symmetric matrix. For example, the overlapping effect of Resources and Activities is negative for Resources, but is positive for Activities.

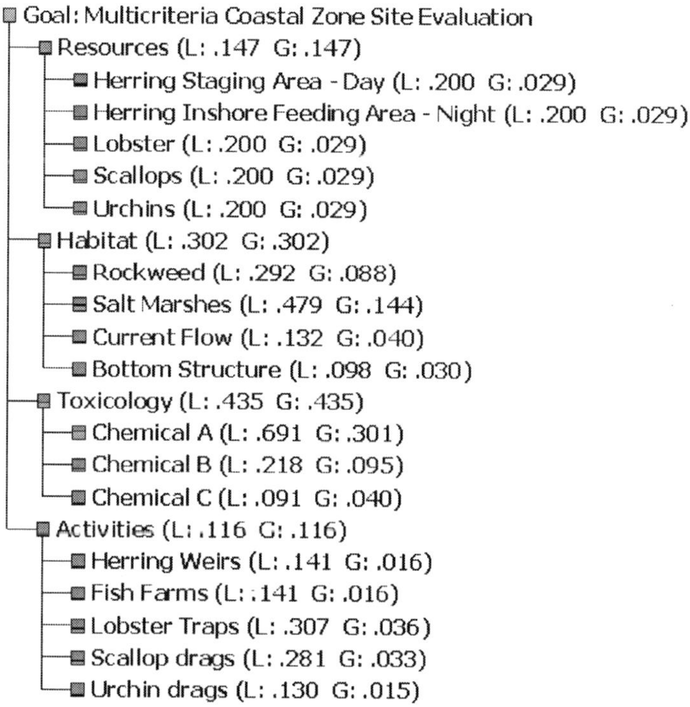

Fig. 16.3 The hierarchical model for the "Local Communities" participants group
Source: Output of problem formulation from *Expert Choice 2000*.

The participants denote the collective of groups or organizations that are the key decision makers in the governance process for coastal marine site use. These are represented by five groups in this framework: (1) Local Communities – residents of the coastal zone who are affected visibly by marine use activities like aquaculture; (2) Federal Scientists – mandated to apply the government's legislation in respect to federal marine use, e.g., the *Fisheries Act of Canada*; (3) Industrial organizations – seeking the rights of marine use for profitable activities; (4) Non Governmental Organizations – seeking to protect environmental concerns; and (5) Provincial Governments – overseeing the local government jurisdiction of the coastal zone that overlaps with federal marine waters. These groups all play a key role in influencing the privileges associated with marine use.

The attributed weights to the ecosystem components and subcomponents by the five participating groups are provided in Table 16.2. For example, Local Communities attribute most importance to the decision on marine site use to the Effluents (0.435) and Habitat (0.302) components. They attach less weight to Activities and Resources. Scientists, however, consider Resources as the most important component with highest weight (0.546) and Activities (exploitation) at the lowest weight (0.075). These assignments are consistent with the scientific mandate for conservation of natural resources. On the other hand, Industrial Organizations attach

Table 16.2 Attributed AHP weights of the 5 participants to the co and sub components

Components and subcomponents	Local communities	Federal scientists	Industrial organizations	Non-governmental organizations	Provincial governments
Resources	0.147	0.546	0.256	0.235	0.226
Herring Day	0.200	0.200	0.079	0.200	0.096
Herring Night	0.200	0.200	0.079	0.200	0.096
Lobster	0.200	0.200	0.276	0.200	0.330
Scallops	0.200	0.200	0.523	0.200	0.330
Urchins	0.200	0.200	0.043	0.200	0.148
Habitat	0.302	0.217	0.124	0.235	0.075
Rockweed	0.292	0.250	0.375	0.375	0.301
Salt Marshes	0.479	0.250	0.125	0.375	0.410
Current Flow	0.132	0.250	0.125	0.125	0.171
Bottom Structure	0.098	0.250	0.375	0.125	0.118
Effluents	0.435	0.163	0.082	0.439	0.185
Chemical A	0.691	0.691	0.691	0.691	0.691
Chemical B	0.218	0.218	0.218	0.218	0.218
Chemical C	0.191	0.191	0.191	0.191	0.191
Activities	0.116	0.075	0.538	0.083	0.514
Weirs	0.141	0.380	0.087	0.200	0.057
Fish Farms	0.141	0.156	0.147	0.200	0.358
Lobster Traps	0.307	0.326	0.264	0.200	0.158
Scallops Drags	0.281	0.095	0.462	0.200	0.158
Urchin Drags	0.130	0.044	0.039	0.200	0.269

most importance to traditional commercial Activities (0.538) and lower weighting to Habitat (0.124) and Effluents (0.082) components. For NGOs, Effluents from Fish Farms are regarded as the most important (negative) component in the system (0.439). The Provincial Government evaluates components in close comparison to the Industrial Organizations except their opposing differences with respect to the ranking of the Effluents and the Habitat components.

Application and Results

In order to test the significance of different attributions to the presence of the ecosystem inventories, specific experiments of selected marine areas around Grand Manan were chosen for evaluation and comparison. These selected areas included a number of different options: (1) evaluation of the total Grand Manan Island study area, (2) comparison of one marine site relative to an "ideal" or target valuations for the ecosystem subcomponents of the site, (3) direct comparison of two different areas, and (4) comparison of the same area with and without designated fish farms. The model's ability to select and explore different marine sites permit a wide rage of comparative results. For illustration purposes, the discussion below presents the comparisons results for two similar sized selected areas, T2 and T3 as shown in Fig. 16.4.

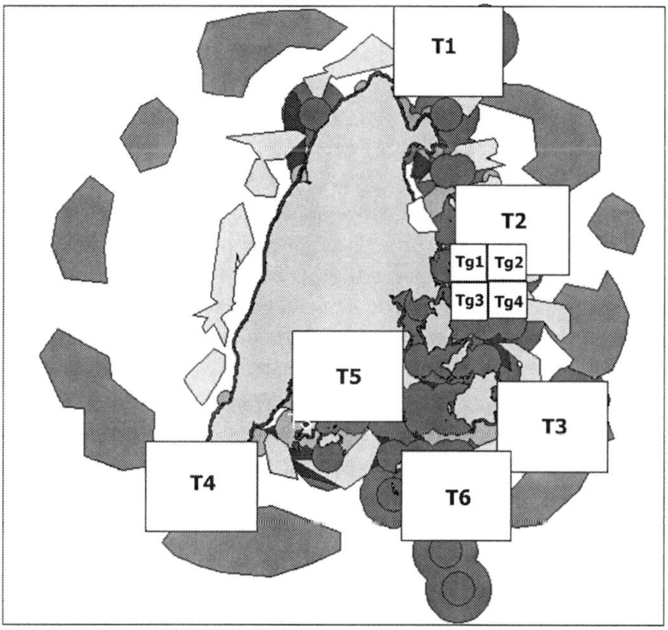

Fig. 16.4 Selected marine areas around Grand Manan Island for evaluation

The status quo position of two selected eastern Grand Manan areas, denoted T2 and T3, of approximately equal size were compared directly in absolute valuation terms. The attributed results provided in Fig. 16.5 demonstrate that the Provincial Government, NGOs, Industrial Organizations and Local Communities attribute more value to Site T2 (light) than Site T3 (dark) although differences vary among groups and with NGOs being indifferent between these two sites. Federal Scientists evaluate Site T2 as less valuable compared to Site T3. Generally

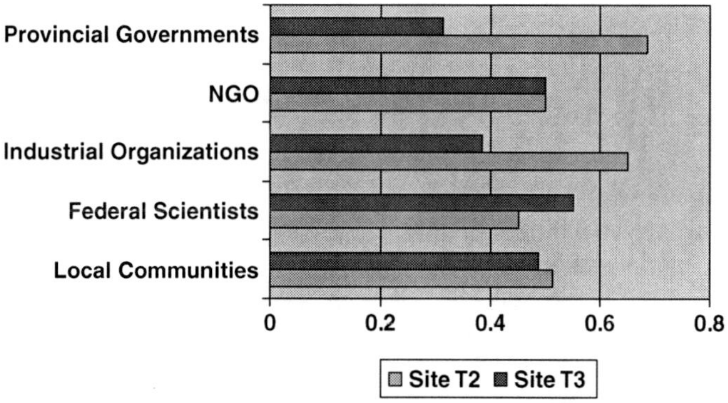

Fig. 16.5 Evaluation summary of the different rankings among 5 participating groups in comparing two selected coastal sites around Grand Manan of equal area

speaking, Site T2 is closer to mainland Grand Manan than Site T3 and therefore, Site T2 is an area in which there is easier access and more beneficial layers of Resources and Habitat that overlap. If pollution and overexploitation activities are avoided, T2 is expected to be a more profitable site than site T3 thereby enhancing the value of Site T2 for commercially oriented groups. Federal Scientists, however, by virtue of their attributed weights independent of the selected areas, are prone to consider the future value of Site T3 over the current value of Site T2 in their evaluations. This divergence on the value ranking of the two sites causes conflicts among participating decision makers. It is important to note this ranking divergence and to balance the interests of each of the different related groups accordingly. Differences between the government groups, federal scientists and provincial governments, are attributed to their fundamental positions. Federal scientists are mandated to protect and conserve the environment first, and are accordingly more concerned with negative impacts, for example, environmental pollution, or losses in resources. Provincial governments are equally concerned with the environment and the promotion of their local economies, e.g., employment and industry. They have more of an affinity in this case with the Industry group. Similarly, NGOs often react as the federal scientists as being more favourable to environmental principles of conservation of resources and mitigation of environmental damage. Local communities on the other hand often exhibit a mixture of support for industrialization and as well as clean environments, much like provincial governments in this case. Overall, the results of the participants' feedback are case dependent. Participants' feedback from the same group for other problem contexts may exhibit other orientations compared to the results illustrated here.

The framework provides the opportunity to explore the valuations of various marine sites under different assumptions, for example, including or excluding certain activities. It does not however provide a "best" solution for ranking selected sites or for choosing sites as fish farms. Rather, it provides support for the development of consensus solutions among the different participants. The use of multicriteria methods of sensitivity analysis also provides insight into why participants are similar and/or different.

Discussion

Attributed pairwise comparison trade-offs are obtained for the various participants in a structured decision framework used to develop priority rankings of selected and characterized Grand Manan ecosystem sites. Marine site valuation analysis assigns relative values to the multiple and integrated components and subcomponents of the ecosystem. These values are designed to be comparable across sites using a yield-value database for the subcomponents including their interactions with other subcomponents. Alterative valuation schemes for subcomponent interactions and cumulative effects are also being explored in support of decision making in this model.

The model provides a means of evaluating integrated marine activity – including or excluding aquaculture fish farms or any other marine activity. As such, the framework is designed to be used to evaluate participants' rankings of sites under various conditions of status quo or under altered conditions for comparison. This procedure permits, for example, the comparison of traditional activities, such as commercial herring weirs with or without fish farms present, compared to other activities. These comparisons are solicited because they are understood to be relevant to the problem context that is designed to model the problem at hand. It is incumbent on the decision analyst to design the appropriate decision hierarchy to contain the relevant and important problem components and to engage the key decision makers to determine their feedback. Alternative decision hierarchies and decision makers can also be explored in a search for robustness of results. However, no single problem design can be judged to be absolute. Results from this approach then are not categorical, but are meant to capture and reflect the positions of all participants in the decision making process. This includes for example, a comparison of the economic benefits attributable to successful aquaculture activities together with the potential negative impacts of reductions in community recreational tourist activities due to the fish farm site's presence in the coastal zone.

Feedback attributed to the different participants in the problem can be identified, and analysed empirically. Representative positions of the groups are constructed form independent feedback from multiple members of each participant group in order to construct a more representative position for the group. Case studies arising from commercial requests for aquaculture sites in the Bay of Fundy provide applied examples of the benefit of this decision support framework.

The broad-based governance environment of the marine coastal zone provides the opportunity to decision makers to understand and explore participants' different perceptions and to explore alternative strategies for developing consensus positions in marine site use. In general, where many active participants exist, as in the marine use coastal zone problem, the formal structure of the governance process leaves something to be desired, that is, the governance processes currently are more exclusive than inclusive, especially in complex settings such as deciding on marine use policy. In many cases, the public is provided with formal opportunities to feedback into a largely government responsible and government-orchestrated system of analyses and judgments. The benefit of the inclusive multicriteria problem framework described here is that it acknowledges the more specific roles of all participating groups and permits them the opportunity to shape the ultimate decisions. The simple idea behind this inclusive approach is that decisions made are not only sanctioned by all participating groups, but they are also, at least in part, responsible for the results, thus claiming ownership and, consequently a smoother implementation of key decisions.

The perspective of this study is that differences among diverse participants are legitimate and need to be accepted, and considered explicitly. The multicriteria analysis described briefly here offers a framework for the sensitivity analysis of the differing positions, and the opportunity to develop positions that provide credence to participants' differing rankings, and alternatives designed to identify options that could be acceptable to all parties. The decision framework is presented as a basis for frank and visual discussions of the positions among all participants toward arriving at mutually acceptable and understood decisions for the sustainable use of the coastal zone.

Acknowledgments Data are obtained with thanks from the St. Andrews Biological Station in St. Andrews, New Brunswick. The Biological Station is part of the federal Department of Fisheries and Oceans and is home to federal fisheries scientists who participate in the monitoring and analyses of aquaculture in the local area including Grand Manan Island. We acknowledge the Station's ongoing involvement in this research with thanks.

References

Allen, P. G., Botsford L. W., Schuur A. M. and Johnston W. E. 1992. *Bioeconomics of Aquaculture*. Elsevier Science Publishers B. V.

Barg, U. C. 1992. Guidelines for the Promotion of Environmental Management of Coastal Aquaculture Development. *FAO Fisheries Techical Paper,* 328, Rome.

Brindley, B. 1991. What is "sustainable"? Some rules for the development road. *Ceres*. 128:35–38.

Canadian Aquaculture Industry Alliance (CAIA). 2003. *Canadian Aquaculture Industry Alliance*. http://www.aquaculture.ca; accessed September 9, 2007.

Department of Finance Canada. 2005. The Budget Plan 2005. Department of Finance. Cat. No.: F1-23/2005-3E, ISBN 0-660-19437-6, p. 194. http://www.fin.gc.ca/budget05/pdf/bp2005e.pdf; accessed September 9, 2007.

Fletcher, K. M. and Neyrey E. 2003. *Marine Aquaculture Zoning: A Sustainable Approach in the Growth of Offshore Aquaculture*. http://www.olemiss.edu/orgs/SGLC/zoning.htm; accessed September 9, 2007.

Lane, D. E. and Stephenson R. L. 1998. Fisheries co-management: organization, process, and decision support. *Journal of Northwest Atlantic Fishery Science* 23: 251–265.

Nath, S. S., Bolte J. P., Ross L. G. and Aguilar-Manjarrez J. 2000. Applications of geographical information systems (GIS) for spatial decision support in aquaculture. *Aquacultural Engineering* 23: 233–278.

Saaty, T. L. 1980. *The Analytic Hierarchy Process*. New York: McGraw-Hill.

Chapter 17
Integrated Systems Analysis for Marine Site Evaluation: Appropriate for the Canadian Marine Farming Industry?

James A. Smith

This commentary considers the issue and solution identified in the Lane et al. paper from the perspective of relevancy to the marine farming industry in Canada. More specifically, commentary is provided that questions whether or not the paper has raised an issue that is truly relevant to the industry, and whether or not the identified solution will support governance models that are relevant to the industry.

Lane's core argument is that decisions regarding marine aquaculture should be made relative not only to a variety of scientific factors about a marine site that might be considered for aquaculture, among other uses, but that decisions must also be made relative to the importance that different stakeholders place on the scientific factors. Lane suggests that his multi-criteria approach to decision making will help to resolve conflicts among stakeholders regarding competition for land and water resources. The decision model is also touted for its capacity to provide insight to decision makers attempting to provide workable solutions to demands for multiple uses of the same site. In this respect, Lane's approach seeks to provide a governance pathway towards consensus, or at least acceptable compromise, between groups with different and potentially conflicting intentions for a marine site. Lane and his colleagues intend that their model be regarded as a practical step forward in the governance of contestable marine resources, and in this respect their efforts are tied to the Government of Canada's responsibility to improve decision making by taking an integrated management approach, as specified in legislation such as the *Oceans Act*.

Within the aquaculture industry, it is widely accepted that effective resolution of disputes regarding marine aquaculture sites is critical to the industry's success. Many are equally convinced that better approaches to dispute resolution are needed, from improved processes for identifying and providing for the participation of interest groups, to convincing demonstration that dispute-resolution processes are fair and adequately transparent. Lane's multi-criteria decision modeling approach provides some avenues for improving the way disputes are resolved. The inclusive

J.A. Smith
Coastal Smith Inc. and Research and Environmental Management Coordinator,
New Brunswick Salmon Growers Association
e-mail: coast@nbnet.nb.ca

© Springer Science+Business Media B.V. 2008

process Lane advocates will no doubt improve decision making. Yet decision making must also be objective and timely for it to have the level of impact, relevance and timeliness that the marine aquaculture industry needs, to make informed business decisions. So if the focus and value of Lane's multi-criteria approach is found in its making the process more inclusive, concentration on this dimension of the governance process might inadvertently undermine the objectivity and timeliness of the decision. There is a trend toward thinking that the more inclusive a decision making process appears from the outside, the process of decision making will be improved internally and will generate a better decision. It may be true that it helps to quell the side-bar controversies that erupt and threaten to derail decision making processes when those who have not been included in the process demand to have their perspectives included. But that may not always be true from the standpoint of making good and timely decisions, particularly if the emphasis on process leads results in decisions that are delayed, unsupported by fact, or fail to meet the needs of those affected by the decision.

Balancing Private and Public Interests

Marine aquaculture in Canada is a private sector business that operates in the public domain. The private aspects of marine aquaculture include the farmer's facilities and equipment, and the organisms that are being grown. Farmers tend to have sole control over all these assets, such that production decisions are planned and timed to meet market needs. With respect to the public dimension of marine aquaculture, marine aquaculture sites are in waters that are regulated and policed by government, and which are typically accessed by public, not private, roads. Farmers recognize their responsibility to operate their facilities in a manner that respects the public property that they are permitted to occupy and use.

Siting and Timing of Aquaculture Operations

Marine aquaculture farmers have to make many decisions that are sensitive to the specific site in which they are working, and these decisions must also address the complex timing demands of aquaculture. The importance of timing raises the concern that conducting integrated systems decision making using Lane's multi-criteria approach will not meet the needs of aquaculturists seeking timely decisions. A year's business activity could be postponed if the decision making process is not timely. Take for example the fact that a salmon farmer begins to make decisions on production when eggs are first laid in the hatchery. About 1.5 years after this the farmer then makes decisions on placing smolt in marine cages. After about one more year the farmer begins to make decisions on harvesting schedules for that crop, a process that could take up to 1 year to complete. The farmer is continually making decisions over a period of greater than 3 years for one crop, with overlapping decision making for sequential crops. Overall the farmer makes these decisions with the intention of obtaining the greatest value in the market. The interaction of siting and time can be illustrated through an example. Consider a typical situation where a salmon farmer is midway through a complete cycle for one crop and production in the hatchery was better than expected, and market conditions are such that high value is expected for the crop in 1 year. This provides a situation where a farmer may want to expand an existing site or develop a new site to capitalize on this production and market opportunity. The process described in the Lane paper could not be completed within the window of opportunity – perhaps 1-4 months – that is available. The farmer must be able to submit applications to federal and provincial agencies, have them reviewed, respond to comments, receive approval and permits, and obtain and place equipment all within a very short time. Notice that nothing in this example conflicts with or denies the need for inclusiveness in marine environment decision making. This example does, however, show that it is doubtful that the process described by Lane would fit into the restrictions on site specificity and timing restrictions faced by marine aquaculture farmers.

Siting, Inclusiveness, and the Facts

Thinking about the shortcomings of Lane's approach in terms of site specificity and timing leads to a broader point about the potential for a trade-off between the kind of inclusiveness that Lane seeks in his process and the effects that inclusiveness might have on the objectivity of the process. From the aquaculturist's standpoint, any decision process which attempts to balance a diversity of perspectives and intentions about a marine site must be associated with a high level of objectivity. The decision making process must have a high level of objectivity so that a farmer can submit targeted applications for detailed site use with a high level of certainty they will be accepted. Again, it is doubtful that the process described in the Lane et al. paper would provide the level of detail needed by the aquaculturist to communicate with the other interested parties in a way which fairly represents the aquaculturist's needs. It is even less likely that the detail needed to make an informed decision about an application for an operation could be provided within the timeframe required by the farmer.

Beyond the complexity of the task involved in getting and communicating evidence, there are issues regarding identification of the relevant scientific facts. The participant groups included, as noted by Lane, all have very different perspectives that may differ considerably on a site-to-site basis. The level of detail required for a proper consultation on any particular site, and the length it takes to conduct the study, may likewise vary a great deal. In addition, there are issues about the industrial scientific expertise needed to conduct a full site use approval. While groups such as environmental non-governmental organizations may have relevant scientific expertise to provide to the site selection process, they frequently do not have legitimate standing in the community or regulatory process. Consequently, the range of contributors to the process, as Lane envisions it, may be limited by the practical constraints imposed by the actual socio-political context in which the consultation is taking place. Finally this process does not allow, within any reasonable time frame, for a farmer-applicant to respond to comments and make alterations to plans to mitigate potential concerns. If a farmer has information that would fill in gaps in the process, answer questions or misperceptions, or resolve a dispute, this information should be heard. It would seem a major shortcoming if a decision making process does not have scientific objectivity highest among the criteria used to guide the process and to evaluate its efficacy.

In addition to concerns related to timeliness and objectivity, it is important to note the importance of confidentiality, which is not among the criteria discussed by Lane. Despite the fact that marine farming takes place in public waters, it is a private sector, private property enterprise that exists in a highly competitive worldwide market place. Farmers must have assurance that plans for their operations remain confidential while in the government approval process. Some revelation of plans is justified as these farms operate in the public domain, and approvals usually involve back and forth discussion between the company and the regulator to get a complete understanding of the application. Some of this activity, which might be integral to the process, cannot be captured in Lane's process. These considerations suggest that subjecting a *full* business application to multi-stakeholder review is not required and may not be consistent with the requirements of the approval process.

A Tidy Process for an Untidy Reality

From the perspective of the marine aquaculturist, the multi-criteria decision making process described by Lane has a place in the overall process of site selection and approval, although more at the planning stage than the decision making stage. Actual aquaculture siting decisions must be made using objective scientific information that may not fit conveniently into the timing and politics of a multi-criteria approach. The process described by Lane could be used to scope potential new farm sites or potential changes to existing sites, and could therefore be used as part of larger marine resources planning exercise in which farmers could apply and receive site approvals. Using a process such as Lane's would result in the transfer of a body of information that would be vetted and ultimately supported by government managers and regulators and distributed to the public. Salmon farmers, and all marine resource users, could use this information to prepare targeted applications. In this respect, this up-front planning would support timely and objective decision making. It would also provide opportunity for the decision making process to respect the interests of and concerns of the public, while simultaneously respecting the confidentiality requirements of the farming business.

Use of a multi-criteria, multi-stakeholder approach in marine resources planning must take into consideration several issues that do not appear to be addressed by Lane. In particular, the described process does not take into account the quality of data or information used, or that use of information must be consistent no matter what the intended resource use. In other words, the balance in Lane's process tips too heavily in the direction of meeting a political agenda of inclusiveness, and too far away from meeting the needs of the industry for science-based decisions made by appropriate officials.

What is *not* being suggested here is that there is a perfect science of marine aquaculture siting for regulators to fall back on, a science that would preclude having to have a process in place to address concerns about aquaculture siting that arise outside of the industry. There needs to be some account taken on the quality of data and information used in planning and/or decision making. It is likely that much of the data and information relevant to aquaculture that describes the marine environment in Canada is out-dated. The data that exists exhibits high variability in temporal and physical scales, different methods of sample collection and analysis, and other aspects of quality control. This does not render the data and information useless. It simply requires that some record of quality is created and stays with the data and information as it is applied. The end result could be a series of flags or footnotes that qualify the outcome of a planning process. Site selection processes must take these factors into account.

Once thinking about regulation of siting reaches a general level of thinking about the relation of confidentiality, evidence, inclusion, and so forth, it is also worth looking across to other national resource sectors to see whether aquaculture is being subjected to a fair regulatory burden. Could the process described by Lane be applied to all resource users, not just marine farming? As aquaculture is a relatively new activity in the Canadian marine environment, it is not unexpected that recent efforts to improve decision making have been developed as the industry has grown, in something near an *ad hoc* manner with too little reference to the longer term and to consistency with governance in other resource sectors and within the marine sector. However marine farming must be looked at in a broader governance structure – planning and decision making – that is consistent with other farming activities, and perhaps activities which rely indirectly on a natural resource. If a marine farm is subject to intensive public review, so should a commercial fishing license, a tourism operation, a ferry service, a port operation, or any other comparable activity. It should be possible to remove any reference to marine farming from the process and replace it with another activity, and vice versa. Whether Lane intended this to be true, or whether marine aquaculture siting is a unique endeavour demanding a special consultation procedure, remains unclear.

In summary, the multi-criteria approach advocated by Lane does address an issue important to marine farmers in Canada. It serves as a useful reminder about how complex a problem siting can be when dealing with multiple stakeholders each of whom has different conceptions about what counts as meaningful data and how the sites current or future use should be valued. The problem is in part one of feasibility – can a multi-criteria approach meet the needs of aquaculturists who need a decision making process that is not going to conflict with the seasonal and biological demands of aquaculture. In addition, the multi-criteria approach might generate more problems than it solves if it raises, rather than answers, questions about the best governance model for scientifically objective, and confidential decision making by the farmer or regulator.

Chapter 18
Models for Analysis and Practical Realities of Marine Aquaculure Siting

Thierry Chopin

Coastal management is a complicated endeavour, perhaps increasingly so as new science and technology creates opportunities for an expanded array of coastal zone uses. It is unsurprising that some researchers and government decision makers would strive to find means to make management a more simplified, perhaps even formalized, procedure to reduce uncertainty in the process and increase the effectiveness of management. One such proposal is offered by Lane and his colleagues who have developed a multicriteria approach to integrated coastal management. Their approach involves a formalized system in which different stakeholders provide input into a ranking system by which decision makers can compare the preferred uses, and the strength of the consultation participants' commitment to those uses. Three concerns appear to motivate Lane and his colleagues. (1) Lane believes that the problem would remain overwhelmingly complex without a formal system. (2) Without a formal system to identify and evaluate preferred coastal uses from a variety of stakeholders, Lane is concerned that the decision process will not be as transparent as decision makers would prefer – particularly those in government making difficult decisions. (3) Lane worries that without a formal system whose architecture is closed while remaining open to whatever input interested parties wish to submit, a decision making process will not have the potential to be maximally inclusive of all views about site use.

In this response to Lane and his colleagues, I wish to focus on three considerations that could be used to enhance their work. The first is to question whether the model in general assumes that aquaculture and other uses of a coastal area are always at odds with one another. An assumption of complementary, rather than conflicting use, might generate a different decision making process out of which there would be different outcomes. Second, a decision making process that involves discussion about marine aquaculture will only be useful if it is realistic. Lane uses Grand Manan Island as a case study, but he does not include some important existing subcomponents on this island which should have important implications for the

T. Chopin
University of New Brunswick, Centre for Coastal Studies and Aquaculture, Centre for Environmental and Molecular Algal Research, P.O. Box 5050, Saint John, N.B., E2L 4L5, Canada
e-mail: tchopin@unbsj.ca

K. Culver, D. Castle (eds.), *Aquaculture, Innovation and Social Transformation*,

© Springer Science+Business Media B.V. 2008

outcomes of their management process. Third, Lane's management process aims to underpin complex decision making processes with a system that will bring about inclusiveness and transparency to the process. But there is as much room, if not more, in his decision making model for a lack of transparency and accountability than an unsupported decision process. In this respect, the decision making model might appear to achieve its goals in idealized circumstances, but in real circumstances its implementation might achieve less than existing decision making processes.

Before turning to a fuller discussion of the three main issues just outlined, it is perhaps helpful to begin with a quite general remark about the decision process model proposed by Lane: it is difficult to comment on the model because it is presented as an interesting concept to assist in the decision making process in integrated coastal management, but, to my knowledge, it remains an untested concept as far as taking it all the way through the decision making process. We do not know if this model has yet "enabled conflicts among traditional and innovative users to be clearly enunciated and explicitly addressed" nor, I would add, led to conflict resolution, for example, in the coastal zone of Grand Manan, which is offered to the reader as a realistic case study.

For example, throughout Lane's discussion, it becomes apparent that aquaculture is considered as it is in its present form in the Bay of Fundy: salmon monoculture. One would hope that innovations will take place and approaches to environmental sustainability, economic profitability and social acceptance will change in the aquaculture sector and with all stakeholders directly or indirectly involved. Aquaculture will evolve, and aquaculture in the Bay of Fundy will diversify from salmon monoculture to integrated multi-trophic aquaculture (IMTA) practices with finfish, shellfish and seaweed cultivation. The model presented by Lane will need to be built with enough flexibility to reflect these changes.

It is, therefore, difficult to evaluate if it will be a useful tool/model for responsible governance. At the same time, I recognize that the authors are in an early stage of project development and we should give them time to apply the concept to real case studies and conclude on its validity. My comments below should, hence, be conceived more as constructive reactions to the reading of the paper than criticisms, as the project may have since evolved, and some of my comments addressed in the meantime.

Irreconcilable Differences?

Towards the end of the chapter, Lane and his colleagues mention that their study focuses on, "a comparison of the economic benefits attributable to successful aquaculture activities together with the potential negative impacts of reductions in community recreational tourist activities due to the fish farm site's presence in the coastal zone." What Lane is assuming is that fish aquaculture and tourism activities are in opposition, and may stand as irreconcilable activities. There are reasons for thinking that this will change. For example, there are fish farmers who make their operations parts of eco-tourism tours or stops in sailing races. Lane might have considered a broader temporal dimension in which to situate the coastal management process, a process which tends to focus on the immediate needs of decision makers. The study of the development of the "écluses à poissons" (since the 11th century) and the oyster tables (since the 15th century) in France, and the weirs (since the 18th century) in Canada, reveals a past with its social troubles and an acceptance over at least 40, 24 and 8 human generations, respectively. In comparison, the salmon aquaculture industry in Canada is just becoming familiar to a second generation of Canadians, many of whom are still struggling with accepting the presence of marine aquaculture in their communities. Marine aquaculture has not yet reach a kind of "traditional" status in many Canadian communities. There is an irony here, since it is a striking fact to see that tourists traveling on Grand Manan Island, or along the coast of New Brunswick, have no problem taking pictures with weirs as part of the "seascape" and will buy postcards with weirs. Yet they avoid taking pictures with salmon aquaculture farms in the "seascape" and, to my knowledge, only one postcard featuring an aquaculture farm is commercially available in the region. On the other hand, magnetic stickers including sketches of weirs and books on weirs are made available to them as souvenirs of a nostalgic era of one type of fishing activity. Is there much visual difference between a weir structure and a salmon cage, especially when looking at the recent "floating weirs" which are made of the same plastic piping? We are probably seeing nothing more here than the difference of a human eye accustomed to one picture for eight generations versus one picture just now becoming familiar to a second generation!

Decisions about marine coast uses must obviously meet the needs of existing people in contemporary communities, but these needs ought to take into consideration longer timeframes to understand that the shock of the new often quickly wears off and becomes yesterday's news. Lane's decision making process has the potential to be locked into the moment in which a decision maker must, for example, approve or reject a siting application from industry. It would be an unfortunate outcome of adopting a model decision making process were it to bind decision makers, and participants in the process, into a narrow timeslice of the evolution of an industry that is elsewhere accepted and may come to be a regular part of Canadians' conception of a "picturesque" viewscape.

Model Realism and Integrated Multi-Trophic Aquaculture (IMTA)

One of the deficiencies of Lane's model is that it considers the present form of aquaculture and does not consider its evolution. The model takes single species aquaculture – that is, salmon aquaculture – as the industry that would be involved in an integrated planning exercise. While Lane does mention "polyculture" in passing, it is not discussed as the serious option it really is for aquaculturists. Polyculture is often thought of as the practice of growing several species of fish together. This has usually meant salmon, cod, haddock or halibut. From an ecological point of view, however, these are all "shades of the same colour" in the sense that this kind of polyculture involves the marine cultivation of fish only. Polyculture can be more sophisticated, in the sense that it can involve true ecological diversification. This means that "poly" means multiple species, but more importantly, it means that there are species cultivated at more than one trophic level (i.e. cultivating finfish along side other groups of organisms such as shellfish, seaweeds, worms, bacteria, etc.). Polyculture systems, with organisms at the same trophic level, perpetuate rather than address environmental issues caused by ecologically unbalanced systems. Integrated multi-trophic aquaculture (IMTA) appears a much more sustainable approach (Chopin et al. 2001; Troell et al. 2003; Neori et al. 2004).

In the marine site components discussion, Lane mentions "the pollution of potentially harmful effluents arising *naturally* from the ecosystem and due to the fish farm wastes". Unfortunately, these subcomponents of the effluent component, arising naturally (?) or not so naturally (!), are not clearly identified and explained in the rest of his discussion. Consequently, it is unclear what they are, and what impact they have in the model. As it stands, it seems that the only remaining type of effluents factored in the model are the fish farm wastes. As portrayed in the model, it is reasonable to expect that the aquaculture industry would be triggered to complain that they are again singled out because they are the "new kid on the block," while in fact, they are not the only contributor of effluents in the coastal zone. Moreover, it may be a misrepresentation of the industry, since if the site is an IMTA one, not a finfish monoculture one, the effluents might be different or reduced.

Lane's Figure 2 gives us the "complete set of 9 overlapping subcomponents currently in use for this study area." It is surprising that dulse is not recognized as a subcomponent, particularly in this area. If harvesting of the brown alga *Ascophyllum nodosum* (rockweed) has been identified, and rightly so, there is also another seaweed being harvested on Grand Manan Island, the red alga *Palmaria palmata* (dulse). This is a significant activity in the region, and specifically on Grand Manan Island (mostly at Dark Harbour and Cheney Passage, parts of the island already coveted by many stakeholders as illustrated by the "busy" Figure 2). This "fishery" has a long tradition and is very much embedded in the social fabric of the island, with a fraction of the population which can be quite vocal at times when the "dulsing" is at stake. At the present time, most of the dulse (90%) still originates from Grand Manan Island, where 50–100 harvesters (and an additional 15–20 people involved in buying, packaging and shipping) harvested approximately 100 dry tonnes (DT) in 1994, between 59 and 68 DT in 1996 (R. Cronk, pers. comm.), and approximately 84 DT in 2002 (T. McEachreon, pers. comm.). The rockweed industry started in 1995 on the New Brunswick side of the Bay of Fundy. Harvested tonnages for the whole of New Brunswick are much higher than those for dulse (the first year landings totalled 703 wet tonnes (WT); they reached 11,801 WT in 2004) but the number of people involved is similar (Chopin and Ugarte 2006). Around 75 harvesters are employed by Acadian Seaplants Limited (http://www.acadianseaplants.com/) along the coast of New Brunswick and Grand Manan Island. The drying and processing plant on the mainland of New Brunswick, in Pennfield, operates with 14 full-time and 15 seasonal employees. Natural beds of dulse have seen quite dramatic changes over the years, with no particular parameters or activities clearly identified as responsible for them (despite no shortage of blaming many culprits!) and the social repercussions have been numerous in the coastal community of Grand Manan Island. Dulse should certainly be a subcomponent to add to the model, but the general point is that a model assumption – that the industrial interest in integrated coastal management is solely in finfish aquaculture – makes the model unrealistic and potentially misleading.

Attributing Value and Achieving Transparency

The aim of Lane's model is to arrive at some measure of valuation of areas using a set of unbiased tools that will combine relevant scientific information with the accurately represented interests of different groups. By engineering a process which is purportedly inclusive and transparent, Lane hopes to mitigate the risk of bias in decision processes as overbearing interests are overrepresented, or skew of results as some interests at stake are only partially represented. In this respect, the process is intended to establish an accountability trail for decision makers who must report back to their superiors and to the public.

Lane has taken upon himself to complete a laudable task, one which is nonetheless extremely difficult to reach in practical terms. There are, however, problems with the process that do not allow the model to overcome all challenges regarding its neutrality, inclusiveness and transparency. Take, for example, what is said about the "selected site specific valuation" process. This is obviously crucial in making decisions about site approvals, however, we are told that for "overlapping subcomponents, 'overlap rules' are determined to deal with these more complex valuation cases." How and by whom these components and rules are established is not specified in Lane's discussion. If they are not established by a process in the model, the model makes no advance on the normal situation in which these valuations are otherwise made by stakeholders. If they are encoded into the model, they ought to be clearly explained to understand the valuation process and agree or disagree with it.

There is another interesting issue about the composition of the participant groups and their selection which could have a significant impact on how participants interact, how interests are represented, and how management decisions are made. Lane says that participants providing their inputs into the Site Comparison and Multi-criteria Group Decision Ranking are represented by five groups in the framework. One of these groups is "Federal Scientists." Are the only scientists from the federal government, or could the group of scientists be broadened to include provincial or municipal public servants, or scientists belonging to academic institutions? If the group really is meant to be federal scientists, it raises questions about the contribution "non-federal scientists" could make. As non-federal scientists are often asked to contribute their perspectives to coastal management decision making processes, it would seem important to mention how their inputs will be included in the framework process. Perhaps a more pressing problem, apart from the composition of scientific expertise drawn into the process, is that "federal scientists" are not necessarily (in fact, not generally) members of the local communities involved in coastal management. While this is sometimes an advantage because it means they are not embroiled in the local politics, it can also be a disadvantage if they are insufficiently aware of the local traditional knowledge and unable to validate and incorporate this knowledge into their own knowledge base.

Suppose, however, that the representation of the scientists *is* from the federal government, since they will be making marine aquaculture siting approval and rejection decisions. One can then think about whether the model allows for the kind of transparency intended. Unfortunately, Lane's explanation of the process raises questions. For example, Lane says that "the attributed weights to the ecosystem components and subcomponents by the five participating groups are provided in Table 2." While all of these numbers are provided in the table, it is unclear how they have been calculated, or from what data they have been derived. Detail about how these numbers have been obtained would be better than providing a table of impenetrable numbers. More important than any individual data set, however, is the need to have some discussion about how the numbers are arrived at, and agreed upon for use in the model by the participants. Where there are ecological systems uncertainties, and disagreements about what and how to measure those systems, the source and reliability of data will be contentious.

How the model will resolve these disagreements is unclear. One can envision a scenario in which federal scientists are called upon to contemplate a case, like Grand Manan Island, for evaluating potential aquaculture sites. Obviously enough, federal scientists come to the table with their interests and agenda, like everybody else and there is nothing unexpected about that. What remains unclear, however, is how scientists would attribute weights independent of the selected areas under consideration. Can one reasonably expect that their criteria will not be closely tied to their duties as federal servants, and thus not totally independent – whatever that could mean. Finally, the model allows a graphic representation of the noted ranking divergence, and provides some explanations as to the origin of the divergence, but does not "balance the interests of each of the different related groups accordingly." So, what is the mechanism for the balancing act, and who does it? The model brings interesting data and information supporting and explaining the basis of the divergence, but does not provide the governance structure to resolve the conflict. Perhaps the problem is that, at root, the model still quests for the Holy Grail of how to take decisions when taking authoritative decisions can be questioned because the idiosyncratic weights attributed to the different participants, the different components and subcomponents are constantly debated, and the Hierarchy in the Analytical Process (the H in the AHP) is the variable dimension!

Conclusion

The framework proposed by Daniel Lane and his co-authors is an interesting tool being developed to contribute to the decision making process, but it is not a decision system *per se*. Perhaps its authors will argue that it was never intended for that purpose anyway. I am sure they will refine the model, including more components, subcomponents, valuation yields, attributed weights, and participant groups and organizations, to make it a valuable tool towards consensus governance in the coastal zone among the many stakeholders who want both to use and preserve this key ecosystem.

References

Chopin, T., Buschmann, A.H., Halling, C., Troell, M., Kautsky, N., Neori, A., Kraemer, G.P., Zertuche-Gonzalez, J.A., Yarish, C., and Neefus, C. 2001. Integrating seaweeds into marine aquaculture systems: a key towards sustainability. *Journal of Phycology* 37: 975–986.

Chopin, T. and Ugarte, R. 2006. The seaweed resource of Eastern Canada: 46 p. In: *World Seaweed Resources. An Authoritative Reference System*. A.T. Critchley, M. Ohno and D.B. Largo (Eds.). A multimedia, interactive DVD-ROM. ETI BioInformatics Publishers, Amsterdam. ISBN: 90 75000 80 4.

Cronk, R. New Brunswick Department of Agriculture, Fisheries and Aquaculture (NBDAFA). Personal Communication.

McEachreon, T. New Brunswick Department of Agriculture, Fisheries and Aquaculture (NBDAFA). Personal Communication.

Neori, A., Chopin, T., Troell, M., Buschmann, A.H., Kraemer, G.P., Halling, C., Shpigel, M., and Yarish, C. 2004. Integrated aquaculture: rationale, evolution and state of the art emphasizing seaweed biofiltration in modern mariculture. *Aquaculture* 231: 361–391.

Troell, M., Halling, C., Neori, A., Chopin, T., Buschmann, A.H., Kautsky, N., and Yarish, C. 2003. Integrated mariculture: asking the right questions. *Aquaculture* 226: 69–90.

Part VII
New practices for Global Competitiveness: Alternate Species, Alternate Uses, and Value-Added Aquaculture

Introduction

Aquaculture can be socially innovative irrespective of how resource intensive it is, whether it is based heavily in science and technology, and whether it is located in developing or industrialized countries. At one extreme, aquaculture is socially innovative just because it is introduced to a context unaccustomed to farming fish. For example, programs in land-locked Malawi teach people how to dig small, rainwater fed ponds in which tilapia grow on kitchen scraps and other low-cost inputs (WFC 2007; Economist 2007). The cultured fish replace declining capture fish stocks in local rivers that cannot keep pace with population growth, and in turn the dredged pond bottoms provide farmers with an excellent supply of sustainable fertilizer for crops. The WorldFish Center estimates that in regions where aquaculture has been adopted, malnutrition in children has dropped by from 45 to 15%. Meanwhile, surplus fish are being sold off-farm, generating badly needed income for some of the world's poorest farmers. What starts as a small-scale, project-by-project, local effort to improve the lives of people through better nutrition finishes in market-driven self-regulation of aquaculture by fish farmers.

Malawian society is being transformed by a socially innovative, extensive rather than intensive aquaculture. The scientific knowledge and technologies underpinning these programs are not drawn from the leading edge of science. What matters most is application of well-established criteria for selecting of tilapia well-adapted to the available feed and grow-out conditions. At the other end of the range of aquaculture innovation are situations where market conditions supply fish farmers with incentives to adopt science and technology-based solutions to production issues in intensive aquaculture. Canada finds itself in this situation with respect to salmon aquaculture. While domestic sales of fresh farmed fish have stayed relatively stable, export markets face considerable competitive pressure from Chilean producers (Brown et al. 2007) supplying a comparable yet cheaper product. Canadian farmers are turning to science and technology to generate innovative ways of producing fish more cheaply and quickly, and to accelerate commercialization of alternative aquaculture species such as Atlantic cod.

Extensive aquaculture innovation in developing countries is directed toward food security, yet in Canada, scientific and technological innovation in intensive aquaculture is sought as a driver of competitiveness. This contrast appears even more pronounced once it is recognized that Canadian emphasis of scientific and technological innovation in aquaculture is not unique to aquaculture. The federal government's science and technology framework (2007) directly connects science and technology *in general* with competitiveness and social and economic wealth, in industries from aquaculture to wireless technologies. Yet the goals of aquaculture innovation need not be so starkly different as those of Malawi and Canada. Nor are the options for governance of aquaculture innovation limited to the experimental, project-based approaches seen in Malawi, or the heavily centralized approach taken by the Canadian federal government. Much work remains to be done, however, to connect alternate uses, species, and practices to appropriate governance practices for sustainable aquaculture contributing to food security and competitiveness according to the capacities and needs of aquaculture countries.

Some of the work needed to match governance systems to aquaculture is undertaken in Jeremy Rayner's contribution. Rayner identifies two obstacles to beneficial use of well-governed innovation, often overlooked in governance and management of innovation. The first is the confusion of an increase in technological intensiveness with an automatic accompanying increase in good governance. In science-dominated government agencies and departments, it is all too easy to treat growth of a science- and technology-intensive industry as the end goal, representing in itself successful governance of an innovation process. The second problem follows directly on the first. Having mistaken science and technology intensiveness for both an end goal and an instance of good governance, managers ignore or forget to explore avenues of good governance connected with the real value of science- and technology-intensive aquaculture. As the means to a well-governed, sustainable industry producing social and economic benefits for citizens, Rayner suggests that Canadian competitiveness will be best served not by higher levels of investment in the already-strong strong science and technology base of the aquaculture industry, but in governance innovation aimed at better better coordination of governmental decision makers and the industry enabled and regulated by them.

Rayner's thesis runs against the common sense view that technological sophistication is the main difference between Canadian and Malawian aquaculture and their readiness for participation in competitive markets. Yet the common sense view has further detractors. Paul Lyon and Mark Burgham make Rayner's case from the standpoint of two civil servants contributing to governance of sustainable aquaculture in Canada. Taking up the example of recent widely-discussed concerns regarding PCB contamination of the flesh of farmed fish consumed by humans, Lyon and Burgham stress the importance of coordination between government, industry and the public to identify and address concerns cutting across food production, trade and safety. Lessons can be learned from cases where coordination has been lacking, and those lessons can be incorporated into improved governance practices – even when the lessons are as simple as the PCB scare's demonstration that coordination is just as important in communication with consumers as it is in coordinating good governance decisions across multiple departments and agencies.

While this outlook sounds encouraging, Colin Barrow expresses reservations about the extent to which conscientious attempts at good governance really meets industry and consumer needs. Barrow suggests that government may *not* be sufficiently nimble to adapt to market conditions as it pursues better coordination of actors. It is worth considering this criticism seriously. After all, Lyon and Burgham's example is about improving governance *after* the fact of a problem in the industry. Without disputing the value of governance, Barrow emphasizes that industry, not government, has the best incentives to react quickly to market conditions. In this respect, coordination of aquaculture might better be industry based, which is to say largely self-regulated. Perhaps surprisingly, this is ultimately where the cases of Malawi and Canada converge. Otherwise worlds apart in terms of their production base, capacity and governance, self-reliance and self-regulation may be the preferred option for socially innovative aquaculture. Whether a world suspicious of aquaculture's promises will give aquaculture a 'social license' to experiment in this sort of self-governance remains to be seen. Yet in an era of seemingly endless regulation, this alternative vision of the future of aquaculture governance deserves consideration as a surprising, yet serious contender.

References

Brown P., Brown J., and A. Rubio. 2007. Salmon supply continues to outpace demand. *Global Aquaculture Advocate* 10: 25.

Government of Canada. 2007. Mobilizing Science and Technology to Canada's Advantage. http://www.ic.gc.ca/cmb/welcomeic.nsf/vRTF/PublicationST/$file/S&Tstrategy.pdf. Accessd August 17, 2007.

The Economist. 2007. Fish versus AIDS. August 30. https://www.economist.com/world/africa/PrinterFriendly.cfm?story_id=9725881. Accessed September 9, 2007.

World Fish Centre (WFC). 2007. Innovative Fish Farming Project for HIV-affected Families in Malawi Doubles Incomes and Boosts Household Nutrition. http:// www.worldfishcentre.org/resource_centre/WorldFish%20Press%20Release_FINAL.pdf. Accessed September 9, 2007.

Chapter 19
Governance for Global Competitiveness: The Future of Aquaculture Policy in a World Turned Upside Down

Jeremy Rayner

> *It may help here to recall Humphrey Lyttleton's rejoinder to someone who asked him where Jazz was going: 'If I knew where Jazz was going, I'd be there already' (Winch 1958: 93–4).*

Introduction

There is no shortage of technical innovation in Canadian aquaculture. There are new farming techniques and new products, and whole new industries are emerging based on breakthroughs in the farming of new species. This chapter, however, is not about innovation *in* aquaculture but about the kind of innovative public policies that are needed to release the potential of aquaculture to contribute to our economy and to our quality of life.

The argument is a simple one. Aquaculture presents special problems of policy coordination that urgently require a more collaborative, less "top-down" approach to policy-making than traditional governing arrangements are able to deliver. Aquaculture is not alone in presenting this challenge to conventional government and there is now considerable, though mixed, experience with alternative "governance" relationships from which we can learn. Among the lessons that have been learned from other resource policy areas is that governance is an increasingly important component of comparative advantage, so getting governance right is central to the present and future global competitiveness of the aquaculture industry in Canada. Getting governance right is a combination of understanding the problems facing the industry (which is relatively easy) and understanding how these problems can best be addressed in the new governance context in which aquaculture and other resource industries must operate (which is rather more difficult).

The chapter is organized as follows. In the next section, I introduce the idea of governance and explain its special relation to comparative advantage in a globalized economy. I then outline the most pressing problems of Canadian aquaculture that

J. Rayner
Department of Political Science, University of Regina
e-mail: jeremy.rayner@uregina.ca

K. Culver, D. Castle (eds.), *Aquaculture, Innovation and Social Transformation*,

© Springer Science+Business Media B.V. 2008

will need to be addressed by new governance arrangements, and identify lessons to be learned from the experiences of other jurisdictions and other resource policy areas. Finally, I want to draw special attention to the new political context in which aquaculture must operate and the implications of this context for how problems are identified and policy solutions formulated and implemented. This new context is characterized by a poorly understood reversal of a number of familiar and important relationships: it is a world turned upside down.

One implication of the argument needs to be stated clearly at once. If the chapter is correct, no amount of scientific and technical innovation can address the fundamental problems of Canadian aquaculture – not abandoning net pens for closed container systems, not polyculture, not submerged offshore mariculture, not new species, not better marketing, more rational planning nor new end uses – all of them will simply reproduce the same governance problems in new forms and some will create new governance problems as well. This chapter is a plea for a better appreciation of how to use the full range of different modes of coordination to turn innovation into comparative advantage.

Governance and Comparative Advantage

"Governance" is a protean term. A recent study (Van Kersbergen and van Waarden 2004) lists no fewer than nine broad contemporary meanings whose very lack of precision indicates the scope of the breakdown in traditional governing mechanisms and their replacement by a coordination mechanism at once much looser and more all encompassing. Governance implies certainly a new and perhaps also a diminished role for governments, while at the same time stressing expanded roles for private corporations, civil society and international actors, whose own "good governance" thus becomes a matter of public concern. In this paper, I mean by governance the effort to sustain co-ordination and coherence among a wide variety of private and public actors with different purposes and objectives (Pierre 2000:3).

In the past, governments aimed to achieve co-ordination and coherence in resource policy largely by exploiting aspects of their authoritative relationship with society. Governments acted either directly by exercising ownership over public lands or other assets, or indirectly through subsidies to private enterprises and authoritative regulations with penalties attached aimed at correcting market failure, particularly the failure to value key elements of the global commons such as air and water. For a number of reasons, still poorly understood, there has been a general loss of confidence in both of these forms of authoritative coordination as the basis for public policy. In part, this loss of confidence stems from a growing realization that authoritative coordination of resource exploitation has not been especially cost-effective, failing to deliver environmental protection and other public goods while simultaneously distorting markets and encouraging rent-seeking behaviour. Right across the developed world, the "regulatory state" is in retreat (Jordan et al. 2005).

It is the deeper reasons for these failures that remain a matter of dispute, but two candidates stand out for their relevance to aquaculture. First, there is the increasing complexity of policy coordination, especially where governments attempt to regulate new activities under old regulations (Bourgault and Lapierre 2000). The regulatory review of Canadian aquaculture conducted by the Commissioner for Aquaculture Development revealed a variety of different policy regimes organized by several federal and provincial departments and agencies operating under different legislative mandates (OCAD 2001: 8). For the targets of regulation, the effects are experienced as a series of ambiguous and contradictory policies "layered" one on top of the other or as policies that seem to "drift", disconnected from emerging problems on the ground (Hacker 2005).

The other reason for regulatory failure is globalization. Canada's aquaculture industry is heavily export oriented and operates in a rapidly changing and highly competitive global environment. It is not at all obvious that governments are best positioned to determine what a particular industry needs to survive and prosper in a global rather than a protected national market. As Peters puts it:

> The most fundamental transformation in the environment of the public sector is that change itself – technological, social, economic – tends to be more rapid and less predictable than in the past . . . governments must find ways of coping with rapidly changing problems and a socio-economic environment that is less predictable than in much of their previous experience. This change in governance will require enhanced flexibility, and with that flexibility comes designs for governance that recognize the modification of preferences through learning and the inadequacies of many technologies for achieving programmatic goals. Decisions that might once have been programmable will, under these circumstances, be more subject to circumstance and opportunities, rather than planning and formalized procedures (Peters 2002: 5).

Hence much of the literature on "governance" stresses the attempt to create some version of a flexible, learning partnership between government and other social and economic actors rather than to renew attempts to steer and coordinate from the top. It is perhaps significant in this respect that the Commissioner's recommendation for an Aquaculture Act that would try to create a more logical regulatory structure to achieve newly specified programmatic goals for the aquaculture sector has been ignored. There is clearly no stomach for this traditional top-down approach to regulating the sector.

More worryingly there is no sign that a new governance structure has emerged to replace the old regulatory framework, either. A very large literature now exists exploring the relationship between governance and comparative advantage (Hall and Soskice 2001). The fundamental premise of most of this work is that, contrary to the scenario laid out in introductory economics textbooks, comparative advantage is (with an unusual but significant exception described below) not something we can assume as a natural attribute but something that itself needs to be explained. And the explanation usually turns on the ability of superior governance arrangements to coordinate and combine the factors of production in more efficient ways than in competing countries with nominally more attractive attributes like soil, climate or natural resources (Hall and Soskice 2003; Watson 2003).

The relationship between governance and comparative advantage has a special significance for Canada, where a long, if somewhat marginalized tradition of historical economics has drawn attention to a recurring pattern of resource exploitation, the production of "staples": raw or semi-finished commodity exports in which the original source of comparative advantage *is* the textbook kind of abundant natural endowment (Innis 1930). As writers in this tradition have emphasized, it is difficult for industries that exploit natural advantages in this way to escape the "staples trap" where they, and the communities whose livelihoods depend upon them, are driven into the ground by a combination of substitution of other, cheaper resources and competition from countries with similar natural endowments but lower labour costs and more cost-effective regulation. The policy options here are by no means reducible to a simple contrast between an innovation strategy that relies more on the flexible application of superior skills and technology *versus* a traditional approach that aims to wring the last ounce of comparative advantage from natural endowments. In cases where the endowments are very significant, for example, Canada's extensive coastline, existing marine infrastructure and seafood processing capacity, it may very well make sense to combine an approach that continues to trade on these endowments with one that plans for an ultimate escape strategy. However, in a country where labour costs tend to be high by world standards, getting the regulatory context right becomes critical both to the success of the traditional staples approach and to providing the breathing space necessary for the standard staples trap escape techniques, such as technological innovation, value added production and product diversification (Hutton 2007).

As we have seen, however, the regulatory context is very complex. In part, this complexity is the result of significant and unique policy legacies, such as the jurisdictional issues created by Canadian federalism, by many decades of government organization and re-organization, often prompted by mutual learning with cognate jurisdictions, and by the continuing uncertainty over the precise extent of aboriginal rights and title. Just as significantly, however, the nature of the policy issues themselves, at the intersection of two highly contested policy domains, environment and food, adds several degrees of complexity of their own. During the era of the regulatory state, aquaculture became embroiled in the febrile atmosphere of contemporary environmental disputes. As a food industry, it was inevitably touched by some of the highly visible failures of the regulatory state, notably the health consequences of BSE in beef cattle and, in the Canadian context, the tainted water episode in Walkerton, Ontario. In the aquaculture context, the predictable outcome has been complex, arbitrary and burdensome regulatory requirements that have simultaneously hobbled the industry without satisfying its critics (Howlett and Rayner 2004). In some similar cases, the difficulties encountered by the regulatory state have promoted innovative and sophisticated responses from civil society, such as certification schemes or monitoring organizations. In others, however, the outcome has been a pattern of "interfering regulation" that is both counterproductive and difficult to change (Knill and Lehmkuhl 2002; Howlett and Rayner 2006).

The Problem Set: Lessons from Other Resource Sectors

Consider just one element of the regulatory context, the environmental and food-safety issues confronting the aquaculture industry in Canada. On the finfish side, environmental issues include the effects of discharges of nutrients and therapeutants into confined bodies of water and the interactions of farmed and wild fish populations, both in the shape of the potential exchange of diseases and parasites and the escape of the farmed fish themselves. On a larger scale, environmentalists have questioned the sustainability of the finfish industry on the grounds that feed typically includes other fish products, with conversion ratios from feed to biomass outside the laboratory that are substantially greater than one to one. For shellfish, the issues include the impacts on bird populations, modifications to benthic communities in the vicinity of lines and rafts, and concerns over exceeding the carrying capacity of enclosed waters. Food safety concerns in finfish have focused on the apparent presence of elevated levels of contaminants such as dioxins in farmed salmon and other fatty fish species (Hites et al. 2004). For shellfish, the issue is usually the concentration of heavy metals in some farmed species in particular locations. BC oysters, for example, are currently excluded from both European and Japanese markets (though not domestic ones) because of elevated cadmium levels (Deal 2005). Running alongside these concerns – all of which are contested in some form or another – are traditional distributive conflicts with other users of the foreshore and nearshore and with traditional capture fishers (Howlett and Rayner 2007).

Devising a governance framework that can adequately address these concerns and include all the various interested parties has been a challenge. Not the least of the challenges has been disentangling cause and effect to uncover the real motivation of political conflict over aquaculture. Are those who raise concerns purportedly about the impact of shellfish farms on bird populations in fact aiming at a quite different purpose, to protect their beach recreation opportunities or views from their waterfront properties, for example? Is the drive to publicize dioxin levels in farmed salmon really motivated by a disinterested concern for human health or by philosophical opposition to the concept of farming fish as an "unnatural" kind of activity? It is here that the experience of other resource policy areas is invaluable, for the governance issues raised by these problems are little different from those of other resource industries, old and new. A recent and already much-cited article (Nie 2003) helpfully summarizes what the author calls the "drivers" of such natural resource-based conflicts. Table one is based on Nie's categories and. anyone familiar with the politics of aquaculture will find the drivers instantly recognizable.

Take the first of these drivers, scarcity. Its surface manifestations, increasing population densities and opposition to resource development by local populations, are common problems in all resource sectors. It may seem on the surface that there is more than enough coastline/ocean/forest/wilderness, to go around if only we could plan carefully enough for everyone. Opposition to resource development looks like a special case of selfishness (NIMBYism) or a general attribute of population growth. But closer examination will soon show that, to take an aquaculture example, the number of potential sites that meet all the requirements of the industry is actually

Table 19.1 Drivers of natural-resource based conflicts (adapted from Nie 2003)

Driver	Selected attributes
Scarcity	Population density NIMBY
Policy surrogates	Strategic venue shifting Lightning rod for larger social and cultural conflicts
The spiritual	Changing public attitudes to nature First Nations strategies
Policy design	Policy legacies Policy "lock in" Institutional incentives
Policy frames	"framing" as a strategy symbolism policy narratives
Scientific uncertainty	Politicization of science wars Attitudes to risk
Electoral politics	Search for "wedge issues" to differentiate parties The "political spectacle" (Edelman 1988)
Group strategies	Organizational maintenance The value of a "crisis" Communication "mirroring"
Constitutional and administrative obstacles	Environment and the division of powers Blame avoidance Discretionary statutes Contradictory mandates
Trust	Generalized lack of trust Adversarialism and litigation Vicious circle: lack of trust is a both driver and a by-product of conflict

rather small and that many of the attributes which draw the industry to them are the very same ones that make them desirable places to live or recreate. Areas that may be highly appropriate from a biological point of view may lack infrastructure or some other key socioeconomic attribute. So the real problem is how to manage a scarce resource and complaining about NIMBYism is beside the point.

Nie argues that, in addition to mistaking the surface manifestations of environmental conflict for the drivers, the peculiar intractability of such conflicts arises from the fact that, in real life, multiple drivers are usually at work. Thus, even if we manage to uncover the real drivers of conflict, a focus on addressing one while ignoring the others is bound to fail. The solution is a more careful attention to all the drivers that "will enable us to match the right resolution strategy to the right driver"(Nie 2003: 35). Nie's relatively optimistic conclusion (and his interesting endorsement of conflict as necessary for a functioning democracy) puts him at odds with the currently fashionable focus on "wicked" resource policy problems. A "wicked" problem is generally defined as one which is inherently perspectival: where you stand really does depend on where you sit. With no right or wrong answers to be found, wicked problems involving many such points of view supposedly fuel conflicts that are so complex and multifaceted that they ultimately resist resolution altogether (Dorsey 1986; Schindler and Cramer, 1999). Nie's claim that more careful analysis will reveal that there are few, if any, genuinely wicked resource policy problems, is ultimately an empirical one. On this view, more careful problem analysis will only occasionally come up against real incommensurability of perspectives or values. More often, clarification of the multiple drivers at work will help the parties to the dispute find common ground.

As I shall explain in more detail below, I think Nie is at least partly right. The vogue for describing policy problems as "wicked" and hence intractable, in addition to fostering the harmless illusion among middle-aged policy analysts that they are right up-to-date with the language of their children, has the more dangerous outcome of encouraging a tendency to stand around gloomily surveying the damage without proposing any positive measures. This attitude, especially when accompanied by a public display of *Schadenfreude*, has often characterized the contribution of social scientists to natural resource conflicts. The comfortable conclusion that such problems are never susceptible of resolution, but only of further study, has fuelled the impatience of practitioners with social science research in this area.

Instead, we shall proceed on the assumption that the kinds of policy problems created by the potential environmental impacts of aquaculture are complex, but not impossibly so – at worst, they bear a momentary resemblance to impossibly complex problems. Unfortunately, their complexity means that they are likely to cross jurisdictional boundaries, both of the vertical kind between different levels of government and the horizontal kind between different departments and agencies of a single government. Even if we had other reasons to be optimistic about the capacity of regulatory approaches to solve social and environmental problems, this kind of complexity would likely defeat them. What kinds of policy innovations are needed?

Three Reversals: The Inverted World of Contemporary Policy-Making

Here we need to understand the new social and political context in which resource-based industries operate and try to resolve the conflicts identified by Nie. And here we encounter the world turned upside down. Drawing on the work of Maarten Hajer, we can identify three relevant reversals of traditional relationships that are features of this new context. They are reversals in the relationship between politics and policy between knowledge and policy and between analysis and action (Hajer 2003a).

The first is the group of changes that has taken place in the *polity*, narrowly defined, the changes that are implicated in the shift from government to governance that has already been noted. Pre-eminent amongst them is the close relationship between the moribund state of traditional political mechanisms, as evidenced by declining voter turnout and party identification, and the lively, perhaps hyperactive state of issue politics, including environmentalism. The effect, as Hajer perceptively notes, is to reverse the traditional relationship between politics and policy. Instead of an electoral debate, followed by the formation of a government and the announcement of policy, there is the formation of a government, the announcement of policy and mobilization of issue activism. Policy becomes the *occasion* for politics, not the outcome of it (Hajer 2003b). The argument here is not that parties no longer run on platforms involving policy promises; rather, that so few citizens pay any attention to these promises or expect them to be fulfilled that policy announcements generally come as a surprise.

This reversal of political relationships has many important consequences, of which two are important for the governance of aquaculture. First, there is the phenomenon that so baffles and enrages those involved in resource industries: they can take part in all kinds of time-consuming planning exercises and comply with every legal requirement in the jurisdiction in which they are operating but still be targeted by activists who challenge the legitimacy of the original plan or regulation. We may huff and puff about the constitutional irregularity of these manoeuvres as much we like but, from Shell and the (non) sinking of the Brent Spar (Lofstedt and Renn 1997; Jordan 1998) to the continuing pressure on the forest industry to certify their forest management to standards more stringent than those imposed by the relevant governments (Cashore 2002), the dispersal of effective decision-making authority beyond governments is a fact of doing business. Second, and even more important, issue activism becomes, by default, the locus for the discovery and enactment of political identity (Harper 2001). That is, while many people have little or no idea what it means to be conservative or liberal today, they do know what it means to be an environmentalist and it is this kind of "issue-focused" identity that seems to make sense for them of what it means to be a citizen. Hence the high symbolic value of such activism noted by Nie as one of the drivers of conflict, a value, incidentally, that is likely to be reinforced rather than dissolved by participation in the usual kind of multistakeholder planning exercises (Rayner 1996).

The second reversal is the one that has taken place between *knowledge* and politics. The golden age when decision-makers took no steps without first consulting their scientific advisors who gave accurate and dispassionate advice on the likely consequences of alternative courses of action was always something of a myth, though one that continues to exert an influence on contemporary statements of the proper relation between science and policy. It is found, for example, in the Department of Fisheries and Oceans' response to declining public confidence in its scientific advice to government, a statement that amounts to an intention to try harder rather than re-examine the whole relationship (Canada 2002). Governments have long engaged in all sorts of activities for which (the claims of economists notwithstanding) no "science" exists to support a decision. It was only relatively recently in the history of modern states, and for a short time at that, that the physical and biological sciences were in a condition to provide useful and uncontested advice. That time has now passed. However unfairly, given the catalogue of scientific achievements over the last century and a half, the reaction of the public to scientific endeavour now seems to be a kind of "mitigated scepticism", not unlike the reaction of seventeenth-century intellectuals to the revelations of Galileo and Descartes (Tuck 1989). That is, if the world is as strange and as counter-intuitive as science seems to tell us it is, and if scientists can disagree so forcefully amongst themselves about it, then many citizens seem more impressed by what we obviously don't know about that world than what we do. This *awareness of unawareness* is, of course, powerfully supported by the periodic failures and crises of science and technology and has given rise to a number of consequences that feed into and mutually reinforce the changes in the polity.

For our purposes, the most important consequence is the way that scientific controversies have become so explosive, in the sense that they cannot be contained within the boundaries of what scientists accept as scientific argument. The BSE scare, which has led to the politicization of food production and animal husbandry with profound consequences for aquaculture, is a classic case in point. It has given another thrust to public scepticism about expert knowledge and further reduced trust in the abstract system of modern food distribution. People have fallen back on experiential knowledge: they want to know more about where their food came and what happened to it on the way to the table, a development that was noted especially in Europe in the wake of BSE. In this case, citizens were observed to be less and less likely to believe that hierarchical regulation underpinned by scientific advice is going to provide them with the level of security that they want (Vogel 2003; Abels 2002). They may be mistaken in this belief, and they may be doubly mistaken that some form of civil society-based certification scheme could replace regulation by states; but their scepticism is going to be, for a long time ahead, part of the reality that the governance of aquaculture must address.

Thus, the response of Canadian aquaculture to these developments can neither be "trust us, we know what we are doing", as the scientific establishment is inclined to say, nor "wait until all the facts are in" as the more radical exponents of the precautionary principle are asserting. Trust has gone and all the facts will never be in (which, of course, fits happily with the political agenda of those making the latter claim). The question of action, of what to do, and more specifically, of what we as a "community of interest" in aquaculture can do, remains.

Thus, when we come to the third theme, that of *intervention*, although the evidence of reversal is less immediately apparent, it is important that our efforts take proper account of the new realities of the governance of natural resources. Governance, as already noted, tries to match the complexity of the problem area by enhancing the ability of policy makers to learn from their mistakes and by giving them the flexibility to respond to the lessons they have drawn. But if we take the governance idea seriously we have to ask, both , "who are the policy makers?" and "who is doing the learning"? The idea that policy makers are a small band of easily identifiable senior public servants and ministers of territorially-based governments belongs to the political science of the day before yesterday. As we have seen, it's not only, or even primarily, technical experts who are doing the learning. The latter are not capable of learning fast enough in a globalized world and the former lack the capacity to respond quickly to what has been learned. Thus, not only are citizens less inclined to accept the decisions of their territorially-based governments about aquaculture development as definitive (the legitimacy problem), they are even less likely to be impressed by the knowledge that the decision has been informed by the advice of a group of self-accrediting experts – which is, of course, why senior public servants and ministers are not especially eager to seek out this advice in the first instance. The reversal of the old world here is that no one stands outside the political arena, looking on from some privileged neutral vantage point offering advice and encouragement to the players, who are ungrateful enough not to take it. Everyone is already part of the game (which the players know perfectly well); not everyone is very good at playing it yet.

The three reversals that Hajer describes provide us, by contrast, with a picture of what coordination by governance will actually entail. First, it will privilege different forms of self-organization over authoritative, hierarchical relationships. Secondly, these forms of self organization will last only so long as those who have created them continue to feel that they are delivering something useful, so that the most permanent kinds will be those where there is some continuing mutual and reciprocal interdependence between the members of the organization. Thirdly, the principal benefit of belonging to such an organization will be participation in learning how to coordinate and steer the organization to its mutually-agreed upon goals so that whatever forms of self-governance emerge will be reflexive learning organizations. The weakness of the European Commission in the area of legislative competence and implementation capacity has led the Commission to experiment extensively with policy networks of this kind but the results have been mixed. There is much we still need to know about network management and the background conditions which allow networks to promote policy learning or to constitute an appropriate governance mechanism (Montpetit 2003).

Knowledge Gaps – and How We Can Fill Them

Two conclusions jump out from this account of coordination via governance. First, the impermanence of governance relationships, and their ultimate dependence on voluntary delegation of authority from the authorising state holding a monopoly on use of coercive power, means that they are unlikely to replace traditional states and the emphasis on coordination through mutual agreement and interaction makes them equally unlikely to replace markets. Instead, they will exist side by side with states and markets as a third major mode of coordination. Second, the ambitious characterization of the goals and methods of governance vividly suggests that governance is as likely to fail as either of the other modes of coordination. As one recent commentator has put it " In part, recent interest in [governance] ... reflects one more turn in the never ending policy cycle whereby disenchantment with one mode of coordination leads to an excessive faith in another – until that too disappoints (Jessop 2002a)."

In fact, as Hajer and many others point out, we do know quite a lot about how contemporary governance is carried on and there are some valuable lessons available about the conditions under which governance succeeds and fails that suggest the kinds of policy innovations needed in aquaculture. First, coordination by governance is actually implemented along a continuum of policies that range from more reliance by government on policy instruments that aim to speed information flows and improve learning at one end to open coordination without the use of the traditional instruments of governing at all at the other (Knill and Lenschow 2004). In the middle lie various modes of self regulation involving more or less appeal to authority as a last resort: from self regulation "in the shadow of hierarchy"(Scharpf 1993) where standards may be set by governments and there are ultimately penalties for non-compliance to pure self-organization around emergent standards like the use of English in business communication or the adoption of Windows-based software. Secondly, the appropriateness of moving towards the self-regulation end of the continuum depends on the capacity of policy actors to regulate themselves in this way. The prevailing metaphor of capacity is the "network," which stresses the mutually recognized interdependence of network members and the exchange relationships – of knowledge, of power derived from positional advantage on particular issues, and so on – that move the network in one direction or another. Governments are network members as are research communities and communities of interest and each, but especially governments, may have significant resources to exchange with other network members. However, there is general agreement that any attempt to govern a network hierarchically is bound to be a failure – and may dissolve the network entirely.

Though empirical studies are lacking (though see Paradis 2003 for shellfish) there is a general concern that the capacity for self organization in Canadian aquaculture is low. Key evidence includes the adversarial relationships that have developed between the industry and other actors, including some levels of government. Especially noteworthy is the tendency to see the appearance and disappearance of temporary "issue networks" that are organized around campaignable themes by environmental NGOs and which draw an *ad hoc* and piecemeal response from other interested parties. In shellfish, the lack of importance of the industry and the small size of firms in many provinces results in an even less sophisticated kind of network based largely on personal acquaintance and informal groups, social networks that "oscillate between latency and activism" and frequently fall apart after occasional bursts of activism (Mayntz 1993:19). In consequence governance failures can be expected – both because of the same failure to grasp the complexities and interdependencies of the sector as a whole that dogged traditional regulation and because of intermittent but recurring gaps opening up between those who are most actively engaged in the networks through information exchange, communication and negotiation and those that they try to represent.

In these circumstances, it is clearly dangerous to imagine that the current interest in governance as a mode of policy coordination will, by itself, prove a panacea for the problems of the aquaculture industry. On the contrary, excessive faith in new governance techniques, such as decentralization, voluntary agreements, and policy networks is likely to prove a cure worse than the disease of traditional regulatory approaches and has already been blamed for a number of high profile policy failures (Moran 2001). And yet, as I have argued, countries that get governance right can reap immediate rewards in the form of vital comparative advantage in fast-moving international markets. The *status quo* is clearly not an option or, at least, represents a kind of upper limit and eventual stagnation of the industry that we are currently seeing with salmon farming in BC.

Four courses of action seem appropriate in these circumstances. First, governments should immediately engage in active support for the creation of more sophisticated aquaculture policy networks, both by actively supporting groups who are vital to the process but currently find it difficult to organize and maintain a presence – a category that would include small farmers – and by better use of information instruments to encourage exchange and the development of network relationships. These "capacity building" ventures are the crucial foundation for any more ambitious use of the governance mode.

Second, the existing networks, perhaps with government support again if we wish to speed up the process, need more active engagement and experience with the tools of self-regulation, including environmental management systems and certification of products. The danger here is that, lacking a complete network, the industries will engage with standards that fail to command any recognition in the market, though even these are useful tools for improvement as the literature on competing sustainable forestry certification schemes suggests (Cashore et al. 2004).

Third, there can be experimentation with some of the tools of open coordination, even if these tools cannot yet stand outside the shadow of hierarchy in the form of regulation. While true self organization is probably not a feasible outcome, as the slow but steady encroachment of regulation into the internet suggests, self-regulation and the use of voluntary agreements backed up by a more focussed regulatory effort for late adopters and backsliders holds out some promise (Gunningham and Sinclair 2002). In particular, it would be worthwhile to devote some resources to understanding the implications of what is rapidly emerging as the (for the moment) favourite tool of open network management, benchmarking. From the attempt by our federal government to use this information instrument to steer the Canadian health care system in the direction of reduced wait times for common surgical procedures (Postl 2006) to the more sophisticated versions found in Europe on immigration (Hajer 2003a) or forestry, this is where the action is in network governance.

Finally, and implicit in the other three suggestions, we need to think about what has been called "metagovernance." As I have argued in this chapter, experimentation with looser kinds of governance arrangements is not a panacea. If these arrangements succeed at all, they will take their place alongside the other main modes of coordinating action: markets and hierarchies. Metagovernance usually refers to the continuing effort to monitor and adjust the relative weights of coordination through hierarchy, markets and networks in ways that monitor the performance of the system as a whole and learns lessons from successes and failures (Jessop 2002b). At its simplest, metagovernance involves the matching of governance capacity with governance challenges and the judicious use of other modes of coordination where governance capacity is weak or absent. All this sounds a tall order but we cannot turn our backs on governance. In this case, at least, we know it is the future for us, because other people are there already.

References

Abels, G. 2002. Experts, Citizens, and Eurocrats – Towards a Policy Shift in the Governance of Biopolitics in the EU. *European Integration online Papers (EIoP)* 6, No. 19 http://eiop.or.at/eiop/local_old.html

Bourgault, J. and Lapierre, R. 2000. Horizontality and Public Management. Montreal: Canadian Centre for Management Development. http://www.myschool-monecole.gc.ca/Research/ publications/pdfs/horizont_e.pdf

Canada 2002. Federal Science and Technology Report. http://www.innovationstrategy.gc.ca/gol/innovation/site.nsf/en/in04880.html

Cashore, B. 2002. Legitimacy and the Privatization of Environmental Governance: How Non State Market-Driven (NSMD) Governance Systems (Eco-labeling Programs) Gain Rule Making Authority. *Governance* 15 (4): 503–529

Cashore, B., Auld, G., and Newsom, D. 2004. *Governing Through Markets: Forest Certification and the Emergence of Non-State Authority*. New Haven: Yale University Press.

Deal, H. 2005. Sustainable Shellfish: Recommendations for Responsible Aquaculture. Vancouver: David Suzuki Foundation. http://www.davidsuzuki.org/files/Oceans/Shellfish.pdf

Dorcey, A.H.J. 1986. Bargaining in the Governance of Pacific Coastal Resources: Research and Reform. Westwater Research Centre, University of British Columbia,Vancouver, B.C.

Edelman, M. 1988. *Constructing the Political Spectacle*. Chicago: University of Chicago Press.

Gunningham, N. and Sinclair D. 2002. *Leaders and Laggards: Next Generation Environmental Regulation*, Greenleaf, UK.

Hacker, J. 2005. Policy Drift: The Hidden Politics of US Welfare State Retrenchment. In W. Streek and K. Thelen (eds.) *Beyond Continuity: Institutional Change in Advanced Political Economies*. Oxford: Oxford University Press: 40–82.

Hajer, M. 2003a. Policy without polity? Policy Analysis and the Institutional Void. *Policy Sciences* 36: 175–195.

Hajer, M. 2003b. A frame in the fields. Policy making and the reinvention of politics. In M. Hajer and H. Wagenaar (eds.) *Deliberative Policy Analysis: Understanding Governance in the Network Society*. Cambridge: Cambridge University Press.

Hall, P. and Soskice D. (eds.) 2001. *Varieties of Capitalism: The Institutional Foundations of Comparative Advantage*. Oxford. Oxford University Press.

Hall P. and Soskice D. 2003. Varieties of capitalism and institutional change: A response to three critics. *Comparative European Politics* 1: 241–250.

Harper, K. 2001. Introduction: The Environment as Master Narrative: Discourse and Identity in Environmental Problems. *Anthropological Quarterly* 74 (3): 101–103.

Hites, R., Foran, J., Carpenter, D., Hamilton, M., Knuth, B., Schwager, S. 2004.Global Assessment of Organic Contaminants in Farmed Salmon. *Science* 303. no. 5655: 226–229.

Howlett, M. and Rayner, J. 2004 "(Not so) smart regulation"? Canadian shellfish aquaculture policy and the evolution of instrument choice for industrial development. *Marine Policy*, 28, 171–184.

Howlett, M. and Rayner, J. 2006 Globalization and Governance Capacity: Explaining Divergence in National Forest Programs as Instances of "Next-Generation" Regulation in Canada and Europe. *Governance* 19 (2): 251–275.

Howlett, M. and Rayner, J. 2007. Caught in a Staples Vise: The Political Economy of Canadian Aquaculture. *Policy and Society* 26: 49–69.

Hutton, T. 2007. Contours of the Post-Staples State: The Reconstruction of Political Economy and Social Identity in 21st Century Canada. *Policy and Society* 26: 9–29.

Innis, H. 1930. *The Fur Trade in Canada: An Introduction to Canadian Economic History*. Toronto: University of Toronto Press.

Jessop, B. 2002a. Capitalism, steering, and the state. Alexander von Humboldt lecture in human geography, Department of Human Geography, University of Nijmegen, Nijmegen, Netherlands. http://www.kun.nl/socgeo/n/colloquium/CapitalismSteeringState.pdf

Jessop, B. 2002b. 'Governance and Metagovernance: On Reflexivity, Requisite Variety, and Requisite Irony.' Department of Sociology, Lancaster University, Lancaster LA1 4YN, UK, at http://www.comp.lancs.ac.uk/sociology/papers/Jessop- Governance-and-Metagovernance.pdf

Jordan, G 1998. Indirect Causes and Effects in Policy Change: The Brent Spar Case. *Public Administration*, 76 (4):713–740.

Jordan, A., Wurzel, R and Zito, A. 2005. The Rise of 'New' Policy Instruments in Comparative Perspective: Has Governance Eclipsed Government? *Political Studies* 53 (3): 477–496.

Knill, C. and Lehmkuhl, D 2002. Private Actors and the State: Internationalization and Changing Patterns of Governance. *Governance* 15(1): 41–63.

Knill, C. and Lenschow, A. 2004. "Modes of Regulation in the Governance of the European Union: Towards a Comprehensive Evaluation." In Jacint Jordana and David Levi-Faur (Eds.). *The Politics of Regulation: Examining Regulatory Institutions and Instruments in the Age of Governance*. London: Edward Elgar.

Lofstedt, R.E. and Renn, O. 1997. The Brent Spar Controversy: An Example of Risk Communication Gone Wrong. *Risk Analysis* 17(2):131–136.

Mayntz, R. 1993. "Governing Failures and the Problem of Governability: Some Comments on a Theoretical Paradigm." In Jan Kooiman (Ed.) *Modern Governance: New Government-Society Interactions*. London: Sage.

Montpetit, E. 2003. *Misplaced Distrust: Policy Networks and the Environment in France, the United States and Canada*. Vancouver: UBC Press.

Moran, M. 2001. Not Steering But Drowning: Policy Catastrophes and the Regulatory State, *Political Quarterly* 72 (4): 414–427

Nie, M. 2003. Drivers of natural resource-based conflicts. *Policy Sciences* 36: 307–341.

Office of the Commissioner for Aquaculture Development. 2001. *Legislative and Regulatory Review of Aquaculture in Canada*. Communications Branch: Fisheries and Oceans Canada. http://www.dfo-mpo.gc.ca/aquaculture/ref/legalreview_e.pdf

Paradis, E. 2003. Industry-Science Collaboration in Shellfish Aquaculture and the Management of the Knowledge Process. Unpublished M.Sc. Thesis. UBC Resource Management and Environmental Studies Program.

Peters, B. Guy. 2002. Governance: A Garbage Can Perspective. Institute for Advanced Studies (Vienna), Department of Political Science Paper Series No. 84, available at http://www.ihs.ac.at/publications/pol/wp_84.pdf

Pierre, J. 2000. Introduction: Understanding Governance. In Pierre, J., (Ed.), Debating governance. Authority, steering, and democracy. Oxford: Oxford University Press.

Postl, B. 2006. *Final Report of the Federal Advisor on Wait Times*. http://www.hc-sc.gc.ca/hcs-sss/pubs/care-soins/2006-wait-attente/index_e.html

Rayner, J. 1996. Implementing sustainability in West Coast forests: CORE and FEMAT as experiments in process. *Journal of Canadian Studies* 31: 82–101.

Scharpf, F. 1993. *Games in Hierarchies and Networks Analytical and Empirical Approaches to the Study of Governance Institutions*. Campus Verlag: Frankfurt am Main.

Shindler, B.A. and Cramer, L.A. 1999. Shifting public values in forest management: making sense of wicked problems. *Western Journal of Applied Forestry*. 14(1):11–17.

Tuck, R. 1989. *Hobbes* Oxford: Oxford University Press.

Van Kersbergen, K. and van Waarden, F. 2004. 'Governance' as a bridge between disciplines: Cross disciplinary inspiration regarding shifts in governance and problems of governability, accountability and legitimacy. *European Journal of Political Research* 43: 143–171.

Vogel, D. 2003. *The Politics of Risk Regulation in Europe and the United States*. Yearbook of European Environmental Law Volume 3. Oxford: Oxford University Press.

Watson, M. 2003. Ricardian Political Economy and the 'Varieties of Capitalism 'Approach: Specialization, Trade and Comparative Institutional Advantage. *Comparative European Politics* 1: 227–240.

Winch, P. 1958 . *The Idea of a Social Science and its Relation to Philosophy*. London: Routledge and Kegan Paul.

Chapter 20
Consumer Confidence, Food Safety, and Salmon Farming

Paul Lyon

Introduction

Aquaculture is conducted in a remarkably complex social and scientific context. While health-savvy consumers turn increasingly to seafood as a source of protein, the commercial capture fishery faces dwindling and depleted wild stocks. What is next for consumers, and for fishers? One answer, offered by the Food and Agriculture Organization of the United Nations, is to consider farming of fish as a critical part of a strategy to meet a potential global fish protein deficit (Tidwell and Allen 2001; Brugère and Ridler 2004).

In the Americas, farmed salmon has become a vital component of two nations' exports: Chile and Canada each compete for a share of the United States salmon market. Alaska's commercial wild salmon fishery, with home field advantage, is also competing for a share of the same salmon market. Yet, as we demonstrate below, it is no simple matter for consumers, the commercial capture fishery, and aquaculture to transition from captured fish to farmed fish. Complex social, economic, and natural forces are in tension, and there is no easy governance recipe for resolution of these tensions.

A consumer's choice of a specific product is motivated by many factors. For salmon, motivating factors include price, convenience, nutritional value, and food safety. When the safety of consuming farmed salmon was challenged in 2004, industry and governments were sent scrambling to respond. Meanwhile, commercial salmon fishery proponents used farmed salmon's bad press to leverage a competitive advantage for wild salmon products in a market where consumers had suddenly altered their purchasing behavior.

The "bad press" was part of an ongoing and very public debate between environmental advocates and the salmon farming industry. Environmental performance of salmon farming has long been at the core of the environmental nongovernmental organizations' (ENGOs) critique of the industry. Beginning in 2002, contaminants were appended to their list of criticisms. The issue of contaminants, and

P. Lyon
Fisheries and Oceans Canada
e-mail: paul.lyon@dfo-mpo.gc.ca

K. Culver, D. Castle (eds.), *Aquaculture, Innovation and Social Transformation*,
© Springer Science+Business Media B.V. 2008

consequently, the safety of eating farmed salmon was encapsulated in a widely publicized study regarding polychlorinated biphenyls (PCBs) in farmed and wild salmon, published in the journal *Science* on January 9, 2004 (Hites et al. 2004).

The following includes a timeline of events with respect to the public debate on contaminants in farmed salmon, the impact the *Science* article had on industry and on public confidence, and considers governance options for responding to misinformation regarding food safety of aquaculture products. This discussion will illustrate of some of the issues discussed by Rayner, together with identification of innovative methods of governance which balance effectively the social and environmental issues surrounding aquaculture practices.

PCBs

On May 17, 2002 a front-page article in the Toronto *Globe & Mail* featured a study supporting the claim that farmed salmon contained sufficiently high levels of PCBs to constitute a health threat (Easton et al. 2002). The study, commissioned by the David Suzuki Foundation in British Columbia was a limited investigation of just four farmed and four wild salmon. The article did not arouse much response from the public.

On January 9, 2004, the issue reappeared in the context of a larger, global contaminant study of farmed and wild salmon, published in the journal *Science*. Hites et al. (2004) asserted that higher levels of PCBs in farmed salmon should result in consumption limits – despite the fact that international standards, including those of the World Health Organization and the United States Food and Drug Administration, were consistent with Health Canada's position that farmed and wild salmon were safe for human consumption (Green 2004; Health Canada 2004). The *Science* article resonated in the media around the globe, echoing in the news for months after publication of the original journal article. This broad coverage may have been obtained because of a reported public relations effort funded by the United States based Pew Charitable Trust (Linklater 2004).

The impact on the salmon farming industry was significant. North American sales dropped immediately and domestic sales in Canada remained consistently depressed (down 30%) for over five months. Consumers perceived wild salmon as the safer choice and responded accordingly. In a June 2004 article in IntraFish (Nielson 2004) it was reported to be clear, according to at least one retailer, that the sale of wild salmon previously lagging behind farmed salmon by a ratio of 1 out of 10 fillets was now on an equal (1–1) footing. The reason cited? Consumers were reacting to the media coverage on the PCB study.

The story proved to be an opportunity for the wild salmon fishery. A mere four days after the PCB study was released in the journal *Science*, Intrafish Media published an article entitled "Northwest seafood chain follows PCB study with wild salmon promotion" (Intrafish 2004). Fearing a weakening consumer confidence in salmon in general, retail outlets attempted to avoid a loss of revenue by positioning wild salmon as a safe alternative to farmed. Nearly simultaneously, the Marine Stewardship Council (MSC) issued a press release suggesting that its eco-labelling certification of wild Alaskan salmon was "a useful tool" for consumers. Indirectly the Council's assertion could have been interpreted by consumers that Alaskan wild salmon was a healthier choice because of their certification – even if, as the MSC official indicated, the MSC had no immediate plans to test for contaminants (Cherry 2004).

The salmon farming sector responded through *Salmon of the Americas*, an organization of salmon-producing companies in Canada, Chile and the United States whose mission is to "improve health, awareness and dining enjoyment of consumers in North America by providing timely, complete, accurate and insightful information about salmon (SOTA 2003)." *Salmon of the Americas* released a separate study showing no significant difference in contaminants found in both farmed and wild salmon. The study received very little media coverage (Brown and Green 2004). Subsequent strategies from *Salmon of the Americas* further emphasised the nutritional and health benefits of salmon, yet public pressure continued, and in response retailers requested that the industry implement an appropriate certification program to respond to consumers' concerns. *Salmon of the Americas* considered standard-based certification systems. In developing its own program, it examined the Safe Quality Food certification of the United States-based Food Marketing Institute (FMI 2005). (Note: The Global Food Safety Initiative was formed in April 2000 by a group of international retailer CEOs whom identified the need to enhance food safety, ensure consumer protection, strengthen consumer confidence, to set requirements for food safety schemes and to improve cost efficiency throughout the food supply chain. The Food Marketing Institute's Safe Quality food certification is recognized by the Global Food Safety Initiative.)

The 2004 Hites study on PCBs in farmed salmon openly challenged international food safety standards, with serious practical effects. In California, for example, the story appeared around the same time as an attempt to ban the sale of farmed salmon, and a lawsuit based on Proposition 65, California's tough anti- toxics law (WSJ 2004). The Environmental Working Group (EWG) and The Center for Environmental Health (CEH) had filed legal notice under Proposition 65 of plans to sue the manufacturers, distributors and retailers of farmed salmon over potentially dangerous levels of PCBs in the fish. They claimed that the businesses failed to warn consumers of dangerous PCBs in farmed salmon and were pushing for a warning label indicating PCB exposure from eating farmed salmon.

Repercussions from this story brought salmon producing nations together, led by the Norwegian Fisheries Minister, to discuss food security and seek out ways to restore consumer confidence. Whatever else may be said about the Hites study, it is clear that it raised the issue that consumers are susceptible to food safety scares despite critical reaction to its conclusions from international food safety agencies and other scientists (Lawrence and Leister 2004; Green 2004).

Many lessons may be learned from this experience. One important lesson for governance is this: some critical negative factors impacting the aquaculture industry and other food producers – such as inaccurate or premature scientific information influencing consumer behaviour – are both impossible to anticipate and beyond the scope of governments to regulate. Governments do not know what research will be published, and in democratic societies are rightly constrained against interference with free expression of scientific research, including research that is strongly disputed by other researchers. This lesson casts doubt on the view sometimes heard that if only the regulatory "balance" were struck properly, salmon aquaculture would be successful by the standards of all stakeholders. Even good governance, in which Canadians are global leaders, cannot predict the path of science or consumers' reactions to media. And even with good governance, industries may face obstacles to their success.

Public Perception

In the wake of the Hites PCB study, Canada's lead federal agency responsible for aquaculture, the Department of Fisheries and Oceans (DFO), took a number of steps to analyse the consequences of the study. In particular a series of focus groups were designed to explore the perceptions, attitudes, and concerns of the average Canadian towards aquaculture. A total of 138 respondents from the general public participated with 22 groups in 11 locations. The study included two groups at each location: Halifax (NS), Saint John (NB), St.John's (NL), Montreal and Quebec City (PQ), Toronto and Ottawa (ON), Vancouver, Prince Rupert and Campbell River (BC) and Whitehorse (YK). It soon became clear that there was a need for balanced information to reach the public, as results from the research revealed negative views on aquaculture that intensified on an east to west continuum, with Eastern Canada tending to be generally positive, and Western Canada generally negative. Beneath this regional variation, the study found that participants shared a generally poor understanding of aquaculture. Probing revealed that respondents had concerns and fears about food safety and the environment, and although aquaculture was not top of mind, or fully understood, they associated fish farming with food safety and environmental sustainability issues (Créatec 2005). There were also indications that Canadians would be receptive to positive information about aquaculture if they were better informed about the industry and how it is regulated, provided their doubts and fears were addressed to their satisfaction. Clearly the public expect government to provide them with reassurances that their food is safe.

> What people expect and want from government on aquaculture is very consistent across the country: reassurance. Overall, there is a desire and need to know the values and principles that guide the government's thinking and policies on aquaculture (Créatec 2005).

The focus group research also suggested that the public would be more supportive of the sector when the aquaculture industry was compared to terrestrial farming practices. Initially, participants' perceived aquaculture as a form of fishing, however following their discussions many believed that aquaculture could be considered a part of the farming industry, even though this was quite a new idea to them. There were three general rationales for their insight. The most frequently mentioned was that aquaculture was done in a controlled area set aside for a particular activity, like farming. The second rational was that the aquaculturists' production driven approach is more akin to terrestrial farming than the hunter/gatherer approach taken by fishers. And thirdly, they perceived farmed fish as a product, like cattle. Their acceptance of aquaculture as a form of agriculture was as much an acknowledgement that aquaculture shares many of the same attributes with agriculture, which largely enjoys social acceptance.

Many recognized the socio-economic benefits of aquaculture and some participants believed that Canada had the potential to be a world leader in aquaculture technology and research. They speculated that aquaculture technologies and intellectual properties developed in Canada could be exported globally, in addition to benefiting the industry domestically. They acknowledged that aquaculture in Canada need not be limited to a few species, like salmon or mussels. They envisioned a role for Canadian researchers in the development of alternate species for the aquaculture industry.

Food Safety

There are many issues and sources at the root of negative attitudes among some Canadians with respect to the aquaculture industry. In British Columbia, for example, a long history of environmental advocacy has predisposed British Columbians to give environmental issues serious consideration. In the case of salmon farming this has resulted in a less favourable perception of the industry than exists in many of the other provinces.

By contrast, the public's fear of PCBs is a product not of longstanding political preferences, but of heightened awareness, whether or not justified. Bioaccumulation of environmental contaminants like PCBs has been studied and publicly debated for years and although much of the science and application of risk assessment remains a mystery to the average person, the public now associates PCBs with a greater risk of cancer for a variety of reasons including, high-profile scientific articles such as the Hites article.

While it remains notoriously difficult to predict how specific risk issues will be received by the public, risk management scholars have defined useful general criteria for identification of the acceptability of specific kinds of risk. Slovic, Fischoff et al., for example, have concluded that for a risk to be perceived as less acceptable it likely meets one or more of the following criteria: involuntary exposure; dreaded consequence; affects children; human origin; high media concern; high symbolism (Butte and Thorne 2004; DHHS 2002). The Hites PCB study met many of these criteria: PCBs are well-known carcinogens which could certainly affect children, the study garnered a great deal of media attention (Butterworth 2004), and farmed salmon served as an easily presented visual symbol of the claimed risk. Where this story of salmon diverges, it does so at the crucial point of involuntary exposure.

Exposure to environmental contaminants is generally involuntary yet in this case, there are steps one can take to reduce the risk. This explains why many consumers took corrective measures to reduce their perceived exposure by responding with a significant reduction in their farmed salmon purchases, even if it meant sacrificing the health benefits of marine fish oils. Faced with a perceived risk which seems to fit in several categories of what risk experts recognize as triggers of risk-avoiding behaviour, and an easy way to avoid the risk altogether, the public did so, preferring risk avoidance over the time-consuming and difficult task of evaluating the quality of the science claiming that consumption of farmed salmon really does pose a risk to human health.

Genetically modified food, and persistent pollutants were raised as a concern in every aquaculture focus group discussion (mercury was raised far more often than PCBs). Frequent reference to avian flu and BSE were indicative of the participants' concern for food safety. As can be seen in participants' concerns regarding those diseases clearly *not* linked to aquaculture, citizens are not prepared to negotiate regarding perceived risks to their wellbeing. While Canadians are willing to accept some level of managed risk on environmental concerns they desire zero risk in the area of food safety. And as we have seen in public response to the PCB-contaminant story accelerated by the Hites article, when food safety is perceived to have been compromised it can elicit a swift public response.

Despite publicised evidence challenging Hites' conclusions and literature that supported the industry's practices, some ENGOs continue to portray salmon farming as an unregulated industry freely using steroids, chemicals and antibiotics without consideration for the end user, the consumer. Repeated messaging and a very effective use of mainstream media by these groups has resulted in these views being widely accepted by some members of the public, regardless of existing science, legislation, regulations and food safety systems that suggest otherwise. By simply linking farmed salmon to PCBs and food safety, environmental advocacy groups have been able to sow doubt on a relatively new form of food production. In part, these messages are accepted because salmon farming does not benefit from the same level of public acceptance demonstrated toward established food production sectors.

Leveraging Credibility

The focus group report noted a general "mistrust of and anger for DFO on both coasts" (Créatec 2005), based in part on issues such as the collapse of cod stocks in the east, predisposing the public to hold similar suspicions regarding DFO's aquaculture policies. The collapse of the cod fishery was raised in nearly every focus group. In British Columbia, there was general dissatisfaction with DFO's management of the Pacific salmon commercial fishery. In the West, the perception that salmon farming is an alliance of industry, the province, and DFO does not help. This perception, although fundamentally inaccurate, appears to have been encouraged by those who seek to discredit the industry and government's oversight and has brought about a general perception of an industry being promoted by governments, an industry that is under-regulated, and an industry where environmental impacts are overlooked.

When asked who they would like to obtain information about aquaculture focus group participants regarded Health Canada, Environment Canada, and the Canadian Food Inspection Agency (CFIA) as credible sources. These results suggest that if DFO were to collaborate more closely with other federal departments on appropriate issues, notably Health Canada, the CFIA and Environment Canada, acceptance of DFO messages would be enhanced.

Evidence from the focus group discussions, comparing aquaculture to terrestrial farming, suggest that it is likely that Agriculture Agri-Food Canada would be viewed by the public as a legitimate conveyor of messages relating to aquaculture. The focus group study concluded that the public might be more receptive to messages about aquaculture if aquaculture was referred to as aqua-farming.

Recognizing Food as a Determinant of Health

Consumers want healthier food options, and product assurance systems guaranteeing the safety of those foods. Focus groups clearly showed us that Canadian consumers "are connected to their food and its impact on their lives on an emotional level (Créatec 2005)." When asked, Canadian consumers were well aware of the dietary value attributed to fish and all were aware of the health benefits of consumption specific marine fish oils (particularly Omega-3 fatty acids).

In light of Canadians' desire for positive, or even added health benefits from particular food, it is no longer sufficient for government to simply respond reactively to safety issues. Governments need to prepare for emerging issues – and even crises of citizens' confidence in particular foods and the food safety system. We can expect some groups will continue to exploit the public's desire for zero risk as it applies to their food. Government can no longer simply respond to legitimate food safety issues, but must also actively engage unsubstantiated claims that may potentially alter consumer behaviour away from safe and beneficial food choices. Risk communication is not just about communicating risk, but it is also about communicating benefits. Recognition that food is a determinant of health is a critical first step.

Conclusion

The PCB story has had serious repercussions for citizens, industry, and government bodies concerned with aquaculture. In the short term, concern regarding PCBs in farmed salmon resulted in: significant losses to the salmon farming industry; public scepticism regarding known and well-established science and global food safety systems; and it caused some consumers to avoid a food associated with certain health benefits.

These repercussions, as suggested, are not all negative. From a positive perspective the debate regarding PCB contamination has assisted industry in moving forward with certification programs, and it has heightened the awareness of government on the need to engage Canadians more fully regarding aquaculture. From a food safety perspective the PCB story has shown that the public needs to see that Canada's food safety network is functioning as it should and it needs to see food safety issues dealt with promptly. At the same time the public are looking for leadership in addressing "food scares." When the food safety system is challenged by erroneous science the government needs to be as proactive with its messages to the public as it would in the case of a legitimate food safety issue, recognising that government has a legitimate role defending foods with significant health benefits. Finally, it should be noted that the PCB story fostered valuable international dialogue, as cold water aquaculture countries gathered to discuss and share information on issues of public concern. As cold water aquaculture countries face the complexities of global markets and savvy consumers, meeting on a common issue is a step toward improved international coordination in response to emerging issues in food production, trade and safety.

References

Brown J. and Green K. 2004. Good News is No News: Media Failure to Cover Salmon Study Skews Public Perception. *CANSTATS Bulletins*. (October 25, 2004)

Brugère, C. and Ridler, N. 2004. Global Aquaculture Outlook in the Next Decades: An Analysis of National Aquaculture Production Forecasts to 2030. FAO Fisheries Circular No. C1001. Food and Agriculture Organization of the United Nations, Rome.

Butte, G. and Thorne, S. 2004. Enabling Well-Informed Decisions and Action on Risks, Powerpoint presentation to Society for Risk Analysis Workshop, June 22–24, 2004 http://www.risktrace.com/sediments/talks/DP.pdf

Butterworth, T. 2004. PCBs Anatomy of a Health Scare in Salmon. STATS. April 4, 2004 http://www.stats.org/record.jsp?type=news&ID=466

Cherry, D. 2004. MSC sees opportunity for eco-labeled salmon in wake of PCB report. Intrafish Media. (June 15, 2004) http://www.intrafish.no/global/news/article16330.ece

Créatec, 2005. Final Report – Qualitative Research Exploring Canadians' Perceptions, Attitudes and Concerns Towards Aquaculture. (February, 2005). http://www.dfo-mpo.gc.ca/misc/focus-aquaculture_e.htm

Department of Health and Human Services (DHHS), United States. 2002. Communicating in a Crisis: Risk Communication Guidelines for Public Officials. Washington, D.C. http://www.hhs.gov/od/documents/RiskCommunication.pdf

Easton, M.D.L., Luszniak, D., and Von der Geest, E. 2002. Preliminary examination of contaminant loadings in farmed salmon, wild salmon and commercial salmon feed. *Chemosphere* 46:1053–1074.

Food Marketing Institute (FMI). 2005. Food Marketing Institute and Salmon of the Americas Announce New Safe Quality Food Certification Initiative. (February 22, 2005) http://www.fmi.org/media/mediatext.cfm?id=718.

Green, K. 2004. The Great Salmon Panic of 2004, *CANSTATS Bulletins*, January 14, 2004

Health Canada. 2004. Fact Sheet: Food safety and PCBs found in Fish (January 12, 2004). http://www.hc-sc.gc.ca/ahc-asc/media/nr-cp/2004/2004_pcb-bpc_e.html

Hites, R.A. et al. 2004. Global Assessment of Organic Contaminants in Farmed Salmon. *Science* 303: 226–229.

Intrafish. 2004. Northwest seafood chain follows PCB study with wild salmon promotion. Intrafish Media. (January 13, 2004). http://www.intrafish.no/global/news/article16282.ece

Lawrence, J. and Leister, B. 2004. Catch of the Day: Choosing the right seafood in a world of scaremongers, fishmongers and scientists trolling for credible answers. *Eating Well* (Spring issue).

Linklater, M. 2004. Thunderer: Answer this: who benefits from the salmon scare? *TimesOnline*, Health News, January 15, 2004. http://www.timesonline.co.uk/article/0,8122-964617,00.html

Nielson, P. 2004. Wild Salmon finding bigger demand on US East Coast. *Cape Cod Times* (June 20, 2004). Intrafish Media. http://www.intrafish.no/global/news/article23197.ece

Salmon of the Americas (SOTA). 2003. SOTA Mission Statement. http://www.salmonoftheamericas.com/sitemap.html

Tidwell, J.H. and Allen, G. L. 2001. Fish as Food: Aquaculture's Contribution, EMBO Reports. 2 (11): 958–963.

The Wall Street Journal. 2004. Eat Your Salmon-The latest PCB scare has nothing to do with science. *The Wall Street Journal*. (February 11, 2004; A.18)

Chapter 21
Aquaculture Policies for Global Competitiveness: An Industry Perspective

Colin J. Barrow

Introduction

Rayner's "Governance for Global Competitiveness" identifies some key governance problems for aquaculture in Canada and proposes some specific courses of action. The paper identifies the need for innovative public policy and a more collaborative approach to governance. Rayner suggests that a rapidly changing environment and globalization are causing the older "authoritative co-ordination" and "regulatory state" models to break down. Rayner also argues that government efforts to reform aquaculture policy have met with little success, at least partly attributable to the difficulty of creating policies to bridge effectively the many jurisdictional boundaries present in the Canadian model of confederation. Additionally, Rayner explores the diverse interests and sometimes hidden agendas of groups such as industry, scientists, aboriginal peoples, and environmentalists, interests and agendas which often conflict and create further uncertainty regarding the possibility of broadly supported policies. Rayner's solution includes the creation of more sophisticated aquaculture policy networks by government, the strengthening of existing networks, and creation of better methods of coordination including use of appropriate benchmarks.

In this commentary I plan to discuss in more detail some of the issues and potential solutions raised by Rayner. These include: (a) the confused regulatory and governance situation; (b) the impact of rapid change and globalization; (c) dealing with the interests of diverse groups; (d) aspects of the solution.

Canadian Governance and Regulatory Situation for Aquaculture

Rayner uses the definition of governance as "the effort to sustain co-ordination and coherence among a wide variety of private and public actors with different purposes and objectives (Pierre 2000)." Inherent in this definition is the understanding that

C.J. Barrow
Ocean-Nutrition Canada, Dartmouth, Nova Scotia
e-mail: cbarrow@ocean-nutrition.com

K. Culver, D. Castle (eds.), *Aquaculture, Innovation and Social Transformation*,

© Springer Science+Business Media B.V. 2008

no matter what governance model is chosen not everyone will be happy with the outcome. Essentially there must be a trade-off between the interests of different groups so that a happy medium is reached. A key to this success is to clearly define what is a successful outcome, or successful happy medium. To have effective governance of aquaculture governments need to create clear aquaculture policy that has regulatory aspects to ensure safety and life-style sustainment, while creating an environment where industry can succeed. Policy needs to be consistent and specific to aquaculture, reducing the tangle of national and provincial regulations and avoiding arbitrary and burdensome regulatory requirements that do not succeed in their purpose.

Rather than having a unified and unambiguous aquaculture policy in Canada there are many federal acts that in some way regulate aquaculture, including the *Fisheries Act* (1985), the *Oceans Act* (1996), the *Canadian Environmental Protection Act* (1999), and the *Food and Drugs Act* (1985), amongst others. In 2001 the Office of the Commissioner for Aquaculture Development drew attention to the shortcomings of the status quo, explaining that:

> Most of the measures in place today were not developed with aquaculture in mind and thus are often applied to the sector in an inconsistent manner….. the aquaculture industry (is) managed as a subset of the traditional fisheries. This is analogous to equating traditional livestock and crop agriculture to the hunting and gathering of animals and plants. (OCAD 2001:1)

Fisheries and Oceans Canada play a leading role in aquaculture policy but must interact with other federal departments such as Health Canada and the Canadian Food Inspection Agency on issues such as food safety and fish feed. However, provincial rather than federal governments are responsible for many issues such as providing licenses and regulating environmental issues, making regulation complex. In British Columbia alone 16 provincial legislative acts and regulations apply to aquaculture (Power 2005).

In summary, I believe that the type of governance model is less important to the success of the aquaculture industry, and of the government's ability to satisfy all stakeholders, than is having a consistent and simplified policy framework. The federal government through the federal Aquaculture Management Directorate (previously The Office of the Commissioner for Aquaculture Development) and Fisheries and Ocean Canada should establish a clear and consistent policy for aquaculture. What industry needs more than regulatory experiments is certainty and predictability in the regulatory environment. To be sure, Canadian regulation must keep pace with competitors, but a large-scale retreat from the "regulatory state" seems premature. There is a pressing need for certainty which needs to be addressed prior to undertaking large-scale renovations of the regulatory system. As the 2001 *Legislative and Regulatory Review of Aquaculture in Canada* explains:

> … it is unclear what rights and obligations aquaculturists have under the existing legislative and regulatory regime, and how these rights and obligations are upheld and enforced. Part of the uncertainty relates to the lack of long-term security for various forms of authorization and licensing of aquaculture activities. This is a deterrent to private investment in aquaculture (OCAD 2001:19).

In summary, innovation in aquaculture needs to be tracked by innovation in governance, but there is some catching up to do in governance before we worry about just how governance innovation should track aquaculture innovation.

The Impact of Rapid Change and Globalization on Aquaculture

Rayner correctly argues that technological, social and economic change tends to be more rapid and less predictable than in the past. This is partly due to globalization, both with respect to markets and with respect to socio-economics. Governance and the regulatory framework need to be clear and consistent, while also enabling enough flexibility to respond to change.

Globalization means that Canadian aquaculture companies compete heavily with those of other countries that have more industry-friendly policies. For example, the Chilean aquaculture industry has been given considerable leeway by the state to maximize profits over environmental sustainability (Barton 1997). The Chinese as they have done in many industries, have targeted success in aquaculture. Chinese government and governance models differ radically from those of westernized countries such as Canada. The Chinese government tends to steer and financially support industries rather than simply regulate them as Canada does (Liu 2005). These differing models add to the difficulty of competing internationally for Canadian aquaculture companies.

Rayner rightly suggests that "it is not at all obvious that governments are best positioned to determine what a particular industry needs to survive and prosper in a global rather than a protected national market." In a free market society, rather more attention needs to be given to the potential for industries to find a way forward themselves, enabled not by picking and choosing of specific industries to support, but by fair and transparent regulation enabling global competitiveness. If regulations are clear and industry understands the policy framework under which it works then industry can use other advantages such as natural resources, quality and technology to overcome certain policy disadvantages. However, this cannot occur in an environment of policy uncertainty as currently exists in Canada. It seems to me that the sort of clarity I advocate would have additional benefits with respect to tensions between the aquaculture industry and other users of Canada's aquatic resources. In my experience members of the aquaculture industry do not resemble in real life the sort of caricatures sometimes seen, presenting the aquaculture industry as a "profit above all" business with little concern for the environment and local communities. Much of the Canadian aquaculture industry has been developed in small coastal communities by members of those communities, including some of Canada's most successful firms such as family-run Cooke Aquaculture of New Brunswick. While these members of the industry want clarity in regulation in order to build their businesses, they recognise the value of Canada's quality of life and the importance Canadians attach to environmental quality as a key component of quality of life. The aquaculture industry does not want to increase international competitiveness by simply reducing environmental standards as Chile has done. That policy is both risky and possibly self-defeating for the long-term health of the industry. What the industry needs is clarity of regulation to enable it to use its comparative advantages in innovation capacity and know-how to overcome competitors' willingness to put short-term gain ahead of environmental quality. Canadian innovation in technology and know-how may enable Canada to have both a high quality of life and comparatively higher environmental standards, but industry needs regulatory clarity in order to pursue this goal while remaining internationally competitive.

In addition to facing the complexities of governance and regulatory overlap amongst federal and provincial jurisdictions, the Canadian aquaculture industry must respond to international and transnational regulations. In particular international accords impact aquaculture decisions and have a different impact on Canada versus other countries, depending upon the desire and efforts of a particular country to implement and enforce the accords. International and trans-national agreements directly or indirectly impacting Canadian aquaculture governance include the United Nations Convention on the Law of the Sea (UNCLOS), the United Nations Conference on Environment and Development (UNCED), the Food and Agriculture Organization of the United Nations (FAO), the North Atlantic Salmon Conservation Organisation (NASCO) and the Network of Aquaculture Centres in Asia-Pacific (NACA). Of these only UNCLOS has a clear means of enforcement, but this only applies to countries that have ratified the agreement.

Dealing with the Interests of Diverse Groups

A variety of groups and individuals are impacted in relation to aquaculture policy and outcomes. Each has quite different objectives. For example, companies involved in the aquaculture industry want to see aquaculture expand, costs decrease, regulations clarified and market access improved. First Nations in general want to see wild fisheries protected and ensure environmental and biological safety, yet this is by no means their universal view. In 1997 a British Columbia Salmon Aquaculture Review concluded that "First Nations almost unanimously remain opposed to salmon farming and the fish farms for the coast of British Columbia (BCEAO 1997)." However, at the recent conference for which this proceeding is based some First Nations members expressed support for limited aquaculture as long as they could participate financially and also environmental and land access issues could be addressed adequately by government and industry. Another interest group closely connected with daily aquaculture operations is the body of working persons wanting to ensure continued or growing employment in the aquaculture industry. Aquaculture can often offer employment in rural communities where unemployment rates are highest.

Less immediately concerned with aquaculture operations yet very concerned with their effects, there are other interest groups in the general public. Specific interest groups range from those wanting to ensure food safety (e.g., The USA-based "Centre for Food Safety": http://www.centerforfoodsafety.org/aquacultur.cfm), environmental safety, protection of beach recreation opportunities, property values, amongst other concerns. Members and groups representative of the general voting public can significantly impact aquaculture governance and public policy through lobbying and other efforts.

The increasing impact of activist groups formed from the general voting public is a major factor in what Rayner refers to as "the inverted world of contemporary policy-making" (Rayner – Chapter 19), where issue activism follows policy and activists challenge the legitimacy of an agreed to plan or regulation even after groups such as industry have spend considerable time and effort in planning exercises and complying with legal requirements. However, this is an inherent outcome of our political system and needs to be accepted as the price of doing business in a democracy. In a global economy where some major economic players such as China do not need to respond directly to a voting public and therefore can focus on setting long-term policies that benefit the country without necessarily pandering to a particular activist group, democracy can be an economic disadvantage for specific industries such as aquaculture. However, it is a disadvantage that must be accepted and worked around as it conveys many other societal advantages. The Canadian aquaculture industry, in my experience, would have it no other way: it is recognised that quality of life cannot simply be subordinated to commercial interests.

As stated in the introductory paragraph and identified by Rayner, the diverse interests and sometimes hidden agendas of groups such as industry, scientists, aboriginal peoples, and environmentalists cause conflict and uncertainty. However, government needs to balance the needs of various conflicting groups and set clear policy specific to aquaculture and stand by the policy only changing when situations change, rather then in the face of activism after the fact. But more importantly, government needs to take hold of aquaculture policy and recognise that setting a clear policy is inherently going to offend some groups, but still persist with setting policy and defining coherent governance.

Aspects of the Solution

Rayner argues that governance will exist side-by-side with traditional states and markets as a mode of coordination, yet suggests that governance is as likely to fail as either of the current methods of coordination. Rayner advocates instead a *balance* between self-regulation by industry and traditional governance, at least partly enabled by strengthening of existing networks and creation of more sophisticated aquaculture policy networks.

From my perspective as a member of the private sector, it is important to ensure that we do not overplay the potential of new governance relations, and recognize equally the importance of market forces in the process of stable transition to any new governance relations. Globalization of markets is certainly changing the way aquaculture business operates, yet beneath this surface appearance of change, there remains the fact that each aquaculture industry sits within a particular state, each industry is producing food for a changing market, and each industry is seeking business and production methods to enable the industry to remain socially acceptable and environmentally sustainable. Each national industry faces a variation on the same set of challenges with respect to its position within its home state and economy: each industry needs regulatory certainty to support investor confidence, and as the measuring stick against which industry performance can be shown to satisfy the standards of good corporate and environmental citizenship. In this context, the key aspects of overall aquaculture coordination are much as they have been in the past: government regulations (a traditional state), market forces, and an overlooked feature of globalisation, self-regulation. A good example of the interactions and impact of market forces and regulations is in relation to food safety and elevated contaminant levels in farmed salmon. Levels of PCBs, dioxins and heavy metals in farmed salmon correlate with levels in fish feed (Hites et al. 2004). European farmed salmon in particular have high contaminant levels due to more polluted oceans from which fishmeal is obtained. Markets are requiring that farmed salmon have contaminant levels below certain maximal levels and to succeed in, or even access, certain markets require decreasing contaminant levels and therefore increasing product quality. To succeed in aquaculture in global markets Canadian companies need to continue to improve fish quality by decreasing contaminant levels. This is also true for shellfish such as oysters, which tend to concentrate contaminants such as heavy metals. In this way markets are governing aquaculture feed practices and also forcing aquaculture companies to adopt better environmental practices to be successful in the marketplace.

Market forces also encourage self-regulation. For example, to achieve acceptable contaminant levels aquaculture companies need to regulate their practices, such as avoidance of over-crowding of fish to avoid parasites. To successfully self-regulate, companies need to form successful networks. An example in the case of fish oil is a network established by Ocean Nutrition Canada under the banner of the Council for Responsible Nutrition (CRN) in the USA. This network included every major fish oil manufacturer and aimed to establish quality standards and analytical methods that would ensure oil supplement safety and consistency. The aim for the companies was to ensure that only good quality oil reached the market and thereby better establish industry credibility in the market place and consumer acceptance of fish oil supplements. This network resulted in the formation of an omega-3 monograph that set quality standards and established acceptable validated analytical methods. (CRN monograph). Adherence to the standards expressed monograph is voluntary, but educated customers now require bulk oil product to be CRN monograph compliant. What is especially noteworthy about this network is its emergence from what some critics would suggest is an unlikely or even impossible source, since fish oil production is a relatively small and fragmented industry, similar to shellfish aquaculture in Canada. Despite these apparent challenges, the creation of a successful network with a clear mandate led to a successful outcome both in terms of self-regulation of product quality, and market growth. Fish oil has been the fastest growing supplement ingredient in North America since the monograph was created two years ago (Nutrition Business Journal 2004, Nutrition Business Journal 2005).

None of this, of course, denies that self-regulation needs to work within government policy framework. The issue at present is that the current aquaculture policy framework in Canada is in disarray. I agree with Rayner that aquaculture policy networks are required to aid government to set policy that helps develop the industry while keeping in mind the interests of other stakeholders. However, these networks will not be successful unless industry can be engaged in the process as active members of the networks. This requires the networks to establish early on a clear mandate and expected outcomes, and also a process to reach the outcomes. Without this the networks will be composed of academic and government groups with agendas not necessarily aligned with industry. Getting industry involved in networks is probably the biggest challenge to the success of networks. Networks can successfully impact policy, but without industry as drivers this policy will not necessarily improve the conditions for the growth of aquaculture in Canada.

References

Barton, J.R. 1997. Environment, sustainability and regulation in commercial aquaculture: The case of Chilean salmonid production. *Geoforum* 28(3–4): 313–328.

British Columbia Environmental Assessment Office (BCEAO). 1997. Salmon Aquaculture Review Final Report – Volume 2: First Nations Perspectives. Victoria, British Columbia Environmental Assessment Office, 69.

Council for Responsible Nutrition (CRN) Omega-3 Voluntary Monograph. http://www.crnusa.org/pdfs/O3FINALMONOGRAPHdoc.pdf

Hites, R.A., Foran, J.A., Carpenter, D.O., Hamilton, M.C., Knuth, B.A. and Schwager, S.J. 2004. Global Assessment of Organic Contaminants in Farmed Salmon. *Science* Vol. 303, Jan 2004, 226–229.

Liu, G.S. 2005. Comparative Corporate Governance: the experience between China and the UK. In *Corporate Governance: An International Review* Volume 13, issue 1, January 2005, 1.

Nutrition Business Journal NBJ 2004 Supplement Web Seminar. http://www.nutritionbusiness.com/.

Nutrition Business Journal NBJ Overview Web Seminar 2005. http://www.nutritionbusiness.com/

Office of the Commissioner for Aquaculture Development (OCAD). 2001. Legislative and Regulatory Review of Aquaculture in Canada. Ottawa: Fisheries and Oceans Canada. 76. http://www.dfo-mpo.gc.ca/aquaculture/ref/legalreview_e.pdf

Pierre, J. 2000. Introduction: Understanding Governance. In Pierre, J., (Ed.), *Debating Governance: Authority, Steering and Democracy*. Oxford: Oxford University Press.

Power, M.D. 2005. Lots of fish in the sea: Salmon aquaculture, genomics and ethics. Paper No. DEG 004, CAE Electronic Working Paper Series, 2005.

Statutes Cited

Fisheries Act, R.S., 1985, c. F-14, s.1.

Oceans Act, 1996, c.31.

Canadian Environmental Protection Act, 1999, c.33.

Food and Drugs Act, R.S., 1985, c. F-27.

Epilogue: Aquaculture, Innovation and Social Transformation

Barry A. Costa-Pierce

> *Aquaculture, not the Internet, represents the most promising investment opportunity of the 21st Century.*
> Peter Drucker, Economist and Nobel Laureate

This pioneering volume is about two things: the poignant clashes of views regarding aquaculture, and the underlying question of how we are to make complex choices regarding use of new technologies to produce food for a hungry planet. As the editors put it, "The debate [over aquaculture] is at least emblematic of, if not literally about, the struggle over the place of technology in human life and the life of the planet." The context of this volume's approach to the debate could hardly be better suited to facing this two-layered question: the waning growth of aquaculture in Canada, a global small-player in aquaculture which nonetheless has huge potential to participate in aquaculture. The particular value of this volume lies in the way Culver and Castle have intentionally engineered a clash of cultures as they have asked the very diverse authors of this volume to take up various parts of a provocative question: "Does Canada – the world's largest ocean nation – want to become an 'aquaculture nation'?"

A tangled set of further questions lies beneath the challenge facing Canadians. What can the world learn from Canadian experience in aquaculture? In Canada, aquaculture is a controversial "new" use of aquatic ecosystems, and while the potential for aquaculture is significant, the sector flounders in a context in which political will is often enough to lead to substantial financial backing of innovation in a sector. What can the world tell Canada about how to face the question of whether, and in what way it can it be, a global leader in aquaculture? What would it mean for other countries were Canada to face the social transformation caused by new technology innovation in an industry that competes, and threatens to replace, a capture fishery imbued with history and mythology about traditional practices? Can a new vision arise where innovative aquaculture businesses evolve into modern, 21st century, knowledge-based operations dedicated to pioneering the development of

B.A. Costa-Pierce
Professor of Fisheries and Aquaculture, Department of Fisheries, Animal and Veterinary Science and Director, Rhode Island Sea Grant College Program, Graduate School of Oceanography, University of Rhode Island, Narragansett, RI 02882-1197 USA

planned, well-integrated, socially acceptable aquaculture systems that have positive multiplier impacts on both natural and social ecosystems?

Aquaculture is nothing new. It arose multiple times in many indigenous societies where coastal population densities of "seafood eating peoples" increased beyond the carrying capacities of natural, aquatic ecosystems to provide for them. There have been many "blue revolutions" throughout history! But modern aquaculture is very different, being technologically complex, and, given a social makeup of a society (urban, rural, rich, poor, etc.) communities can either embrace such dramatic change, and accommodate the necessary social transformations, or reject such change and continue their social, cultural, and economic evolution without aquaculture. One of the fundamental problems of aquaculture development is the lack of understanding and planning for the social/community transformations wrought by aquaculture innovations, and the lack of knowledge on how to govern aquaculture innovations. The divergent and fascinating First Nations perspectives on aquaculture reviewed herein are a case in point, one that has fascinated me for a long time, namely, the evolution of aquaculture in ancient societies, and the use of that traditional knowledge to evolve an alternative path for coastal societies worldwide (Costa-Pierce 1987, 2002). Yet no matter the sources of the knowledge we use to chart a course for aquaculture – whether traditional knowledge, science, or some blend, well planned, transparent, participatory processes are required; ones that may drag out for longer than decision-makers want, but in the end will lead to some shared visions of more common futures.

In this regard, the nation-state has a particularly difficult time dealing with such long, oftentimes contentious local/regional participatory processes. Aquaculture planning issues are invariably tied to people's visions of their communities, their past, and their ideas of the future of a working waterfront. Welcome the stark contrast in the social/community acceptance of salmon aquaculture in Newfoundland versus British Columbia where different regional visions of the past and the future drive the acceptability of innovations such as aquaculture. Matters become all the more complicated when community identities and aspirations for the future have to interact with and decide to use or step away technologies of uncertain merits, often under the pressure of advice from people who are not members of the affected community. Culver and Castle point out that Canadians' favorite public value of consensus works poorly at the nation-state level, and hardly any better within the country's diverse regions. Furthermore, aquaculture innovations are so diverse that no one pathway for coordinating social acceptance with technological innovation is possible. It may be the case that in each instance of technology uptake into the aquaculture industry, active social engagement – a learning community of sorts – needs to be developed to take stock of the wide diversity of aquaculture innovations.

There is an intriguing contrast visible in the difference between global opinion and what authors in this volume represent as an often negative Canadian public opinion of aquaculture. Most global analysts agree that a rapid acceleration of aquaculture – the "blue revolution" – is not only needed, but is a forgone conclusion given global demand for sustainable sources of food protein. In addition, any rigorous comparisons of aquaculture with other food protein production systems reveal that aquaculture is a very efficient mass producer of animal proteins for a crowded, coastal planet, and also has great potential to transform coastal societies. Comparisons against an array of capture fisheries and terrestrial agriculture systems confirm that aquaculture – even as currently practiced (e.g. in its "current state of evolution") is superior in terms of financial and social impacts (profit per ha; social multiplier effects), and energy and production efficiencies.

Aquaculture could mitigate the effects of the disastrous trajectory for terrestrial meat production and the capture fisheries. FAO (2007) estimate that 77% of all global fish stocks are either fully exploited with no room for further expansion; or are overexploited, depleted, or recovering from depletion owing to excess fishing pressure. FAO (2007) project that capture fisheries production from 2004 to 2030 will remain stable at 86–87 million metric tones, while aquaculture will almost double from 45 to 83 MMT during the same time. Much of the pressure to increase aquaculture production is derived from considerations of efficiency. Goodland & Pimental (2000) state, "In the worldwide effort to increase food production, aquaculture merits more attention than raising grain-fed cattle." Production efficiencies of edible mass for aquaculture range from 2.5 to 4.5 kg dry feed/kg edible mass compared with 3.0–17.4 for conventional terrestrial animal production systems. Beef cattle require over 10 kg of feed to add 1 kg of edible weight, whereas catfish can add a kg of edible weight with less than 3 kg of feed. By comparison, wild, "capture" fisheries are less efficient from both trophic and energy efficiency perspectives.

Coastal and oceanic ecosystems have trophic efficiencies of 10–15%, and mean trophic levels of 3.0–5.0 (Ryther 1967), whereas mean trophic levels in aquaculture systems range from 2.3 to 3.3, with highest aquaculture trophic levels in North America and Europe (Pullin 2001). Pullin (2001) state that "most ocean fish consumed by humans have trophic levels ranging from 3.0 to 4.5 (Pauly et al. 1998); that is 0–1.5 levels above that of lions." Vernon (2007) completed an analysis of salmon ranching versus salmon farming in the north Pacific Ocean, and concluded that salmon ranching as practiced currently in British Columbia, Alaska, and Japan, introduces 250 times more smolts and consumes 70 times more feed to produce less than 10 times the volume of marketable salmon than salmon farms. To produce 1 kcal of catfish protein about 34 kcal of fossil fuel energy is required; lobster and shrimp capture fisheries use more than 5 times this amount of energy. Energy costs for even the most intensive salmon cages are three times less than lobster and shrimp fishing, and are comparable to beef production in feedlots.

In light of the superior efficiency of aquaculture, the sometimes negative public view of aquaculture in Canada appears for a moment to be a strange anomaly – until we bring to mind again the observation of Culver and Castle that the debate over aquaculture is simultaneously a debate about the place of technology on our coasts and in our lives. If aquaculture is to be a vital part of the knowledge-based, innovation economy for coastal societies in Canada, and worldwide, aquaculture must be better integrated into social life. Aquaculture is technology intensive and rapidly evolving, and so those who wish to use it must keep follow the pace of change. What is required is an entirely new view of what the blue revolution is not, to clear away false expectations and misconceptions, but also what it can be:

a new pathway to sustainable development for aquaculture – an alternative path to wasteful food production techniques – and the birth of new institutions to help direct and support the massive social transformation of some coastal societies dealing with new aquaculture proposals. It is crucial to remember when thinking of coastal societies that like the rest of us, most are not keen to become historic theme parks where fishermen's wharves are devoid of fishermen. Coastal societies are typically intent on both preserving their glorious past, and securing the future of their working waterfront.

The Green Revolution is Not a Model for the Blue One

In my opinion, aquaculture's future growth as projected by FAO will not occur in Canada or elsewhere unless aquaculture developers choose a route other than the failed "green" revolution with its adverse environmental, ecological, community, social, and ethical impacts.

The Green Revolution raised agricultural yields, especially in India, but did not strive to involve agribusiness and multinational food corporations with communities, farmers, or local governments to build, invest in, and nurture change in local institutions, communities and cultures. As a result, accelerated production did not alleviate poverty or eliminate hunger, and production gains damaged the environment. Numerous multidisciplinary studies, especially in Mexico and India, have shown that the Green Revolution's expensive seed, fertilizer, pesticide and irrigation "packages" favored a minority of economically-privileged farmers (Rosset 2006). Environmentally, the green revolution produced a well-documented litany of "externalities": accelerated pollution of rivers and water tables, widespread soil degradation, losses of biodiversity, and occupational pesticide poisonings. In India, the Green Revolution's "technology packages" required irrigation; so the government subsidized the digging of tens of thousands of wells which pumped many water tables dry, forcing vast areas to return to traditional, dryland farming systems. Most of India's grains were exported, so national elites profited mightily. The local, hydrological result of the Green Revolution's technology packages was "the sacrifice of India's ancient aquifers to the international grain trade" (Rosset, 2006).

My Crystal Ball is Blue...If Viewed Through the Right Lens

Aquaculture developments are rarely planned as social transformations. Modern aquaculture is still new to our policy institutions, governments and universities. Aquaculture is rarely considered as a part of social design and transformation; and is rarely thought of as an exciting new field of multidisciplinary scholarship having its own merits in the context of innovation, cultural evolution, knowledge systems, knowledge acquisition and management. Discussion of advances in the field of aquaculture is similarly at an early stage of development, largely dominated by technological and natural science findings while the relatively few social researchers interested in aquaculture struggle to keep up with the sheer volume of the science and its interaction with social issues. Aquaculture developments today are commonly considered as "industrial primary production." Metrics of "success" rarely include anything other an exports of tons of fish produced. In the face of what is presented as "success" we are now facing concerns that can no longer be ignored: charges that aquaculture's current mode of industrial expansion is leading to a failed pathway of "boom and bust" cycles. Such cycles have plagued the development of shrimp, tilapia, and salmon aquaculture worldwide. Why have these occurred?

The relative narrowness of aquaculture research is unfortunately paralleled in many governments by equally narrow expertise and responsibility for aquaculture. Aquaculture impacts many sectors of society, but aquaculture is rarely considered comprehensively by government. Decision-making about its future is currently made principally by the minds of a few specialists in governments who oftentimes are untrained in aquaculture; these power-brokers are routinely ensconced in the bowels of government agriculture or fisheries departments. As a result, "boom times" get much attention by important people; "bust times" are those where a few specialists are left to clean up the mess.

Inadequate training and knowledge are often at the root of poor planning and exaggerated claims regarding the merits of aquaculture. Too often, aquaculture comes into coastal communities with great promises of innovation amid projections of positive impacts on the local economy. Aquaculture's extraordinary potential for generating large scale multiplier impacts are rarely achieved, however, or if they are, they are achieved as poorly planned afterthoughts, not as well planned, collaborative developments that have at the outset included a broad range of local stakeholders, policy-makers and allied supply-side businesses. According to Dicks et al. (1996), aquaculture production in the USA accounted for only 16,500 jobs and just 8% of the income. Aquaculture goods and services (feeds, fertilizers, processing, transport, equipment, supplies, etc.) accounted for 92% of the income and about 165,500 jobs.

Multidisciplinary Aquaculture Social Science Needs

In my view, we have yet to take two crucial steps needed to develop the alternative path for aquaculture development as social transformation: (1) industry commitment to implementation of an ecosystem approach to aquaculture (EAA); (2) development of alternative institutions that would team government, universities and industry, whose mission would be to multidisciplinary learning, social transformation, and the aquaculture "innovation portfolio." University involvement is essential as a "keeper of the flame" since neither government nor industry can effectively play the role of an honest broker. What is required is a neutral convener and purveyor who can help lead the integration of an EAA with institutions committed to social change who pay close attention to market and societal transformations (such as progress towards: internationalization and multiculturalism, sustainability, bioregionalism, carbon footprints/trading/food miles). Societal transformations need time and stable funding bases to foster unique partnerships. Such type of leadership is fully underway in agriculture, and could be an important model for aquaculture. I hope and anticipate that this seminal volume will, in and of itself, go a long way towards developing further a cadre of multidisciplinary scholars in aquaculture; and additionally stimulate formation of new institutions committed to developing cadres of science-policy-innovation experts.

An Ecosystem Approach to Aquaculture (Natural and Social Ecology)

Implementing an ecosystem approach to aquaculture (EAA) is a major global trend (FAO, 2007). At an FAO Workshop on EAA in Mallorca, Spain in 2007, an EAA was defined as "a strategy for the integration of the activity within the wider ecosystem such that it promotes sustainable development, equity, and resilience of interlinked social-ecological systems."

Soto (2008) has defined an ecosystem approach, like any systems approach to management, as accounting for a complete range of stakeholders, spheres of influences and other interlinked processes. Applying an ecosystem-based approach must involve physical, ecological and social systems in the planning for community development, while also taking into account stakeholders in the wider social, economic and environmental contexts of aquaculture. Soto (2008) developed the following three principles and key issues of an EAA at the different scales:

Principle 1: Aquaculture should be developed in the context of ecosystem functions and services (including biodiversity) with no degradation of these beyond their resilience capacity. The key issue here is to define and estimate resilience capacity or the limits to "acceptable environmental change." A range of terms has been used to characterize the appropriate limits, including "environmental carrying capacity," "environmental capacity," "limits to ecosystem function," "ecosystem health," "ecosystem integrity," "fully functioning ecosystems," all of which are subject to a specific social, cultural, and political context.

Principle 2: Aquaculture should improve human-well being and equity for key stakeholders. This principle ensures that aquaculture provides equal opportunities for development, and that the benefits from aquaculture are properly shared while not resulting in detriment to society, especially the poor. Both food security and safety are key components of societal well-being.

Principle 3: Aquaculture should be developed in the context of other sectors, policies and goals. This principle recognizes the important interactions between aquaculture and the larger ecosystem, in particular, the influence of surrounding natural and social environments on aquaculture. Aquaculture does not take place in isolation from other activities, and, in is usually not the only human activity impacting water bodies that are otherwise more severely affected by agriculture, industrial activities, and human wastes. This principle acknowledges the opportunity to couple aquaculture activities with other production sectors in order to promote material and energy recycling. Aquaculture sites are not only economic engines of primary production that work so long as they comply with local regulations. Aquaculture sites can also be sources of innovation and pride if they can be designed, community-based, farming ecosystems. Regular reviews of progress towards an ecosystems approach to aquaculture are necessary to inspire planners and environmental decision-makers (at many societal scales – national, regional, local) to use new, innovative approaches. Sophisticated site planning of aquaculture can occur so that farms "fit with nature" and do not displace or disrupt invaluable aquatic ecosystems or conservation areas, and also contribute to the local economy and society.

In as much as EAA calls for technological solutions that fit aquaculture technology into ecological constraints, it also requires social change for the acceptance and wise use of innovations. All too often, aquaculture comes into coastal communities so poorly planned that its extraordinary potential for generating large scale multiplier impacts is recognized only as an afterthought and by accident, not as the result of well-planned, collaborative developments involving from the outset a broad range of policy-makers and allied supply-side businesses. Authors of this book make a strong case that aquaculture's future in Canada, as an example nation, is tied to this "innovation portfolio." Aquaculture's future is much more than a simple technological exercise—it is an exercise in multidisciplinary, multi-institutional, environmental scholarship—requiring aquaculture developers to devise planned, systems approaches that include communities and their traditional knowledge systems to insure their "coastal futures."

Felt in this volume points out "it all depends on the lens, b'y." Indeed, at the community level, science facts may mean little to policy decisions; communities can find it difficult to separate societal values from science. The precise meaning ascribed to key science and ecosystem terms is usually not a battle over science: ultimately, it is a debate over which policy option is selected. The debate over what might appear to be semantic nuances is really a surrogate debate over values and policy preferences. Words such as "degraded," "damaged," "healthy," and the relative importance of natural versus altered environmental conditions are calibrated by societal values and preferences, not only science. One person's "damaged" ecosystem is another's "improved" one. A "healthy" ecosystem can be either a swamp infested with mosquitoes, or an intensively managed fish pond.

Ryan in this volume states that "The pro/anti approach is frustrating for resource managers to have to wade through rhetoric instead of spending precious resources on progressive innovation". Such debates are rarely sophisticated enough, however, since the art and science of "the process" – of the proper design to air completely opposing views and bring them together (Martin 2007) – is rarely considered important enough to invest in. For example, we routinely review the "environmental impacts of aquaculture," but fail to examine fully the environmental benefits of aquaculture (Rensel and Forster 2007), or the notion of aquaculture ecosystems (Costa-Pierce 2003), similar to the popular ideas of "agroecosystems" (Gliessman 2006). As a more provocative example, there are reports of land-based salmon aquaculture being not economically feasible but there are few visionary, multidisciplinary, social-ecological, economic feasibility studies on how it to make it viable. Scientists also say that other industries release more nitrogen to the BC marine environment than aquaculture (Cubitt et al., Chapter 9), but we never discuss these other industries, and their plans for the future. Continuing on, we know that salmon escapees of non-indigenous salmon species seed the marine environment with an exotic species that Mother Nature never intended to be there, but we rarely discuss how the use of indigenous salmon species in aquaculture may be a higher risk to the genetic integrity of native salmon (Fleming 2005). Plus, we know that billions of salmon smolts are seeded into the Pacific Ocean from salmon hatcheries – can these produce more detrimental food chain impacts than aquaculture (Vernon 2007)? All of these examples support the argument of Antle and Wagenet (1995) who emphasized, over a decade ago, that systematic frameworks are required, so that economic data and data from other, interacting disciplines can be collected and integrated using common units of measurement.

Aquaculture knowledge is fluid, international and innovation is at its heart. The fascinating topic of "intangible assets" in this book reminded me of how poorly we plan for innovations in aquaculture. Much of aquaculture's true worth is not in fish or clams; it is in the intellectual capital that moves freely across borders. Culver calls these "national and supra-national systems of innovation," and applied aquaculture scientists/outreach specialists know exactly what he is talking about. Western scientists have lived for years in the major aquaculture growth regions of the world and have brought with them not only their brains, but also their attitudes, habits, families, Many stayed, "went bush," and became essential parts of the cultural evolution of aquaculture in those societies. "Outsiders" became "insiders" and grew to know as much (or more) than many native-born about the local, cultural development of aquaculture, aquaculture families and communities in those societies.

Institution-Building

Alternative paths must be nurtured by alternative institutions capable of translating research into action and governance of action. Fostering an alternative path to aquaculture development will require the design and development of alternative institutions committed to developing the "innovation portfolio" and charting the future development of socially and ecologically responsible aquaculture industries that are well integrated into coastal societies. Globally, we do not have these institutions yet, and where they made baby steps forward, they have ultimately failed. The causes of institutional failure in aquaculture need further investigation before we can say with certainty what went wrong. Yet that need should not stop us from trying to be clearer about the conditions for success. Sustainable aquaculture needs well-designed, facilitative, collaborative (industry, government, universities, NGOs) institutions working with adequate funding, using multidisciplinary approaches to setting research and outreach priorities, and assessing research and outreach impacts. This "triangulated" approach to the problems, and some of the pitfalls, is usefully displayed in this book's model of contested exchange between interested and affected stakeholders in aquaculture.

There are several lessons to be drawn from this experiment with a model of contested exchange, but perhaps one of the most important is this: As aquaculture continues to expand, the role of research and development will be key and will perhaps increase, contrary to the "once and done" attitudes which sometimes plague bureaucrats looking for single-step innovation which will then provide a durable socio-economic base for some society. Globally, aquaculture development has been hampered by three factors relating to the fundamental requirements of research and development: insufficient funds, lack of core research staff, and weak research infrastructure (FAO 2007). An increase in the quantity and quality of human resources focusing on aquaculture innovation is essential in the search for new aquaculture opportunities. Successful human resources development will trigger the development of more efficient aquaculture-related technology, legislation, and management (FAO 2007).

The International Agriculture Research Centers allied under the CGIAR network (Consultative Groups in International Agriculture Research) are good models to consider, since these institutions develop research priorities that are driven not only by science needs but also economic efficiency, equity, internationality and sustainability (Kelley et al. 1995). Such characteristics are very important to support the development of an ecosystem approach to aquaculture, especially in a time of intense competition for scarce funds. It is important when laying out research priorities that anticipated benefits flowing from additional investments as well as the opportunity costs are made explicit. Kelley et al. (1995) state how ICRISAT (International Crops Research for the Semi Arid Tropics) provided clear criteria for establishing choices among competing research activities, that were analytically rigorous, drew on scientists' empirical and intuitive knowledge base, and were transparent and interactive. ICRISAT developed themes for collaborative, multidisciplinary research that were impact-oriented, and projected clear milestones against which progress was measured and evaluated. In the US, the Land Grant and Sea Grant College Programs is another model, not only for aquaculture, but also for many diverse aquatic/ocean/coastal disciplines with which it interacts. Sea Grant connects universities to industry and government in a collaborative manner to address integrated, multidisciplinary, multi-stakeholder problems; and is a proven commodity; but also is vastly underfunded (OSB/NRC 1994).

Collaborative research designs are needed to meet the standards of good science while also satisfying the standards of accountability and policy relevance. As scientists become involved in the priority setting process, they become aware of the ways that their research can contribute to the mission of publicly funded research. Proper valuation of the economic, environmental, and human health impacts of aquaculture production systems is essential for priority setting and impact assessment; multidisciplinary collaboration is also a critical element of the valuation stage of impact assessment.

The coastal zone management (CZM) and site selection chapters by Lane et al., Smith and Chopin emphasize the many aspects of this practical need for proper valuation. Smith calls CZM "a tidy process for an untidy reality." Much discussion over the impacts of aquaculture is based on its siting, and any process that determines siting is, by definition, multi-issue, multi-stakeholder, multi-institutional, and multi-jurisdictional. Smith questions if a multi-criteria approach could ever meet the needs of aquaculture when it involves multiple stakeholders, each of whom have different concepts about what counts as meaningful data, and how the sites current and future use should be valued. Chopin et al. (2004), the pioneer of the Integrated Multi-trophic Aquaculture concept where disparate aquaculture systems are integrated into a designed ecosystem, has brought great attention to this as an alternative model of aquaculture development which could include recreational, tourism, and other stakeholders. Chopin's view of the complementarity of uses and the importance of scientific validation of that complementarity resonates with a future of aquaculture as a vital part of coastal communities that are proud of its new, exciting, knowledge-based future.

Funding for innovation is always a major issue, but fostering the long-term development of such innovative institutions for aquaculture is impossible on competitive funding alone. Huffman and Just (1994) argue that with respect to the development of agricultural experiment stations at US universities, directed ("formula") funding was more productive than competitive grant funding, and that "heavier administrator involvement... enhanced productivity of applied research where approaches and products were specified in advance." One take home message could be that the demise of AquaNet in Canada leaves a vacuum that cannot be filled with many disconnected, short-term projects.

A larger take home message must, however, return to the two-layered problem with which Culver and Castle began. This volume's authors demonstrated convincingly that it is a serious mistake to take a one-dimensional view to aquaculture, whether that view sees aquaculture as a tool of regional development, as a threat to ecosystem health, or something else. Aquaculture is itself multi-dimensional and making the best use of its potential to contribute to global food security will require far more than the knowledge and skills of natural scientists. Once the multi-dimensional nature of aquaculture is recognized and widely accepted, it is much easier to understand the second layer of this volume's motivating problem. Aquaculture is just one among many complex technologies vying for support in national systems of innovation, and there are lessons to be learned by aquaculture advocates from understanding national systems of innovation, and advocates and operators of those systems can learn about the adequacy of those systems from the way they handle scientifically and socially complex topics such as aquaculture. It's all in the lense, as Felt argues here, and we need to work together – aquaculture advocates, global food security advocates, national science, social science, humanities and other researchers – to develop a point of view we can all live with into the future.

References

Antle, J. M. and Wagenet, R. J. 1995. Why scientists should talk to economists. *Journal of Agronomy* 87: 1033–1040.

Chopin T, Robinson S, Sawhney M, Bastarache S, Belyea E, Shea R, Armstrong W, Stewart and Fitzgerald P. 2004. The AquaNet integrated multi-trophic aquaculture project: rationale of the project and development of kelp cultivation as the inorganic extractive component of the system. *Bulletin of the Aquaculture Association of Canada* 104(3): 11–18.

Costa-Pierce, B.A. 1987. Aquaculture in ancient Hawaii. *BioScience* 37: 320–331.

Costa-Pierce, B.A. 2002. *Ecological Aquaculture: The Evolution of the Blue Revolution*. Oxford: Blackwell Science.

Costa-Pierce, B.A. 2003. Use of ecosystems science in ecological aquaculture. *Bulletin of the Aquaculture Association of Canada* 103(2): 32–40.

Dicks, M. R., McHugh, R., and Webb, B. 1996. *Economy-Wide Impacts of U.S. Aquaculture*, Oklahoma Agricultural Experiment Station, P. 946.

FAO (Food and Agriculture Organization of the United Nations). 2007. *State of the World's Fisheries and Aquaculture*. Rome: FAO.

Fleming, I. 2005. Fugitive fish: Risk associated with farmed fish escapees. Available at: http://www.eurekalert.org/pub_releases/2005–02/nsae-fff021605.php (accessed 27 November 2007).

Gliessman, S. 2006. *Agroecology: The Ecology of Sustainable Food Systems*. Atlanta: CRC Press.

Huffman, W. E. and Just, R. E. 1994. Funding, structure, and management of public agricultural research in the United States. *American Journal of Agricultural Economics* 76: 744–759.

Kelley, T. G., Ryan, J. G., and Patel, B. K. 1995. Applied participatory priority setting in international agricultural research: making trade offs transparent and explicit. *Agriculture Systems* 49: 177–216.

Martin, R. 2007. *The Opposable Mind*. Cambridge, MA Harvard Business Press.

OSB/NRC (Ocean Studies Board, National Research Council). 1994. A Review of the NOAA National Sea Grant College Program. Washington, DC: National Academy Press.

Paul, D., Christensen, V., Dalsgaard, J., Froese, R., and Torres, F. 1998. Fishing down marine food webs. *Science* 279: 860–863.

Pullin, R. S. V. 2001. Integrated agriculture – aquaculture and the environment. In *Integrated Agriculture – Aquaculture: a Primer*. FAO Fisheries Technical Paper 407: 16–17.

Rensel, J.E. and Forster, J.R.M. 2007. Beneficial environmental effects of marine finfish mariculture. NMFS, NOAA, Washington, DC http://www.wfga.net/documents/marine_finfish_finalreport.pdf

Rosset, P.M. 2006. *Food is Different: Why We Must Get the WTO Out of Agriculture*. London: Zed Books, London.

Ryther, J.H. 1967. Occurrence of red-water off Peru. *Nature* 214: 1318–1319.

Soto, D. (2008). Aguilar-Manjarrez, J., Hishamunda, N. and Deudero S. (eds). Building an Ecosystem Approach to Aquaculture (EAA). Rome: FAO Fisheries Proceedings.

Vernon, B. 2007. Salmon ranching examined. Pacific pastures under siege? *Northern Aquaculture* September/October 2007: 14.

Workshop Report

Aquaculture, Innovation and Social Transformation

Workshop Proceedings
Montreal, Quebec
August 18 to 20, 2004

Prepared for:

David Castle
Department of Philosophy
University of Guelph

Keith Culver
Department of Philosophy
University of New Brunswick

FINAL REPORT – October 25, 2004

Prepared by:

Jacques Paynter
AMEC Earth and Environmental,
A division of AMEC Americas Limited
25 Waggoners Lane
Fredericton NB E3B 2L2

AquaNet
Aquaculture, Innovation and Social Transformation Workshop
Final Report
October 25, 2004

Table of Contents

AquaNet
Aquaculture, Innovation and Social Transformation Workshop
Final Report
October 25, 2004

1.0 Introduction

1.1 Introduction

This report presents the results of the Aquaculture, Innovation and Social Transformation Workshop convened in Montreal on August 18–20, 2004.

The objectives of the workshop were to:

- Investigate the state and future of Canadian aquaculture from the perspective of policy-oriented social science and humanities research.
- Comprehensively identify emerging social issues arising because of innovation in aquaculture.
- Develop a forecast of responsible governance models for aquaculture innovation.

The workshop was co-sponsored by the Canadian Program on Genomics and Global Health and AquaNet, with additional support from the Department of Philosophy at the University of Guelph, the University of New Brunswick's Centre for Social Innovation Research and McGill University's Centre for Intellectual Property Policy.

The workshop was subdivided into seven sessions. For each session, two research papers were presented, followed by a critique and commentary by a designated participant. These presentations were then the subject of discussion oriented to exploring the issues in greater depth.

1.2 Participants

Attendance at the workshop was comprised of a broad representation from: University of Glascow; University of Guelph; Aquaculture Innovation; University of New Brunswick; University of British Columbia; Fisheries and Oceans Canada; University of Ottawa; Fish Farming International; Norwegian Institute for Fisheries and Aquaculture; Malaspina University-College; Institute on Governance; McGill University; Carrier Sekani Tribal Council; Ocean Nutrition Canada; Networks of Centres of Excellence; and, New Brunswick Salmon Growers Association.

AquaNet
Aquaculture, Innovation and Social Transformation Workshop
Final Report
October 25, 2004

amec

1.3 Structure of the Report

The remainder of this report is structured as follows:

Chapter 2.0	Summary of Discussions on Animal Welfare in Aquaculture
Chapter 3.0	Summary of Discussions on Knowledge Management and Intellectual Property Issues
Chapter 4.0	Summary of Discussions on Environmental Sustainability of Aquaculture
Chapter 5.0	Summary of Discussions on Interaction Between Traditional Knowledge and Modern Aquaculture
Chapter 6.0	Summary of Discussions on Messages, Consumers, and Aquaculture
Chapter 7.0	Summary of Discussions on Integrated Coastal Zone Management
Chapter 8.0	Summary of Discussions on New Practices for Global Competitiveness
Chapter 9.0	Conclusions

2.0 ANIMAL WELFARE IN AQUACULTURE

2.1 Discussion Results

- Definitions of Animal Welfare and Suffering

The need for a common definition of animal welfare was discussed. Specifically, most definitions relating to animal welfare fall into three broad categories:

- Feelings-based definitions
- Function-based definitions
- Nature-based definitions

Feelings-based definitions depend upon the controversial point that animals have conscious feelings, whereas function-based definitions are based on the assumption that an animal whose physiological systems are poorly functioning is suffering. Nature-based definitions rely on the view that animals suffer if they cannot meet their behavioural needs.

As it is difficult to define animal welfare, an alternate approach is to identify the conditions that must be fulfilled to ensure that animals have a good quality of life (and death). These conditions fall under five "domains of welfare" which include the following:

- Domain 1: Water and food depravation, malnutrition
- Domain 2: Environmental challenge
- Domain 3: Disease, injury and functional impairment
- Domain 4: Behavioural/interactive restriction
- Domain 5: Mental and physical suffering

- Measurement of Welfare

Natural responses of animals to adverse conditions, particularly changes in physiology, health, and behaviour, can be used to assess their welfare. Although links between these conditions and actual welfare are considered complex and none of these assessment methods can provide all the answers, welfare researchers tend to collect data on all three as they monitor health, physiology, and behaviour of their subjects. This provides a broad view, however it also presents the challenge of how to combine information from different biological systems.

- Fish Capacity for Suffering

Some information presented indicates that fish have limited behavioural capacities and are not capable of complex mental processes, and therefore tend to behave in a reflex-based manner. As such, it is not anticipated that fish can experience pain and fish welfare is a meaningless concept.

In opposition to the above, it was discussed that fish do indeed have pain receptors (nociceptors) that react in a similar fashion to those of humans and respond to a range of harmful stimuli. Their brains are organized differently than those of other vertebrates, but it is still possible to identify many of the same structures. Fish behaviour is highly complex and flexible as they have a capacity to learn and remember. For these reasons it was suggested that pain can be experienced by fish, and therefore fish welfare is meaningful.

- Fish Welfare Interest

It was expressed that responsible farmers are concerned for the well-being of their stock as a matter of professional integrity; good welfare results in good production. Farmers and regulators are also responding to the increasing and legitimate public concern for the welfare of farmed fish. It is anticipated that welfare status will become an additional aspect of product quality from which responsible farmers will stand to benefit.

- Fish Welfare in Light of Production Systems

It can be argued that farmed fish experience better conditions than wild fish as they are continually supplied with high quality food, are protected from predators, and inoculated against disease. On the other hand, they are often disturbed by human activities, held at high densities that restrict space, are sometimes fed clumped food, which leads to aggression and competition among the caged fish, and are ultimately slaughtered. These activities may impair welfare, so it is important to consider methods conducive to responsible farming and alleviating stress of the fish. Simple changes to production systems and farm practices that reduce the adverse effects of the farm environment on fish welfare were considered. These include lower stocking densities for species where high stocking densities impair welfare and the selection of slaughter methods that reduce stress.

- A Response to Fish Welfare Issues

The response to the above-mentioned issues was provided from several perspectives:

Research approach – Is circumstantial evidence and analogy a sufficient motivator to attend to animal welfare?
Production approach – Does good welfare improve production or quality?
Ethical approach – Evidence of some degree of consciousness is not an indication of pain as suffering. What is the moral balance between fish interests before human interests?
Evolutionary approach: – What evolutionary advantage would be gained by fish in developing an ability to suffer? Fish have a "fear and flight" mechanism that is not learned from past experience, cannot be shared and is not hereditary.
The application of animal welfare issues to the business of selling fish is considered to be dubious because there is no evidence to suggest the consumer really cares. To the contrary, the data indicates that farmed salmon sales are negatively affected by product quality concerns, and not welfare. Thus fish welfare does not appear to be a marketable trait unless it directly impacts on product quality.

Suggested actions include: regulations based upon actual experience developed in consultation with industry and other stakeholders; regulations based on sound knowledge of the species and environments concerned; practical tools for on-farm assessment of welfare collaboratively developed by scientists and industry; and, industry initiated monitoring, overseen by regulators.

3.0 KNOWLEDGE MANAGEMENT AND INTELLECTUAL PROPERTY ISSUES

3.1 Discussion Results

- Performance Measurement

The need for aquaculture to demonstrate an acceptable balance between environmental impact and high-value economic output was discussed. To achieve this balance, innovation in products and processes is required as well as research and development (R&D) efforts beyond the private sector capacity.

- Evaluation of Intellectual Property

Arguments were provided for a shift from regarding intellectual property and patents as the measure of aquaculture R&D organizations' contribution to the Canadian system of innovation. It was suggested that organizations operating on short-term funding and evaluation systems should emphasize the value of the rivalrous non-intellectual property intangible assets they create, in human capital capable of solving emerging problems, and the variety of ideas and know-how which can be employed to benefit Canada without protection of intellectual property. This emphasis should require aquaculture R&D organizations to actively study their own contributions to innovation and to promote increased sharing of knowledge and transfer of innovations near to market to end users prepared to commercialize those innovations. This will require collaboration between governance and research elements of the national system of innovation, even possibly requiring routine secondments between governance and research elements to create a cadre of science-policy-innovation experts.

amec

- Salmon Aquaculture Forum

A Salmon Aquaculture Forum has been created to resolve contentious issues surrounding industry by bringing together diverse interests. Creation of the Forum arose from the two-sided debate over salmon aquaculture in British Columbia (BC), with one side identifying salmon farming as a source of economic and food security and the other side typifying salmon farming as one of the greatest threats to the diversity and abundance of natural resources on the coast of BC (in particular to wild salmon stocks).

Numerous attempts to settle the debate in the Province through public consultation unfortunately were not successful. The Salmon Aquaculture Forum was proposed as a means of engaging people with different perspectives and backgrounds, including First Nations, stakeholders from industry, the environmental sector, and others, in an attempt to build common ground, implement best practices, and develop a consensus about the future of the aquaculture industry.

The Salmon Aquaculture Forum: Discussion Paper on Practices and Findings and *The Salmon Aquaculture Forum: Briefing Note to Ministers* are two reports produced by the Forum that provide recommendations to the two levels of government on the ways and means of creating a commission-style forum. To date, the governments have not implemented the Forum. However, participants discussed the need for such a forum, as salmon aquaculture appears to be here to stay while the debate surrounding industry continues.

The role of the Forum in consensus building within the Forum body and conflict resolution within the general public was also discussed. To accomplish this, the Forum must earn the trust and respect of a large and diverse stakeholder community, the general public, and the media.

4.0 ENVIRONMENTAL SUSTAINABILITY OF AQUACULTURE

4.1 Discussion Results

- Management for Survival vs. Management for Profit

Participants discussed farming approaches as management for survival or management for profit. Management for survival was considered to be production based on semi-intensive and extensive systems (i.e. providing finfish for rural markets and subsistence in developing countries); management for profit was considered to be intensive rearing (i.e. finfish such as salmonids for profit). It was suggested that to adequately address the issue of aquaculture sustainability, it is important to recognize and address the diversity of farming practices and economic and social factors influencing development.

It was also suggested that the concept of sustainable aquaculture must also address environmental, economic, and social issues, with each of these having equal importance.

- Goal of Environmental Sustainability

The goal of environmental sustainability was presented as the maintenance of the current level of environmental assets to ensure that future generations benefit from the same level of environmental goods and services presently available. In regards to aquaculture, accumulation of waste and pollution of aquatic habitats were identified as some of the major negative environmental impacts. Dependence on wild seed to establish new generations was also identified as a pressure point on already threatened stocks.

- Approaches to Successful Environmental Sustainability

The environmental sustainability of aquaculture is dependent upon intrinsic factors including water quality, species characteristics, availability of artificial and natural feeds, seed supply, and location of facilities, and extrinsic factors including natural hazards, climate change, pollution, the market, sociocultural conditions of culture sites, and legislative control. In consideration of these factors influencing successful environmental sustainability, adoption of an integrated coastal management (ICM) scheme was suggested. The goal of the ICM would be to identify all economic activities within a coastal resource system in order to manage them in a sustainable manner to mitigate negative cumulative environmental impacts.

It was also suggested that the ICM be combined with Environmental Impact Assessments currently conducted for sites to form an Integrated Environmental Impact Assessment (IEIA). The IEIA would take into account all present and future economic development activities to ensure the sustainability of the coastal resources in a specific area.

Suggestions were made regarding proper planning and management policies on provincial and federal levels, whereby national policies should encourage sound practices, central government should increase awareness of environmental concerns, building capacity, and research and technical assistance, and local governments should develop an ICM specific to their environmental and economic activities.

5.0 INTERACTION BETWEEN TRADITIONAL KNOWLEDGE AND MODERN AQUACULTURE

5.1 Discussion Results

- Perception of Aquaculture and Aboriginal Resistance

A discussion was held regarding the perception of aquaculture by First Nations peoples as an extension of the historical process of colonialism. It was opined that salmon farming results from political processes involving traditional chiefs and new forms of authority that both integrate and clash with existing forms of Native self-government.

Main concerns of the Kwak'wakawakw (BC coast First Nations group) included the transfer of disease to wild stocks, interference with salmon and ooligan runs, and the emission of contaminants onto clam beds. Much of the strongest opposition to the industry comes from hereditary chiefs.

The coastal areas now considered under-developed and suitable for salmon farming were once considered areas of great wealth and prosperity and were used by aboriginal people prior to their relocation. Native economies are pressured by a new form of territorial expansion that brings with it few settlers, technology, exotic species, and extractive capacity.

- Government and Industry Perception

Government and industry claims have focused on past mistakes with respect to resource management. The perception was that the aquaculture industry was not sufficiently efficient, productive, and modern. This has lead to the provision of leases to salmon farming companies in areas considered underdeveloped Native territories. This is viewed as a means to extend the network of bureaucratic control on the Pacific coast that began a century ago with the commercial fishery.

- History as a Part of the Present Approach

It has been suggested that an understanding of the cultural energy and sometimes individual heroism on the part of the hereditary chiefs (who continue their ancestors' struggles against settler encroachment on tribal lands and water) in their resistance to salmon farming, will help reveal the true extent of the barriers to a sustainable aquaculture sector in Canada in regards to current salmon farming practices and research on the Pacific coast.

- First Nation's Panel on Fisheries

A First Nation's Panel on Fisheries [http://www.fns.bc.ca/pdf/FNFishPanelReport0604.pdf] was established in BC to work independently to articulate a vision of a viable post-treaty fishery that addresses both fisheries management and allocation challenges. Goals for the Panel also include providing a description of a workable framework for fisheries management and principles to guide negotiations of both allocations and fisheries management between First Nations and Canada with a focus on migratory fish stocks that are subject to multiple users.

Meetings were held to determine the main issues for the First Nations in the current fishery. The main concerns included food and societal access, aboriginal rights, economic access, government policies and programs, licensing and quota systems, recreational fisheries and tourism, habitat protection and stewardship. Food and societal access, economic access, local management, coordinated management and treaty process were the issues of priority.

Following a review of these concerns, the Panel recommended the following priorities: stewardship of food fish, transfer fish for economic purposes, improve intertribal relations, provide major government commitment, and balance management interests.

Environmental sustainability, use of traditional resources, social stability, and economic well-being were identified as key aquaculture issues. The following effects of consolidation were identified: densely located farms, maximized fish densities in cages, automated equipment and methods that reduce employment, and perception of foreign ownership.

In regards to Aquaculture initiatives, increased opportunities for information sharing and dialogue have been conducted with two Aquaculture and Environment Summits held in 2002 and 2003. Continued environmental research and monitoring has been conducted with collaborative research on contaminant levels in marine species from fish farm wastes and environmental monitoring, planning and strategy development led by First Nations.

6.0 MESSAGES, CONSUMERS AND AQUACULTURE

6.1 Discussion Results

- Biotechnology and Public Acceptance

It was discussed that the greatest uncertainty associated with successful commercial uptake of new biotechnology is public acceptance. Advance inquiries with potential end-users of new biotechnology was suggested as a means of reducing this uncertainty, however proactive public involvement is not an integral part of Canadian agricultural biotechnology innovation. Participants felt that earlier and more earnest public consultation might have reduced existing tensions regarding agricultural biotechnology in Canada. The lack of public involvement was seen as a contributing factor to market failure of some products or nearly total opposition to the technology.

No genetically modified salmon have been approved for consumption in Canada to date. Given this, the opportunity exists to identify reactions of potential consumers to the introduction of specific animal product biotechnologically developed prior to its commercialization.

Participants suggested that the development of transgenic salmon would address Canada's need for a rapid-growing, cold-water tolerant fish to be competitive with other countries where fish are farmed in warmer, more productive waters. A transgenic salmon which grows more rapidly than non-transgenic Atlantic salmon is currently under regulatory review in Canada and the United States.

- Consumer Perception and Potential Reception in Canada

Results from a recent survey conducted in eight Canadian cities indicated that the attitudes regarding salmon do not change significantly when information about biotechnological application is communicated. Although results clearly showed that consumers were not fully accepting of the introduction of transgenic salmon, responses were equal if not enhanced when disclosure occurred. This is thought to be a result of lack of confidence to evaluate the significance of the modification, however with disclosure, consumers with more awareness and understanding can do so. Results also indicated that consumers were equally if not more accepting if sufficient information about the technology was communicated prior to the introduction of the product.

Results indicated that upwards of 40% of consumers, especially women, would not accept biotechnology in the near future. This finding is hypothesized to be a function of the traditional domestic role occupied by women who continue to make the majority of the decisions regarding food purchases.

It was suggested that consumers should have the ability to obtain information they feel they require in order to make informed choices, especially regarding new technologies. Results from the survey seem to indicate that the provision of information does not decrease acceptance if an appropriate level of information regarding risks and benefits is provided. It was suggested that government and industry should take this into consideration when deciding best methods to inform consumers and when targeting messages to male and female audiences.

- Consumer Perception and Potential Reception in Three European Markets

Two additional surveys were conducted in three European markets (France, Germany, and Spain) to determine critical factors related to future market trends for farmed salmon and farmed salmon products. The survey involved professional purchasers for supermarkets, wholesale organizations, and restaurants (of varying quality). The findings of these surveys suggest that the general image of the Atlantic salmon product is positive and well received. However potential environmental threats of widespread aquaculture activity and problems associated with genetically modified organisms have hampered some views.

The growing concern about food safety that has led to increased demand for quality guarantees and high consumer awareness on these issues was discussed. To meet future consumer demands, the industry should be encouraged to develop consistent production and processing strategies. Strong negative opinions are felt for the use of genetically modified ingredients in salmon feed by the trade and seafood business. Environmental effects, fish welfare, and genetically modified organisms remain the foremost concerns for the development of future farmed salmon. As such, increased communication and improved labelling of the fish and fish products will be needed.

7.0 INTEGRATED COASTAL ZONE MANAGEMENT

7.1 Discussion Results

- Integrated Coastal Zone Management Approach

It was suggested that a more integrated approach is required to promote sustainable development in the coastal zone and to identify solutions to the potentially negative impacts. The integrated coastal zone management approach is intended to be an integrated, multi-criteria decision support system for the evaluation of marine sites by multiple participants in support of governance and decision-making processes in the coastal zone.

- Integrated Coastal Zone Model and Framework

Marine site valuation analysis assigns relative values to the multiple, integrated components of the selected study area ecosystem. These values are designed to be comparable by using a yield-value framework. It is recognized, however, that alternative valuation schemes can be used in support for decision making in the generated model. The model provides a means of evaluating integrated marine activity and the procedure permits the comparison of traditional activities with other activities. Participants suggested that the objective of this work is to provide the opportunity for discussions among all parties involved to arrive at mutually acceptable decisions for the sustainable use of the coastal zone and its resources.

8.0 NEW PRACTICES FOR GLOBAL COMPETITIVENESS

8.1 Discussion Results

- Fundamental Problems Facing Aquaculture

Participants expressed the opinion that governance is the fundamental problem of Canadian aquaculture. It was suggested that to remedy these governance problems, an understanding must be gained of both the problems facing the industry and how these problems can best be addressed in the new political context in which aquaculture and other resource industries must operate.

- Suggestions for Proper Governance

A discussion was held regarding opportunities for appropriate governance in the promotion of sustainable aquaculture. Two proposals were presented to attempt to achieve promotion of sustainable aquaculture through proper governance. The first proposal included consideration of implications of new developments for governance, attempting to understand how such developments would be perceived by various sectors of the network, and identifying the risk of disturbing formal and informal arrangements, the soft law, and covenants and agreements that give the network its shape.

The second proposal suggested providing resources to understand the implications of the rapidly emerging tool of network management and benchmarking.

- Gaining Consumer Confidence

According to a recent study (Hites, R. A., et al., Global assessment of organic contaminants in farmed salmon. *Science* (303) 2004, 226–9), consumers believe farmed salmon to be high in toxins (including elevated levels of PCBs) that pose a risk to their health. Negative repercussions from these beliefs are loss of consumer confidence, weak markets in Canada and the United States, and increased pressure on wild product retailers.

Following a review of the study, it was determined that consumers feel betrayed by government and that government's reactions are insufficient to calm fears. Future steps proposed to gain consumer confidence include the following:

- study of risks by intervening parties;
- messages should be prepared in advance;
- Health Canada and Canada Food Inspection Agency need to show credibility;
- establish connections for complimentary efforts; and
- utilize resources to establish credibility and distribution channels.

9.0 CONCLUSIONS AND RECOMMENDATIONS

9.1 Research Recommendations

Awareness and understanding of Innovation and social transformation in aquaculture has progressed considerably in recent years and continues to evolve. However, much more work needs to be undertaken. Although not a specific topic of the workshop, several participants requested that recommendations for future research should be gleaned from the presentations and the open discussion. The following research recommendations, provided for the industry, academe and government sectors, have been identified by the facilitator and do not necessarily represent the views of the workshop participants.

9.1.1 Industry

Industry is interested in research on the following:

- Consumer perceptions regarding how farmed fish are cared for and managed – do they really care? There is a need for better understanding of the consumer purchase process, gender factors, credibility of the source and information distribution channels.

- Enhanced practices, based on emerging research on fish sentience, to manage farms in a manner that would boost productivity, lessen vulnerability to disease and lower costs.
- Means to implement and operationalize Integrated Coastal Zone Management, in the context of the need for site-specific data, and the multiplicity of stakeholder interests.
- Responsible farming methods and practices that would alleviate stress of the fish and reduce the adverse effects of the farm environment on fish welfare.
- Development of consistent production and processing strategies to meet future consumer demands.
- Investigate alternate antibiotic methods, utility and extent of buffer zones, impacts of antibiotics on the environment, results of feed testing and bioaccumulation/ biomagnification.

9.1.2 Academe

Research projects potentially suitable for academe include:

- Does academic research actually reach industry and the general public and do they understand it? How could research dissemination be improved?
- Greater understanding of fish sentience and welfare, and the implications for aquaculture practices.
- Consumer attitudes, preferences and purchase decision criteria (specific to regions, gender, etc.) related to aquaculture methods and products.
- Innovation in products and processes oriented to balancing environmental impact and high-value economic output.
- Means to increase sharing of knowledge and transfer of innovations amongst aquaculture R&D organizations, and industry and government.
- Development of practical guidelines, protocols and methods to promote and implement Integrated Environmental Impact Assessment (IEIA) in an aquaculture context.
- Elaboration of the concept of critical natural capital.

9.1.3 Government

Government research should focus on:

- Better understanding and management of citizen / consumer issues related to general food safety, inspection adequacy, and other environmental concerns.
- Better methods of risk communication. Better intelligence on "food scare" reporting in the media, including issues scanning to identify upcoming controversies and media concerns (what is media currently researching),

identifying what consumers are reading, and how to effectively communicate real meaning.
- Improved policy and regulatory frameworks for harmonizing (provincial and federal levels) the planning, management, environmental assessment practices.
- With respect to First Nations, develop better understanding related to food and societal access, economic access, local management, coordinated management, intertribal relations, and treaty processes.
- Develop best methods and media relations practices for informing consumers generally and dealing with crisis issues in particular about aquaculture food products, as well as considering gender biases. Develop methods to gain and enhance consumer confidence.
- Improved communication and labelling of the fish and fish products.
- Understanding of the implications of information in the public domain; how to be transparent yet maintain confidentiality of proprietary business data.
- The Environmental Impact Assessment process, particularly accepted siting criteria and effective public and stakeholder consultation.

9.2 Governance

Although one of the objectives of the workshop was to identify opportunities for appropriate governance in the promotion of sustainable aquaculture, this topic did not receive much attention or discussion from participants. Limited discussion took place regarding sensitivities of existing institutions to the possible disturbance of current formal and informal arrangements and network management.

It is evident that additional resources and research regarding adequate, workable and acceptable governance models is needed. Research is also required to develop new and appropriate structures to interface with First Nations; the traditional models (usually fraught with western culture biases) are not effective. Governance oriented research should take into account the current and future situation related to Health Canada and CFIA reporting procedures, enhanced and direct citizen engagement, the Aquaculture Policy Framework – Office of Sustainable Aquaculture; and the Responsible Aquaculture Industry Program – Global Aquaculture Alliance.

9.3 Going Forward

The Aquaculture, Innovation and Social Transformation Workshop was successful from several perspectives:

- The participants have been afforded the opportunity to come together, increase their understating of current activities and discuss common concerns and potential solutions.
- All workshop participants expressed a desire to continue deliberations in the future through appropriate mechanisms.

Printed in the United States
127210LV00005B/89/P